Guillermo González-Moraga

Cluster Chemistry

Introduction to the Chemistry of Transition Metal
and Main Group Element Molecular Clusters

With 232 Figures

Springer-Verlag
Berlin Heidelberg GmbH

Professor Dr. Guillermo González-Moraga
Departamento de Química, Facultad de Ciencias
Universidad de Chile, Casilla 653
Santiago, Chile

ISBN 978-3-540-56470-6

Library of Congress Cataloging-in-Publication Data
González-Moraga, Guillermo, 1943–
Cluster chemistry : an introduction to the chemistry of transition metal and main group element molecular
clusters / Guillermo González-Moraga.
Includes bibliographical references and index.
  ISBN 978-3-540-56470-6   ISBN 978-3-642-85926-7 (eBook)
  DOI 10.1007/978-3-642-85926-7

1. Metal crystals.   I. Title.

© Springer-Verlag Berlin Heidelberg 1993
Originally published by Springer-Verlag Berlin Heidelberg New York in 1993

Typesetting: Macmillan India Ltd, Bangalore 25.

51/3020 – 5 4 3 2 1 0 – Printed on acid-free paper

Preface

The vertiginous development experienced by inorganic chemistry in the last decades has signified not only great advances in the knowledge about traditional problems but also the development of new concepts. Cluster chemistry or chemistry of the atomic conglomerates is an important aspect in this Renaissance of inorganic chemistry.

A great amount of work has been carried out lately and numerous research papers and reviews are already available in this new field of inorganic chemistry. However a lack of material appropriate for disseminating and specially for teaching this subject is still apparent.

The aim of this work is to offer, in a simple and systematized form, some fundamental features of the chemistry of molecular clusters. The interdisciplinary nature of this chemistry appears as a very appropriate subject for a course permitting the student to integrate concepts previously learned and to get an integral vision of chemistry. This book is principally directed at chemistry students at either the senior or graduate level and to researchers who want to obtain an introductory and general view of the field. Abundant literature references address the lecturer to original work and principal reviews on the theme.

The present book would not have been possible without the assistance of a number of people and institutions. I am specially indebted to my teacher Prof. Dr. E. Fluck who not only encouraged me to assume this task, but also supported its achievement with valuable discussions and by providing me access to the facilities at the Gmelin Institute, Frankfurt. A fellowship of the Alexander von Humboldt Foundation provided me with a great deal of the time I needed for the preparation of the manuscript. I also wish to thank many of our students and Faculty colleagues, specially to M.A. Santa Ana, N. Yutronic, and C. Diaz, for useful discussions and encouragement. Thanks are also due to the various authors and editors who kindly permitted the reproduction of numerous illustrations. Finally, I specially thank the assistance of Claudia Pereira and María Luz Peña for preparing tables and illustrations.

Santiago, August 1993.   Guillermo González-Moraga

Table of Contents

Chapter 1

Current Concepts in Modern Chemistry

The aim of this book is to describe some relevant features of a branch of chemistry which deals with compounds made up of atomic conglomerates known as clusters. Although much of the knowledge about this kind of species has been developed only in the last 30 years it is apparent they are widely represented in the chemistry of a great number of elements. In most cases homonuclear clusters may be considered as a small fragment of an element constituting a nucleous stabilized as it occurs with the central atoms in mononuclear species. This new chemistry can be considered therefore as a bridge between traditional chemistry and that of the pure elements. This feature implies an attempt of applying already disposable concepts to the interpretation of the abundant experimental information on this new class of compounds. Since current concepts have been mostly elaborated separately for both limit cases, simple molecular species and infinite arrangements of the elements, the understanding of the fundaments of both existence and architecture of clusters is a very interesting but also difficult task.

In order to facilitate the discussion of some aspects of cluster chemistry – especially those related with bonding and its relationship with chemical and structural properties of the compounds – it is convenient to review briefly the fundaments of current bonding concepts. Some germane aspects of the atomic and molecular structure will be therefore summarized in the next sections. As an application of such concepts and also as an introduction to metal clusters, the discussion of some aspects of the metal-metal multiple bonding in dinuclear compounds is also included in this Chapter.

1.1 Atomic Structure

Current chemical concepts are essentially based on the fact that matter is fundamentally constituted by nucleus and electrons and that such systems follow the laws of quantum mechanics.

According to quantum mechanics, any system as an atom or a molecule is described by a function of both the coordinates of the particles constituting the system and the time. They are normally referred to as wave functions, ψ, because of their mathematical expressions which correspond to those traditionally used for describing undulatory processes. ψ itself does not have physical meaning, it is only a mathematical function that can also have negative value or be an imaginary number. However, the square of its absolute value has the meaning of probability. Thus for instance in the case of an electron $|\psi|^2$ integrated over a given volume of space is proportional to the probability of finding the electron in this space.

A particular case specially interesting for chemists are the functions dealing with stationary states, i.e. states that are stable in time and not changing spontaneously without external influence. In these cases the functions are state functions.

The Schrödinger equation is a fundamental relationship – analogous to the Newton equation in classical mechanics – to be applied in the description of quantum systems as the atomic ones.

The Schrödinger equation is commonly written in the form

$$H\psi = E\psi$$

where H is the Hamilton operator that represents the general form of the total energy of the system, i.e. the sum of the kinetic and potential energy. E is the numerical value of the energy for any particular ψ. The Schrödinger equation is a complex differential equation which can be exactly solved only in the case of very simple systems – for example the hydrogen atom. Because of the relationship of ψ with probability mentioned above any solution must satisfy the following boundary conditions:

1. to be always a finite function,
2. to have only one value in each and every point,
3. to be normalized, i.e. to satisfy the relationship

$$\int_0^\infty \psi\psi^* d\tau = 1$$

which establishes that the probability of finding the particle somewhere in space must be one.

1.1.1 Wave Function for the Hydrogen Atom

Considering the boundary conditions above the solution of the Schrödinger equation for the hydrogen atom formulated in spherical coordinates yields wave functions of the form

$$\psi = R_{n,l}(r)\,\Theta_{l,ml}(\theta)\,\Phi_{ml}(\varphi)$$

The numbers n, l and m_l known as *quantum numbers* occur as a result of boundary conditions. The quantum numbers with their allowed values may be summarized as follows:

Principal quantum number $n = 1, 2, 3, \ldots$

Angular quantum number $l = 0, 1, 2, \ldots, n - 1$

Magnetic quantum number $m_l = 0, \pm 1\ \pm 2, \ldots, \pm l$

The function $R_{nl}(r)$ is the *radial part* of the wave function and its square is related to the probability of finding the electron at any distance r of the nucleus. The functions $\Theta_{l,\,ml}(\theta)\,\Phi_{ml}(\phi)$ are the *angular part* of the wave function which is associated with the quantum numbers l and m_l.

The value of quantum number l is often replaced by a letter:

To the number $l = 0$ 1   2   3   4   5   ... corresponds

the letter s p d f g h ...

Thus wave functions are often specified by giving its quantum numbers in the shorthand notation n, l, m_l.

Each of this total wave functions defining a determined state of the electron in hydrogen atom is commonly referred to as an *orbital*. A complete set of orbitals for hydrogen through n = 3 is given in Table 1.1.

The representation of atomic orbitals may be achieved by plotting the radial and angular parts of the wave functions separately.

Figure 1.1 shows typical plots of the orbital radial functions. However, in order to interpret this information as electron density distribution it is often more useful to consider the *surface density functions* S(r) plotted in Fig. 1.2

$$S_{n,l}(r) = R^2_{n,l}(r)\,r^2$$

which correspond to a spherical surface density function for a sphere of radius r centered at the nucleus.

The most common representation of angular functions is to use boundary surfaces of orbitals outlining the region within which there is a large probability

Table 1.1. Orbitals of a hydrogen-like atom

Quantum numbers n l m	General symbol	Radial wave function[a]	Angular wave function
1 0 0	$1s$	$2(Z/a_0)^{3/2}e^{-\rho}$	$(2\pi^{1/2})^{-1}$
2 0 0	$2s$	$2^{-3/2}(Z/a_0)^{3/2}(2-\rho)e^{-\rho/2}$	$(2\pi^{1/2})^{-1}$
2 1 0	$2p_z$	$2^{-1}\cdot6^{-1/2}(Z/a_0)^{3/2}\rho e^{-\rho/2}$	$2^{-1}\cdot3^{1/2}\pi^{-1/2}\cos\theta$
2 1 1	$2p_x$	$2^{-1}\cdot6^{-1/2}(Z/a_0)^{3/2}\rho e^{-\rho/2}$	$2^{-1}\cdot3^{1/2}\pi^{-1/2}\sin\theta\cos\phi$
2 1 -1	$2p_y$	$2^{-1}\cdot6^{-1/2}(Z/a_0)^{3/2}\rho e^{-\rho/2}$	$2^{-1}\cdot3^{1/2}\pi^{-1/2}\sin\theta\sin\phi$
3 0 0	$3s$	$2\cdot81^{-1}\cdot3^{-1/2}(Z/a_0)^{3/2}$ $\cdot(27-18\rho+2\rho^2)e^{-\rho/3}$	$(2\pi^{1/2})^{-1}$
3 1 0	$3p_z$	$4\cdot81^{-1}\cdot6^{-1/2}(Z/a_0)^{3/2}(6\rho-\rho^2)e^{-\rho/3}$	$2^{-1}\cdot3^{1/2}\pi^{-1/2}\cos\theta$
3 1 1	$3p_x$	$4\cdot81^{-1}\cdot6^{-1/2}(Z/a_0)^{3/2}(6\rho-\rho^2)e^{-\rho/3}$	$2^{-1}\cdot3^{1/2}\pi^{-1/2}\sin\theta\cos\phi$
3 1 -1	$3p_y$	$4\cdot81^{-1}\cdot6^{-1/2}(Z/a_0)^{3/2}(6\rho-\rho^2)e^{-\rho/3}$	$2^{-1}\cdot3^{1/2}\pi^{-1/2}\sin\theta\sin\phi$
3 2 0	$3d_{z^2}$	$4\cdot81^{-1}\cdot30^{-1/2}(Z/a_0)^{3/2}\rho^2 e^{-\rho/3}$	$4^{-1}\cdot5^{1/2}\pi^{-1/2}(3\cos^2\theta-1)$
3 2 1	$3d_{xz}$	$4\cdot81^{-1}\cdot30^{-1/2}(Z/a_0)^{3/2}\rho^2 e^{-\rho/3}$	$2^{-3/2}\cdot30^{1/2}\pi^{-1/2}\sin\theta\cos\theta\cos\phi$
3 2 -1	$3d_{yz}$	$4\cdot81^{-1}\cdot30^{-1/2}(Z/a_0)^{3/2}\rho^2 e^{-\rho/3}$	$2^{-3/2}\cdot30^{1/2}\pi^{-1/2}\sin\theta\cos\theta\sin\phi$
3 2 2	$3d_{x^2-y^2}$	$4\cdot81^{-1}\cdot30^{-1/2}(Z/a_0)^{3/2}\rho^2 e^{-\rho/3}$	$4^{-1}\cdot15^{1/2}\pi^{-1/2}\sin^2\theta\cos2\phi$
3 2 -2	$3d_{xy}$	$4\cdot81^{-1}\cdot30^{-1/2}(Z/a_0)^{3/2}\rho^2 e^{-\rho/3}$	$4^{-1}\cdot15^{1/2}\pi^{-1/2}\sin^2\theta\sin2\phi$

[a] Zr/a_0;   Z nuclear charge.

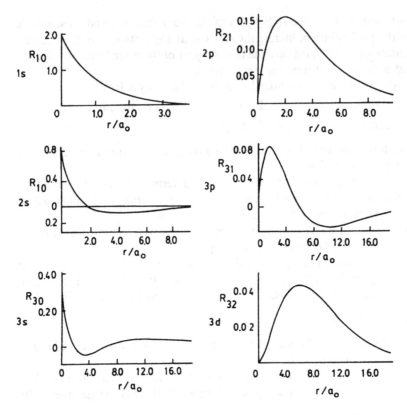

Fig. 1.1. Radial functions $R_{n,1}(r)$ of *1s, 2s, 3s, 2p, 3p* and *3d* hydrogen orbitals (r in atomic units)

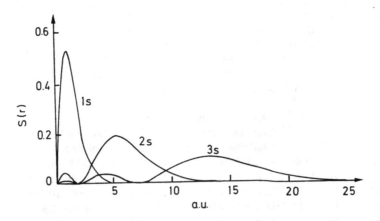

Fig. 1.2. Surface density functions, $S_{n,1}(r) = R^2(r)r^2$, for the *1s, 2s,* and *3s* atomic orbitals (r in atomic units)

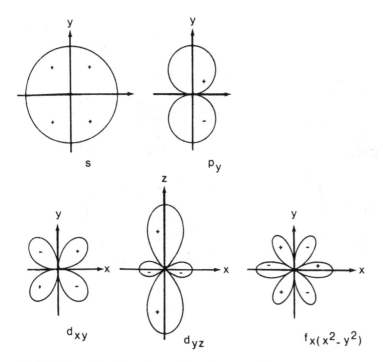

Fig. 1.3. Conventional boundary surfaces of *s*, *p*, *d* and *f* atomic orbitals

of finding the electron. Some examples of this kind of representation is illustrated in Fig. 1.3.

Two different forms of displaying a more complete picture of the structure of chemical orbitals given simultaneously information about shape, nodes, and electron density values are those illustrated in Figs. 1.4 and 1.5.

The spacial wave function already analyzed is however not enough for totally describing the state of an electron. It is necessary to have an additional function containing information about the spin of the electron. This function is characterized by a *spin quantum number m_s* which takes the values $\pm 1/2$. Thus the state of an electron in an atomic system is defined by the quantum numbers n, l, m_l, and m_s.

1.1.2 Multielectron Atoms

The quantum mechanical description of a multielectron system is in principle the same than that of hydrogen atom. However the complexity of differential equations arising from building a Hamilton operator appropriate for multielectron systems prevents us from obtaining exact solutions. Nonetheless approximated methods, based fundamentally on reformulations of the problem

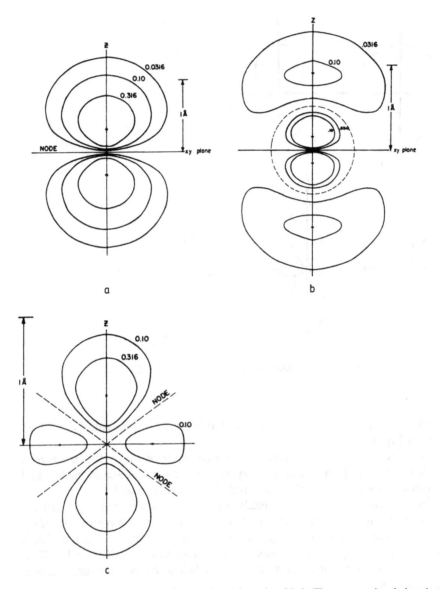

Fig. 1.4. Contours maps for (**a**) *2p*, (**b**) *3p*, and (**c**) *3d* atomic orbitals. The cross-sectional plane is any plane containing the z-axis. Reproduced with permission from E.A. Ogryzlo EA, Porter GB (1963) J. Chem. Ed. 40:256

leading to multielectron system with solutions similar to those for the hydrogen atom, have been developed. Such approximations permit us then to make use of orbitals appropriate for multielectron atoms. Although the shapes of such orbitals are similar to those of the hydrogen atom, substantial differences in the radial functions and thus in their energies are observed.

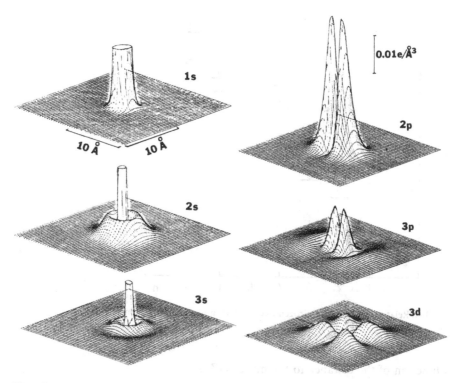

Fig. 1.5. Transversal cross-sectional plots of hydrogen orbitals. Diagrams represent the orbital as a function of the coordinates of a given plane. Elevation correspond to the function value in each point of sectional plane. Reproduced with permission from Fluck E (1989) Allgemeine und anorganische Chemie, 6th edn. Quelle and Meyer Verlag, Heidelberg, Wiesbaden

Hydrogen orbital energy depends exclusively on principal quantum number n, i.e orbitals with different l but the same n are energetically degenerated. In multielectron atoms such orbital degeneracy disappears since orbital energy depends on both n and l quantum numbers. This feature may be easily understood by considering the concept of effective nuclear charge. The plots of the surface density functions of hydrogen orbitals illustrated in Fig. 1.2 show that each orbital is characterized by a determined charge distribution leading to characteristic pattern of penetration into other orbitals. The average potential energy of a given electron in hydrogen atom expressed by

$$V = -e^2 \int \frac{Z}{r} S(r)\,dr$$

depends only upon its own wave function since nuclear charge experimented by it is constant ($Z = 1$). In a multielectron atom, however, the electron average energy depends not only on its own hydrogen-like wave function but also upon the screening effect of internal electrons. Z in this case is not a constant but

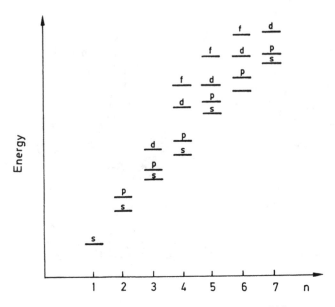

Fig. 1.6. Relative energies of hydrogen-like orbitals in multielectron atoms

a function of the distance to the nucleus (Z^*).

$$V = -e^2 \int \frac{Z^*(r)}{r} S(r)\,dr$$

The scheme in Fig. 1.6 shows schematically the energies of the orbitals in neutral multielectron atoms.

1.1.3 Electron Configurations in Multielectron Atoms

In addition to the conditions for the electronic structures of multielectron atoms established by the monoelectronic wave functions and their relative energies mentioned above, other restrictions should also be considered. One of them is the Pauli principle stating that no two electrons can have the same quantum numbers. Thus one orbital can "contain" a maximum of two electrons provided they have different spin quantum numbers. Other practical rules or restrictions refer to the influence of interelectronic interactions on the electronic structures established by Hund's rules. The electrons with the same n and l values will occupy first orbitals with different m_l and the same m_s (paired spins).

The *aufbau principle* is used to build up the electronic structures of multielectron atoms by adding protons and electrons to the hydrogen atom. According to this principle electrons are placed in hydrogen-like orbitals filling them in the order of decreasing stability (energy of the orbital and interelectronic

Table 1.2. Electron configurations of atoms in their normal states

1	H	$1s$	36	Kr	$3d^{10}4s^24p^6$	71	Lu	$4f^{14}5d6s^2$
2	He	$1s^2$	37	Kb	$[\text{Kr}]5s$	72	Hf	$4f^{14}5d^26s^2$
3	Li	$[\text{He}]2s$	38	Sr	$5s^2$	73	Ta	$4f^{14}5d^36s^2$
4	Be	$2s^2$	39	Y	$4d5s^2$	74	W	$4f^{14}5d^46s^2$
5	B	$2s^22p$	40	Zr	$4d^25s^2$	75	Re	$4f^{14}5d^56s^2$
6	C	$2s^22p^2$	41	Nb	$4d^45s$	76	Os	$4f^{14}5d^66s^2$
7	N	$2s^22p^3$	42	Mo	$4d^55s$	77	Ir	$4f^{14}5d^76s^2$
8	O	$2s^22p^4$	43	Tc	$4d^55s^2$	78	Pt	$4f^{14}5d^96s$
9	F	$2s^22p^5$	44	Ru	$4d^75s$	79	Au	$[4f^{14}5d^{10}]6s$
10	Ne	$2s^22p^6$	45	Rh	$4d^85s$	80	Hg	$5f^{14}5d^{10}6s^2$
11	Na	$[\text{Ne}]3s$	46	Pd	$4d^{10}$	81	Tl	$6s^2$
12	Mg	$3s^2$	47	Ag	$4d^{10}5s$	82	Pb	$6s^26p^2$
13	Al	$3s^23p$	48	Cd	$4d^{10}5s^2$	83	Bi	$6s^26p^3$
14	Si	$3s^23p^2$	49	In	$4d^{10}5s^25p$	84	Po	$6s^26p^4$
15	P	$3s^23p^3$	50	Sn	$4d^{10}5s^25p^2$	85	At	$6s^26p^5$
16	S	$3s^23p^4$	51	Sb	$4d^{10}5s^25p^3$	86	Rn	$6s^26p^6$
17	Cl	$3s^23p^5$	52	Te	$4d^{10}5s^25p^4$	87	Fr	$[\text{Rn}]7s$
18	Ar	$3s^23p^6$	53	I	$4d^{10}5s^25p^5$	88	Ra	$7s^2$
19	K	$[\text{Ar}]4s$	54	Xe	$4d^{10}5s^25p^6$	89	Ac	$6d7s^2$
20	Ca	$4s^2$	55	Cs	$[\text{Xe}]6s$	90	Th	$6d^27s^2$
21	Sc	$3d4s^2$	56	Ba	$6s^2$	91	Pa	$5f^26d7s^2$
22	Ti	$3d^24s^2$	57	La	$5d6s^2$	92	U	$5f^36d7s^2$
23	V	$3d^34s^2$	58	Ce	$4f^26s^2$	93	Np	$5f^57s^2$
24	Cr	$3d^54s$	59	Pr	$4f^36s^2$	94	Pu	$5f^67s^2$
25	Mn	$3d^54s^2$	60	Nd	$4f^46s^2$	95	Am	$5f^77s^2$
26	Fe	$3d^64s^2$	61	Pm	$4f^56s^2$	96	Cm	$5f^76d7s^2$
27	Co	$3d^74s^2$	62	Sm	$4f^66s^2$	97	Bk	$5f^86d7s^2$
28	Ni	$3d^84s^2$	63	Eu	$4f^76s^2$	98	cf	$5f^{10}7s^2$
29	Cu	$3d^{10}4s$	64	Gd	$4f^75d6s^2$	99	Es	$5f^{11}7s^2$
30	Zn	$3d^{10}4s^2$	65	Tb	$4f^96s^2$	100	Fm	$5f^{12}7s^2$
31	Ga	$3d^{10}4s^24p$	66	Dy	$4f^{10}6s^2$	101	Md	$5f^{13}7s^2$
32	Ge	$3d^{10}4s^24p^2$	67	Ho	$4f^{11}6s^2$	102	No	$5f^{14}7s^2$
33	As	$3d^{10}4s^24p^3$	68	Er	$4f^{12}6s^2$	103	Lw	$5f^{14}6d7s^2$
34	Se	$3d^{10}4s^24p^4$	69	Tm	$4f^{13}6s^2$	104	Rt	$5f^{14}6d^27s^2$
35	Br	$3d^{10}4s^24p^5$	70	Yb	$4f^{14}6s^2$	105	Ha	$5f^{14}6d^37s^2$

interactions) and taking into account the Pauli principle. The electronic configurations of the atoms in their normal state are shown in Table 1.2.

The *Periodic Table* constitutes the best representation of the properties of the elements where the tight relationship between their chemical and electronic properties is clearly stated. A version of the Periodic Table with group enumeration as recently accepted by IUPAC is reproduced in Fig. 1.7.

1.1.4 Electronegativities

Electronegativity, defined as a measure of the tendency of an element to hold electrons when it forms a compound, is one of the more widely used concepts in chemistry. Mulliken defined the electronegativity of an atom as an average between the energy associated to the loss of an electron (ionization potential I_A)

PERIOD   GROUP

	1	2	3	4	5	6	7	8	9	10	11	12	13	14	15	16	17	18
1	1 H																	2 He
2	3 Li	4 Be											5 B	6 C	7 N	8 O	9 F	10 Ne
3	11 Na	12 Mg											13 AL	14 Si	15 P	16 S	17 Cl	18 Ar
4	19 K	20 Ca	21 Sc	22 Ti	23 V	24 Cr	25 Mn	26 Fe	27 Co	28 Ni	29 Cu	30 Zn	31 Ga	32 Ge	33 As	34 Se	35 Br	36 Kr
5	37 Rb	38 Sr	39 Y	40 Zr	41 Nb	42 Mo	43 Tc	44 Ru	45 Rh	46 Pd	47 Ag	48 Cd	49 In	50 Sn	51 Sb	52 Te	53 I	54 Xe
6	55 Cs	56 Ba	57* La	72 Hf	73 Ta	74 W	75 Re	76 Os	77 Ir	78 Pt	79 Au	80 Hg	81 Tl	82 Pb	83 Bi	84 Po	85 At	86 Rn
7	87 Fr	88 Ra	89** Ac	104 Rt	105 Ha													

*Lanthanide Series	58 Ce	59 Pr	60 Nd	61 Pm	62 Sm	63 Eu	64 Gd	65 Tb	66 Dy	67 Ho	68 Er	69 Tm	70 Yb	71 Lu
** Actinide Series	90 Th	91 Pa	92 U	93 Np	94 Pu	95 Am	96 Cm	97 Bk	98 Cf	99 Es	100 Fm	101 Md	102 No	103 Lr

Fig. 1.7. Common long form of the Periodic Table of the elements

and that associated to the gain of an electron (electron affinity E_A).

$$A \xrightarrow{I_A} A^+ + e^-$$

$$EN = 1/2 \, (I_A + E_A)$$

$$A + e^- \xrightarrow{E_A} A^-$$

This definition may be applied not only to atoms but also to any atomic state likely to obtain its I_A and E_A values, e.g. for species with any electron configuration, or directly for atomic orbitals (orbital electronegativity). Some Mulliken orbital electronegativities are shown in Table 1.3. Electronegativities are indeed directly related with the effective nuclear charge experimented by the electron in these orbitals.

It is noteworthy that, as shown in Table 1.4, atom electronegativities obtained by different methods are linearly related to each other.

Table 1.3. Mulliken orbital electronegativities (eV)

H						
s 7.2						

	Li	Be	B	C	N	O	F
s	3.1	di^2 4.8	tr^3 6.4	$di^2\pi^2$ 10.4, 5.7	$di^3\pi^2$ 15.7, 7.9	$tr^5\pi$ 17.1, 20.2	*s* 31.3
p	1.8	te^2 3.9	te^3 6.0	$tr^3\pi$ 8.8, 5.6	$tr^4\pi$ 12.9, 8.0	$di^2\pi^4$ 19.1	*p* 12.2
				te^4 8.0	te^5 11.6	te^6 15.3	

	Na	Mg	Al	Si	P	S	Cl
s	2.9	di^2 4.1	tr^3 5.5	$di^2\pi^2$ 9.0, 5.7	$di^3\pi^2$ 11.3, 6.7	$tr^4\pi^2$ 10.9	*s* 19.3
p	1.6	te^2 3.3	te^3 5.4	$tr^3\pi$ 7.9, 5.6	$tr^4\pi$ 9.7, 6.7	te^6 10.2	*p* 9.4
				te^4 7.3	te^5 8.9		

	K	Ca	Ga	Ge	As	Se	Br
s	2.9	di^2 3.4	tr^3 6.0	$di^2\pi^2$ 9.8, 6.5	$di^3\pi^2$ 9.0, 6.5	$tr^4\pi^2$ 10.6	*s* 18.3
p	1.8	te^2 2.5	te^3 6.6	$tr^3\pi$ 8.7, 6.4	$tr^4\pi$ 8.6, 7.0	te^6 9.8	*p* 8.4
				te^4 8.0	te^5 8.3		

	Rb	Sr	In	Sn	Sb	Te	I
s	2.1	di^2 3.2	tr^3 5.3	$di^2\pi^2$ 9.4, 6.5	$di^3\pi^2$ 9.8, 6.3	$tr^4\pi^2$ 10.5	*s* 15.7
p	2.2	te^3 5.1	te^3 2.2	$tr^3\pi$ 8.4, 6.5	$tr^4\pi$ 9.0, 6.7	te^6 9.7	*p* 8.1
					te^5 8.5		

Values can be computed only for orbitals holding 1 electron. For the carbon and nitrogen families it is possible to have both hybrid and π atomic orbitals half-filled.
di = digonal $\equiv$ sp hybrid, tr = trigonal $\equiv$ sp^2 hybrid; te = tetrahedral $\equiv$ sp^3 hybrid

1.2. Chemical Bonding

The description of a molecular system with more than one nucleus and two or more electrons is essentially similar to that of an atom: A function satisfying the Schrödinger equation

$$H_{mol} \psi = E\psi$$

should be found. There H_{mol} is the Hamilton operator containing all possible interactions between electrons and nuclei as well as interelectronic and internuclear interactions. The description of this problem for the hydrogen molecule H_2 is illustrated in Fig. 1.8.

Inclusive for the simple H_2 molecule the solution of the wave equation is rather more complex than for atomic systems, so the exact solution of this problem has been impossible so far. Therefore approaches leading to the formulation of approximate wave functions capable of being improved by iterative methods must be used.

There are two methods for finding such approximate solutions, namely the Valence Bond Theory (VBT) and the Molecular Orbital Theory (MO-theory).

Table 1.4. Electronegativity scales

Authors	Equation[a]	Interconnection equations	Ref.
Pauling (X_P)	$X(A) - X(B) = k(\Delta_{AB})^{1/2}$ [b]	—	1, 2
Mulliken (X_M)	$X(A) = 1/2\,[IP(A) - EA(A)]$ IP = ionization energy EA = electron affinity	$X_P = 0.34 X_M - 0.2$	3–5
Allred–Rochow (X_{AR})	$X(A) = e^2 Z^*(A)/r^2(A)$ Z^* = effective nuclear charge r = covalent radii	$X_P = 0.36 Z^*/r^2 + 0.7$	6
Sanderson (X_S)	$X(A) = \bar{\rho}/\bar{\rho}_i$ $\bar{\rho}$ = Average electron density (total charge/atomic volume, Z/V).	$X_P = (0.21 X_S + 0.77)^2$	7, 8

[a] X(A) = Electronegativity of atom A
[b] Δ_{AB} = Bond energy (D) difference: D(A–B) – 1/2[D(A–A) + D(B–B)]

References

1 Pauling L (1960) The Nature of the Chemical Bond. Cornell University Press. Ithaca N.Y.
2 Allred AL (1961) J. Inorg. Nucl. Chem. 17:215
3 Mulliken RS (1935) J. Chem. Phys. 2:782 and 3:573
4 Hinze J, Jaffe HH (1962) J. Am. Chem. Soc. 84:540
5 Hinze J, Jaffe HH (1963) J. Phys. Chem. 67:1501
6 Allred AL, Rochow E (1958) J. Inorg. Nucl. Chem. 5:264
7 Sanderson RT (1965) J. Inorg. Nucl. Chem. 21:989
8 Sanderson RT (1967) J. Chem. Ed. 44:517

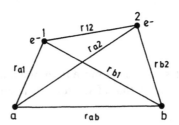

$$H = \left[-\frac{h^2}{8\pi^2 m} (\nabla_1^2 - \nabla_1^2) - \frac{e^2}{r_{a1}} - \frac{e^2}{r_{b1}} - \frac{e^2}{r_{a2}} - \frac{e^2}{r_{b2}} + \frac{e^2}{r_{12}} + \frac{e^2}{r_{ab}} \right]$$

Fig. 1.8. Coordinates and Hamilton operator for the hydrogen molecule

Both methods suppose that the orbitals of involved atoms are known independently of the approximation degree with which they have been obtained. Both approaches lead in general to satisfactory descriptions of main molecular properties. The advantages of one or the other in most cases depend more upon the problem to be solved and the type of information being required than on the methods themselves.

1.2.1 The Valence Bond Theory

From the point of view of the Valence Bond Theory (VBT) the conception of the chemical bonding is practically the same as that of R.N. Lewis. VBT may indeed be considered as representing a rationalization in terms of the wave mechanics of the idea of bonds by sharing electron pairs.

When two atoms with appropriate orbitals – i.e. orbitals of not very different energy and of the same symmetry, and each with one electron – approach one another closely, electron pairing will be produced. In this model electron pair sharing is mathematically represented as a product of the atom wave functions.

$$\psi_{AB} = \psi_A(1)\,\psi_B(2)$$

This product is associated with the probability of finding the electron of both atoms A and B in the space between the nuclei. Because of the indistinguishability of electrons (1) and (2) two wave functions ψ_I and ψ_{II} should be considered.

$$\psi_{cov} = c_1\psi_I \pm c_2\psi_{II}$$

where

$$\psi_I = \psi_A(1)\,\psi_B(2) \qquad \text{and} \qquad \psi_{II} = \psi_A(2)\,\psi_B(1)$$

However, this solution may be considerably improved by considering that there is also a finite chance that both electrons will be occasionally associated to the same atom A or B:

$$\psi_{ion} = c_3\psi_A(1)\,\psi_A(2) \pm c_4\psi_B(1)\,\psi_B(2).$$

Thus the description of real atoms can then be obtained in an acceptable approximation by the hybrid function ψ_{AB}.

$$\psi_{AB} = c_5\psi_{cov} + c_6\psi_{ion}$$

The coefficients c_i, besides their role in the normalization of the functions, determine the contribution of the different canonical structures to the mixed function. These structures, also known as *resonance structures*, state, therefore, the type and magnitude of bond polarity.

The energy calculated for a given system decreases with increasing number of the resonance structures considered in the hybrid wave function used in its description. The difference between the energy of the mixed state and that of the most stable of the resonance structures is called the *resonance energy*. The greater the coefficient c_i of a given canonical structure in the hybrid, the higher its contribution to the resonance energy. The scheme in Fig. 1.9 shows the resonance description of the benzene molecule.

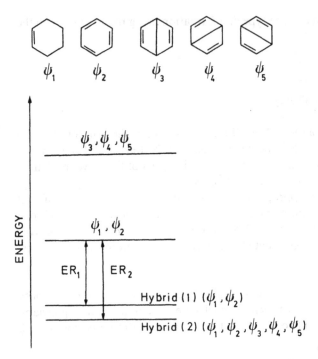

Fig. 1.9. Resonance energy in benzene molecule. Comparison of resonance energies by considering only Kekulé structures $\psi_1-\psi_3$ (RE$_1$) and Kekulé plus Dewar structures $\psi_1-\psi_5$ (RE$_2$)

1.2.2 Directional Atomic Orbitals

One of the more useful features of the model of bonding by sharing electron pairs described by the VBT is the idea of directional bonding. This concept implies that the angular parts of the atomic wave functions involved in the description of a given molecule should have a symmetry according with the molecular structure. In methane for instance no pure hydrogen-like carbon orbitals fulfill the requirements stated by its tetrahedral geometry. Nevertheless one of the properties of Schrödinger wave functions is the feature that linear combinations of a set of solutions is also a solution of the same equation, i.e. from the original hydrogen-like atomic orbitals it is always possible to obtain a new set of atomic orbitals, called hybrid atomic orbitals, that fulfill predetermined symmetry requirements. Analytical as well as pictorial descriptions of some common hybrid atomic orbitals are shown in Table 1.5. Examples of molecules with central atoms displaying these hybrid atomic orbitals are also given in Table 1.5. The application of these sets of equivalent hybrid orbitals is however restricted to very symmetrical species. However when molecular symmetry does not require the equivalence of the hybrid other functions should be built up. Thus for instance for $CHCl_3$ three equivalent and a fourth different hybrid orbitals are necessary.

Table 1.5. Hybrid atomic orbitals. Diagram and description of most common hybrid orbitals

Hybrid	Hybridization	Analytical Forms
$S_p(D_{\infty h})$		$di_1 = (\tfrac{1}{2})^{1/2}(\phi_s + \phi_z)$ $di_2 = (\tfrac{1}{2})^{1/2}(\phi_s - \phi_z)$
$sp^2(D_3)$		$tr_1 = \frac{1}{3^{1/2}}(\phi_s + (2)^{1/2}\phi_x)$ $tr_2 = \frac{1}{3^{1/2}}[\phi_s - (\tfrac{1}{2})^{1/2}\phi_x + (3/2)^{1/2}\phi_y]$ $tr_3 = \frac{1}{3^{1/2}}[\phi_s - (\tfrac{1}{2})^{1/2}\phi_x - (3/2)\phi_y]$
$sp^3(T_d)$		$te_1 = \tfrac{1}{2}(\phi_s + \phi_x + \phi_y + \phi_z)$ $te_2 = \tfrac{1}{2}(\phi_s + \phi_x - \phi_y - \phi_z)$ $te_3 = \tfrac{1}{2}(\phi_s - \phi_x + \phi_y - \phi_z)$ $te_4 = \tfrac{1}{2}(\phi_s - \phi_x - \phi_y + \phi_z)$
$dsp^3(D_3)$		$tr_1 = (\tfrac{1}{3})^{1/2}[\phi_s + (2)^{1/2}\phi_x]$ $tr_2 = (\tfrac{1}{3})^{1/2}[\phi_s - (\tfrac{1}{2})^{1/2}\phi_x + (3/2)^{1/2}\phi_y]$ $tr_3 = (\tfrac{1}{3})^{1/2}[\phi_s - (\tfrac{1}{2})^{1/2}\phi_x - (3/2)^{1/2}\phi_y]$ $di_4 = (\tfrac{1}{2})^{1/2}(\phi_z + \phi_{z^2})$ $di_5 = (\tfrac{1}{2})^{1/2}(\phi_z - \phi_{z^2})$
$d^2sp^3(O_h)$		$oc_1 = \frac{1}{6^{1/2}}[\phi_s + (2)^{1/2}\phi_{z^2} + (3)^{1/2}\phi_z]$ $oc_2 = \frac{1}{6^{1/2}}[\phi_s - (1/2)^{1/2}\phi_{z^2} - (3/2)^{1/2}\phi_{x^2-y^2} + (3)^{1/2}\phi_x]$ $oc_3 = \frac{1}{6^{1/2}}[\phi_s - (1/2)^{1/2}\phi_{z^2} - (3/2)^{1/2}\phi_{x^2-y^2} + (3)^{1/2}\phi_y]$ $oc_4 = \frac{1}{6^{1/2}}[\phi_s - (1/2)^{1/2}\phi_{z^2} + (3/2)^{1/2}\phi_{x^2-y^2} - (3)^{1/2}\phi_x]$ $oc_5 = \frac{1}{6^{1/2}}[\phi_s - (1/2)^{1/2}\phi_{z^2} - (3/2)^{1/2}\phi_{x^2-y^2} - (3)^{1/2}\phi_y]$ $oc_6 = \frac{1}{6^{1/2}}[\phi_s + (2)^{1/2}\phi_{z^2} - (3)^{1/2}\phi_z]$

From the energetic point of view, the formation of hybrid orbitals and their populating with the electrons required for bonding is an endothermic process. It implies a promotion energy. However such energetic cost is compensated by an increase of electron density in the bonding direction which leads to better orbital overlapping and hence to higher bonding energies.

As shown in Table 1.3 in Sect. 1.1.4, hybridization also implies variations in the orbital electronegativity because of changes in the corresponding nuclear effective charges. This will be reflected in the bonding polarity.

1.2.3 Molecular Orbital Theory of Bonding

This theory endeavors to describe the molecule by a method intrinsically similar to that used for obtaining atomic orbitals but considering multicenter wave functions. Thus this approach consists of finding the best functions for describing the state of one electron in a field formed by the totality of the nuclei placed in their equilibrium positions. These monoelectronic molecular wave functions may be obtained according to the MO-theory by a linear combination of atomic orbitals (LCAO).

The hydrogen molecule ion, H_2^+, with two nuclei and one electron is the most simple imaginable molecular system. It can be considered as a prototype for homonuclear biatomic molecules just as hydrogen atom is for atoms in general.

According to MO-theory, the wave function for the system H_2^+ may be formulated as a linear combination of two hydrogen atomic orbitals 1s. The meaning of these two LCAO-MOs can be appreciated in both analytical and pictorial descriptions illustrated in Fig. 1.10. The electron density distribution associated to the orbital ψ_{AB}^b shows that the charge concentration between the nuclei is indeed greater than the simple sum of contributions of two separate orbitals since $(\psi_A + \psi_B)^2 < \psi_A^2 + \psi_B^2$. That results in binding the nuclei together. ψ_{AB}^b is therefore a *bonding molecular orbital*. Contrarily in ψ_{AB}^* there is a nodal plane bisecting the internuclear axis A–B being there the electron density lower than in the separated orbitals ψ_A and ψ_B. It results therefore a high energy state in which there is a considerable repulsion between the nuclei. ψ_{AB}^* is an *antibonding molecular orbital*.

The energies of ψ_{AB}^b and ψ_{AB}^* relative to that of an isolated hydrogen atom are shown in Fig. 1.11.

The nature of the state described by ψ_{AB}^b and ψ_{AB}^* may also be visualized in the plot of the potential energy of the system H_2^+ as a function of the internuclear distance shown in Fig. 1.12. Only when the electron is in the state described by ψ^b there is an energy minimum at the internuclear distance R_e. According to the diagram in Fig. 1.11, the binding energy in this case is $-E_{AB}$.

The energy level diagram derived for the hydrogen molecule ion may also be used for building up the electronic structure of some other chemical species just as the *aufbau principle* used for building up the atom electronic configurations. The addition of one electron to H_2^+ leads to a hydrogen molecule H_2 which according to the energy of the MOs and the exclusion principle has the electron configuration $(\psi_{AB}^b)^2$. The binding energy will be approximately $-2E_{AB}$. Analogously, the electron configuration of the molecule HeH is $(\psi_{AB}^b)^2 (\psi_{AB}^*)^1$ with a binding energy of $-E_{AB}$. However when two He atoms come together there

Bonding

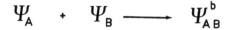

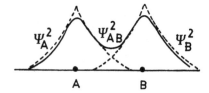

Antibonding

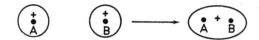

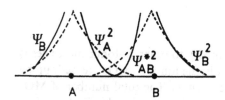

Fig. 1.10. Electron density in bonding and antibonding orbitals in the hydrogen molecule-ion.
(a) Analytical and (b) pictorial descriptions. (c) Electron density distributions along the nuclear axis

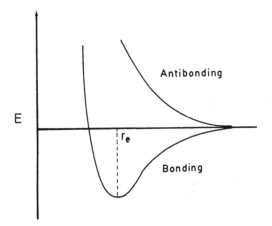

Fig. 1.11. Potential energy diagram for the hydrogen molecule ion in bonding and antibonding states

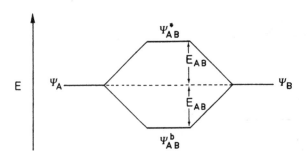

Fig. 1.12. Molecular orbital energy diagram for the H_2 molecule

are four electrons to be housed in the two MOs and the binding energy will be precisely zero.

1.2.4 Homonuclear Diatomic Molecules

The application of the Molecular Orbital theory to other diatomic molecules is essentially similar to that describing the molecule ion H_2^+. MOs are built up by linear combination of appropriate atomic orbitals. The total number of MOs formed is equal to the total number of participating atomic orbitals. According to their energies in respect to those of non-interacting atomic orbitals, molecular orbitals are considered as bonding, antibonding and, in the case of no appreciable changes, to non-bonding MOs respectively. Only the linear combinations of AOs with adequate size (energy) and symmetry produce appreciable changes in the energy of the system. That is frequently expressed in terms of the overlapping integral S:

$$S = \int \psi_A \psi_B \, d\tau \quad \begin{array}{ll} S < 0 & \text{antibonding} \\ S = 0 & \text{non bonding} \\ S > 0 & \text{bonding} \end{array}$$

As illustrated in Fig. 1.13a the overlapping between orbitals of different symmetry is null. Poor energy matching, i.e. great differences in size or in energy between interacting orbitals, leads to low absolute values of S. In general the overlapping between orbitals with symmetry δ is poorer than that with symmetry π and this than that between orbitals with symmetry σ (vide infra). Molecular orbitals of different symmetry as those occurring simultaneously in multiple bond are therefore not equivalent.

Both MOs described in Fig. 1.10 have cylindrical symmetry about the internuclear axis. MOs of this type are called sigma (σ) MOs. Since the internuclear axis is usually defined as the z axis p_z atomic orbital has a σ-symmetry about this axis and thus it can be combined with other orbitals of the same symmetry ($p\sigma$ orbitals). p_x and p_y have however other symmetry. Rotation about the internuclear axis is not symmetric and there is a nodal plane containing this axis. These kind of orbitals are said to have π-symmetry. The linear combination of π atomic orbitals leads to bonding (π^b) and antibonding (π^*) MOs. Analogously atomic orbitals with symmetry δ, i.e. two nodal planes containing the internuclear axis, may be combined to give MOs with the same symmetry. That will be discussed separately in Sect. 1.3. The formation of dinuclear molecular orbitals with two different classes of symmetry is illustrated in Fig. 1.14.

When two atoms containing s and p valence orbitals are combined into a homonuclear diatomic molecule a set of molecular orbitals with shapes and symmetry properties as those already described arises. The relative energies of

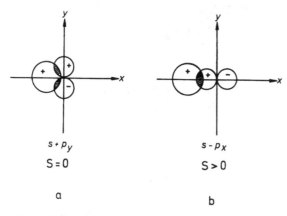

$$s + p_y$$
$$S = 0$$

$$s - p_x$$
$$S > 0$$

a

b

Fig. 1.13. Orbital overlapping. (**a**) Overlapping of orbitals of different symmetry ($S = O$). (**b**) Overlapping between two orbitals of the same symmetry ($S > 0$)

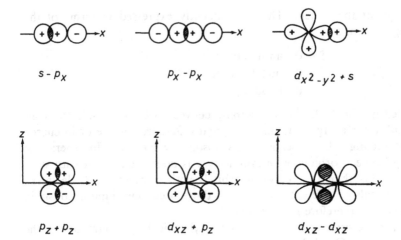

Fig. 1.14. Examples of orbital overlapping leading to molecular orbitals with symmetry σ and π. For examples of molecular orbitals with symmetry δ see Fig. 1.29

these MOs have the order shown in any of the diagrams illustrated in Fig. 1.15. π-orbitals formed by combination of p_x and p_y orbitals have the same energy and they are called *degenerate orbitals*. Energy diagrams (a) and (b) in Fig. 1.15 differ in the relative energies of the π and σp MOs. This feature is related to the

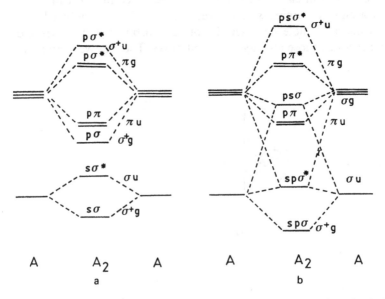

Fig. 1.15. Qualitative orbital energy diagrams for homonuclear diatomic molecules in the limiting approaches (**a**) without and (**b**) with s–p mixing

Table 1.6. Valence electron configurations of diatomic molecules of second period elements

Molecule	Valence electron configuration	Bond order
Li_2	$(\sigma_s^b)^2$	1
Be_2	$(\sigma_s^b)^2(\sigma_s^*)^2$	0
B_2	$(\sigma_s^b)^2(\sigma_s^*)^2(\pi_x^b)(\pi_y^b)$	1
C_2	$(\sigma_s^b)^2(\sigma_s^*)^2(\pi_{x,y}^b)^4$	2
N_2	$(\sigma_s^b)^2(\sigma_s^*)^2(\pi_{x,y}^b)^4(\sigma_z^b)^2$	3
O_2	$(\sigma_s^b)^2(\sigma_s^*)^2(\pi_{x,y}^b)^4(\pi_x^*)(\pi_y^*)$	2
F_2	$(\sigma_z^b)^2(\sigma_s^*)^2(\sigma_z^b)^2(\pi_{x,y}^b)^4(\pi_{x,y}^*)^4$	1
Ne_2	$(\sigma_s^b)^2(\sigma_s^*)^2(\sigma_z^b)^2(\pi_{x,y}^b)^4(\pi_{x,y}^*)^4(\sigma_z^*)^2$	0

coupling of both σ MOs, σs and σp. Since in a diatomic molecule all σ MOs have the same symmetry, the mixture of both σp and σs must be considered. That can produce an inversion in the position of the levels σp and π in the energy diagram. The importance of this effect increases when the energy gap between s and p atomic orbitals is small. Since the effective nuclear change experimented by valence electrons increases along a period, such effect is more important for the lighter elements of the series. Table 1.6 shows the electron configurations of the diatomic molecules of the second period elements established by theoretical and experimental studies. Electron configurations for molecules from the lighter elements follow a pattern similar to that in Fig. 1.15b. While those for heavier elements show a normal pattern (Fig. 1.15a) not reflecting thus the *contamination* of σs and σp orbitals.

In Table 1.6 the *bond orders* associated to the different electron configurations are also indicated. In the MO theory bond order is defined as the number of bonding electrons minus the number of antibonding electrons divided by two.

Strongly associated with the occupation of the molecular orbitals is the concept *frontier molecular orbitals* which is very useful for connecting the electronic structure of chemical species with their reactivity. Oxidation-reduction as well as Lewis acid-base properties of chemical compounds are determined by their capability of accepting or donating electron density. Thus, symmetry and energy of both, *Higher Occupied Molecular Orbital* (HOMO) and *Low Unoccupied Molecular Orbital* (LUMO), of a given molecule determine its ability for interacting with other chemical species.

1.2.5 Heteronuclear Diatomic Molecules

Treatment of heteronuclear biatomic molecules by LCAO-MO theory is similar to that of the homonuclear species. The values of mixture coefficients c_1 and c_2 in the molecular wave equation

$$\psi_{AB} = c_1\psi_A \pm c_2\psi_B$$

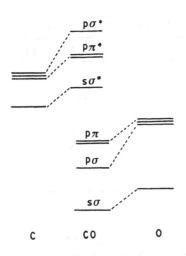

Fig. 1.16. Qualitative molecular orbital energy diagram of carbon monoxide

reflex the degree of asymmetry respect to a plane perpendicular to and bisecting the internuclear axis occurring in the MOs in heteronuclear species. Such an asymmetry is in turn determined by energy differences between interacting AOs. If the energies are vastly different the mixture will be poor and the MOs will behave nearly as pure orbitals ψ_A and ψ_B.

Figure 1.16 illustrates the qualitative MO-diagram of carbon monoxide, CO. Oxygen AOs are somewhat more stable (higher orbital electronegativity) than those of carbon, so they contribute more to the bonding MOs than the carbon orbitals. The latter in turn contributes more to the antibonding MOs. Thus, the HOMO in the CO-molecule is an orbital localized mainly on carbon atom agreeing with the Lewis-base character of this molecule (vide infra). The LUMO in turn is an antibonding π orbital of relatively low energy which actually acts as an acceptor orbital as it occurs in the back-bonding in metal carbonyl species (vide infra).

1.2.6 Polyatomic Molecules

LCAO-MO theory is specially appropriate for describing polyatomic molecules in which multiple bonds occur. Strict application of the theory often demands very complex calculations in which all the valence orbitals in a field formed by the nuclei and the rest of the electrons should be considered. However there are some simplifications that make its application easier.

Most bonds in a great number of molecules may be supposed to be two center/two electron bonds. In a first approximation it may be considered they do not interact with the rest of the molecule so the MO-calculations can be restricted to multicenter delocalized bonds. Thus, for instance, in the description of the NO_2 molecule the molecular skeleton ONO may be viewed as built up by two binuclear σ-bonds arising from the overlapping, for instances, of sp^2 atomic hybrid orbitals. It remains therefore one p-orbital perpendicular to the molecular plane on each atom. By combining these three p-orbitals three molecular orbitals extended over the entire molecule are obtained. Approximately shapes of these MOs are illustrated in Fig. 1.17. In the case of the NO_2 molecule there are 17 valence electrons. Four electrons form the σ-framework; 10 electrons are located as lone pairs in the atomic hybrid orbitals; and the remaining three electrons are therefore placed in the three π molecular orbitals described above. The electron configuration of the NO_2 molecule is $(\psi^b)^2 (\psi^n)^1 (\psi^*)^0$ therefore it is a paramagnetic species. The chemistry of NO_2 agrees with this description. It can for instance, act, as an oxidant forming the species NO_2^- with an electron configuration $(\psi^b)^2 (\psi^n)^2 (\psi^*)^0$, or undergo dimerization giving the diamagnetic species N_2O_4.

In Fig. 1.18 a bonding scheme for the cyclo-pentadienil anion, a species widely used as ligand in organometallic chemistry, is illustrated. Each carbon is assumed to use its s, p_x, and p_y orbitals and three of its electrons to build up the σ-bonded framework. Remaining p_z orbitals form various MOs extended over the planar ring. The stability of this anion and the strongly oxidant properties of the neutral hydrocarbon agree with both orbital distribution and electron configuration schemed in the figure.

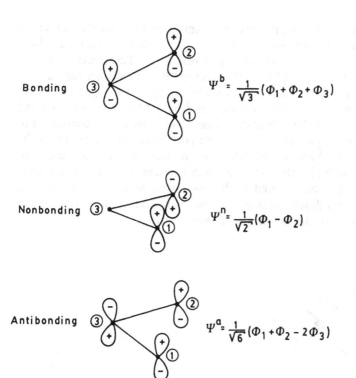

Bonding $\quad\quad \psi^b = \frac{1}{\sqrt{3}}(\Phi_1 + \Phi_2 + \Phi_3)$

Nonbonding $\quad\quad \psi^n = \frac{1}{\sqrt{2}}(\Phi_1 - \Phi_2)$

Antibonding $\quad\quad \psi^a = \frac{1}{\sqrt{6}}(\Phi_1 + \Phi_2 - 2\Phi_3)$

Fig. 1.17. π-MO sketches for the NO_2 molecule. σ-bond skeleton is assumed to be formed by two-electrons bonds

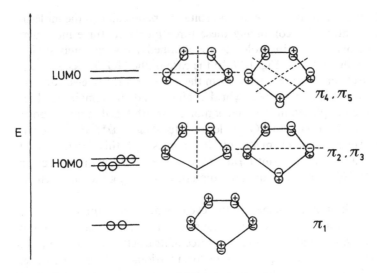

Fig. 1.18. π-molecular orbital sketches and corresponding energy diagram for the cyclopentadienyl anion

Multicenter bonds are not privative of systems with π symmetry. MO theory is indeed very useful in the description of electron deficient systems as those frequently found in borane and carborane chemistry. The structure of the diborane B_2H_6 is described in Fig. 1.19. Terminal B–H bonds are normal two center/two electron bonds, but the bonds with bridging hydrogen atoms are three center/two electron bonds, i.e. electron-deficient linkages. As described schematically in Fig. 1.20 these three center bonds in B_2H_6 can be considered to arise from the combination of two hybrid boron orbitals and the orbital 1s of the hydrogen atom. These hybrid orbitals one on each boron atom could be a mixture of an sp^2 hybrid with a p-orbital. In this system the energy minimum is reached with two electrons located in the low laying MOs of the scheme on Fig. 1.20. Further MO descriptions of *closo* boron hydrides will be analyzed in Chapter 2 in relation to the bonding in cluster species.

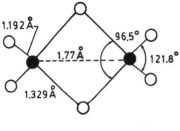

B_2H_6

Fig. 1.19. Schematic structure of diborane

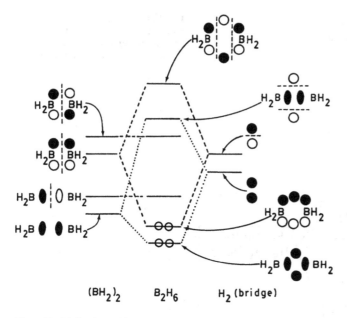

Fig. 1.20. Molecular orbital energy diagram and molecular orbital sketches for diborane

1.2.7 Transition Metal Complexes

Molecular orbital theory may also be successfully used for qualitative bonding descriptions of mononuclear metal complexes, for instance, of species of type $[ML_n]^m$ in which the central atom is a transition metal ion and L are neutral or charged donor ligands. There are nine valence metal orbitals that can be combined with the ligands orbitals according to their symmetries to form bonding, non-bonding, and antibonding molecular orbitals. Figure 1.21 describes a qualitative molecular orbital diagram for an octahedral species ML_6 where L are electron pair donors. In principle this scheme is very similar to those described above. However the notation used for identifying the molecular orbitals is a little more complex. This notation arises from the symmetry properties of the system. Analytically, delocalized molecular orbitals correspond to linear combinations of both metal and ligand atomic orbitals. However for a pictorial description it is more convenient to think of these combinations occurring in two steps. In the first one, ligand atomic orbitals are linearly combined with each other leading to a new set of delocalized orbitals. These orbitals can be ordered and labeled using the irreducible representation symbols of the symmetry point group of the molecule, in this case O_h. Although these symmetrized atomic orbitals may be clearly differentiated by their symmetry properties they are in the complex at a distance great enough to be considered to have practically the same energy (negligible orbital overlapping) as the original orbitals. Figure 1.22 describes the orbitals arising from the combination of

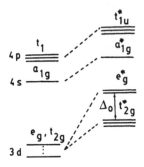

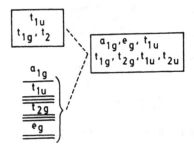

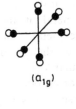

Fig. 1.21. Qualitative MO-energy diagram for a complex ML_6 where M is a transition metal ion and the ligands L are both σ and π donors

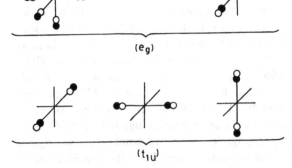

Fig. 1.22. Sketches of a set of symmetrized $p\sigma$-ligands in complex ML_6 with octahedral (O_h) molecular symmetry

a symmetry-equivalent set of six p ligand atom orbitals in the same octahedral arrangement they will have in the complex. In a second step the metal valence orbitals in the central atom are combined with the symmetrized atomic orbitals as exemplified by the schemes in Fig. 1.23.

Returning to molecular orbital diagram in Fig. 1.21, a series of features of special relevance for understanding metal transition complex chemistry are apparent:

1. Because of their relatively large energy the contribution of s and p metal valence orbitals to metal-ligand bonding stabilization is relatively poor, hence the bonding arises principally from t_{2g} and e_g molecular orbital interactions. That implies that populations of the t_{2g}^* and e_g^* will normally affect negatively the metal-ligand bond energy.
2. Low-lying molecular orbitals always have ligand character, and in turn practically all those metal-like in character are antibonding molecular orbitals. For a central atom with electronic configuration d^n HOMO as well LUMO have metal character so the complex chemistry will normally occur at least initially at the metal atom.
3. Color and magnetic properties which are characteristics of the chemistry of transition metals depend fundamentally on the energy gap between the t_{2g}^* and e_g^* levels normally called Δ. This gap which normally has a magnitude of 25–100 kcal mol^{-1} directly affects spectral absorption frequencies e.g. in the electronic spectra of metal complexes with high-spin (vide infra) electron configurations d^1, d^4, d^6, and d^9 and molecular symmetry O_h only a single d^*–d^* band is observed and it has an energy Δ. In the assignment of the ground state configuration in transition metal complexes with configurations

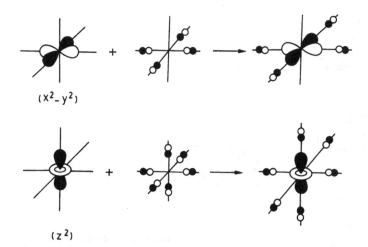

Fig. 1.23. Examples of symmetry-allowed combinations between metal and symmetrized ligand orbitals in an octahedral transition metal complex

d^n with n = 4, 5, 6, 7, or 8 the value of Δ is also critical. If Δ is fairly small, electron repulsions will be minimized according to Hund's rule assigning electrons to the e^* molecular orbitals and thus giving *high spin* electron configurations. On the other hand if Δ is large enough, the lowest energy configuration results in a *low spin* electron configuration by filling first the low-lying t_{2g}^* molecular orbitals.

The d orbital splitting Δ is affected by any factor which alters the metal-ligand bonding. Among them the π-acceptor properties of the ligands is very important. Ligand with low-lying empty orbitals with a symmetry π respect to the metal-ligand axis can experience back-donation of electron density from the metal into these orbitals. As can be observed in the molecular orbital diagram shown in Fig. 1.24, such an interaction involves specially the t_{2g} molecular orbitals affecting thus directly the magnitude of Δ. Typical examples of π-acceptor ligands are the phosphine and arsines with their empty d-orbitals or many organometallic ligands. Among the latter the most important is doubtless carbon monoxide in which the π-acceptor function is carried out by a low-lying C–O antibonding orbital. Electron density transfer from metal into the ligands

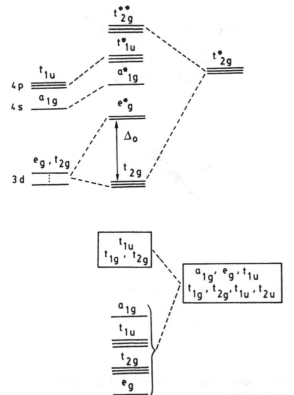

Fig. 1.24. Qualitative MO-energy diagram for an octahedral transition metal complex ML$_6$ when L are π acceptor ligands

is the determining factor for stabilizing metal low oxidation states being therefore of great relevance in the chemistry of transition metal clusters to be discussed in the next Chapter.

1.3 Bonding in Condensed Phases

1.3.1 Molecular Solids

In the preceding section the bonding between atoms was analyzed by different theories or models. There attention was focused mainly on building up molecules which have been considered as discrete aggregates of atoms having determinated chemical properties independently of their compound aggregation state. Moreover, it was systematically supposed such properties refer to species in an infinite diluted gaseous phase. This is an approximation that often finds justification because for many of these compounds the intermolecular interactions in condensed phases as solids, liquids, or solutions are weak in comparison with the strength of the bonding within the molecule.

Solids formed by discrete molecules are called molecular solids. Since all atomic valence orbitals are used in molecule bonding or occupied with non bonding electrons, the solid is held together only by van der Waals forces. Molecular solids have in general low melting temperatures thus reflecting the low bonding energy in the crystal.

1.3.2 Non-Molecular Solids

Besides the molecular solids discussed above, there are numerous and very important solid substances which are formed by extended arrays rather than by molecular units. Indeed, many of the elements are found under normal conditions as non-molecular solids built up as infinite atomic arrays.

Three limiting types of non-molecular solids may be distinguished:

1. Metallic solids. Infinite arrays of closely packed atoms held together by delocalized electrons. Each atom in a metal has a high coordination number, often eight or twelve. Physical properties of metals are quite different from those of other solid substances. They are good conductors of heat and electricity, have high reflectivity, and special mechanic properties such as strength and ductility.
2. Non-metallic network solids. They are infinite arrays of atoms held together by strong covalent bonds. No discrete molecules can be distinguished but the entire solid can be considered as a giant molecule. In general they are substances with very high melting temperatures and with poor properties as conductors of heat and electricity. Typical examples of this class of solids are

carbon in diamond (sublimation temp. $> 3550\,°C$) and SiO_2 (m.p. $1610\,°C$) and SiC (infusible). The coordination number of the atoms in the arrays is rather low being determined by their capacity of forming two center/two electron localized covalent bonds with their neighbors.

3. Ionic solids. They consist of infinite arrays of ions in which each ion is surrounded by a certain number of ions of the opposite sign; coordination numbers are often high, six or eight. In general ionic crystals are hard solids with high melting temperatures and behave as electric insulators (vide infra).

There are also other types of solids, for instance glasses and polymers, that cannot be classified into the groups defined above. These substances may be considered to be intermediate cases between the limiting situations described above. Specially in the cases of both non-metallic and ionic solids the instances for which such classifications match totally are very few. In most cases the behavior of actual solids may be better described by assuming an hypothetical mixture of these limiting cases.

The discussion on the properties of the different kinds of solids above and the analysis of the bonding models used for explaining these properties below are specially relevant for our task of understanding the chemistry of cluster compounds.

Cluster compounds – defined as a group of atoms directly linked to each other leading to a polyatomic nucleus that can be found either isolated or associated with a given number of ligands – should be considered to lie between the classical atomic and molecular compounds on the one hand and the macroscopic condensed non-molecular species on the other. In other words they may be considered as a small piece of an element that has been stabilized by the presence of ligands. Electronic structures and many other properties of these new kind of compounds should be also intermediate between properties of the pure elements and those of the classical mononuclear compounds.

1.3.3 Electrostatic Bonding

The third group in the classification mentioned above corresponds to a vast number of solid compounds which can be considered as aggregates of positive and negative particles interacting in an electrostatic manner.

Most ionic solids are formed by a combination of elements with very different electronegativities, for instance, between halide or oxide ions and cations from electropositive metals as those of the groups 1, 2, or from the transition series.

By considering perfectly ionic materials and determining the arrangements of the ions in the lattice it is possible to calculate in a relatively simple way the potential energy of the system.

The formation of ions, involving the ionization energy (IP) and the electro-affinity (EA) of the electropositive and electronegative atoms respectively, is an

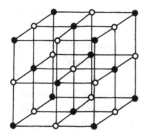

Fig. 1.25. Unit cell for NaCl

endothermic process that should be compensated by the anion-cation electrostatic interactions.

$$M(g) \xrightarrow{\text{PI}} M^+(g) + e^- \qquad M = Na \quad \Delta E = 495.91 \text{ kJ}$$

$$X(g) + e^- \xrightarrow{-\text{EA}} X^- \qquad X = Cl \quad \Delta E = -348.29 \text{ kJ}$$

In the energy calculations for ionic solids at least three factors should be considered: 1. The electrostatic energy determined mainly by the charge of the ions and the interatomic distances, 2. the arrangement of the ions in the solids, and 3. the interelectronic repulsive interactions between the ions at the bonding distances.

In a first approximation a hard-sphere electrostatic model, i.e. ions spherical in shape, incompressible, and having a defined size, can be considered. Thus, when the ions are brought together in large numbers, according to the compound stoichiometry, they will arrange themselves spontaneously trying to minimize the potential energy of the system, i.e. the minimum volume for reaching the least electrostatic energy.

In the case of compounds 1:1 with the two ions relatively similar in size to those in sodium chloride, NaCl, the best arrangement may be described as two interpenetrating face-centered cubic lattices. As observed in Fig. 1.25 which represents a unit cell for NaCl, in this structure each ion is surrounded octahedrally by 6 counter-ions.

1.3.4 Metal Bonding

Metallic elements are normally found forming close packed lattice structures of types body centered, face centered, or hexagonal close packed. All of them display high coordination numbers, either eight or twelve nearest neighbors. Furthermore metals show a number of other characteristic physical properties such as metallic luster and high thermal and electrical conductivities. Theories dealing with metals must therefore be able to explain these properties.

High coordination numbers prevent the application of models based in normal covalent electron pair bonding since there are neither sufficient orbitals nor sufficient electrons for the interaction of each atom with its neighbors. Ionic contributions also have to be discarded.

The most successful model of the electronic structure of metals is the theory known as the *band theory* of metals. This theory can be actually considered as an extended form of the molecular orbital theory.

For instance in the case of any alkali metal the interaction of two atoms, each with one valence electron ns[1], leads to a dinuclear molecule that can be described by a molecular orbital diagram similar to that of the hydrogen molecule with one bonding and one antibonding molecular orbital. If a third atom is incorporated a diagram with three molecular orbitals containing a non bonding MO should be considered. And so succesively, the higher the number of atoms considered in the molecule, the greater the number of molecular orbitals and the smaller the difference in energy between them. This process on going from a dinuclear molecule to an enormous one with practical infinite number of atoms is described in Fig. 1.26. Thus in a crystal of a metallic element so many lattice orbitals are involved that the band of orbital energies of the valence electrons of the metal atoms approaches a continuum. The *energy gap* between bonding and antibonding molecular orbitals and hence the band width depend on the overlap of atomic orbitals at the interatomic distance in the lattice. Relatively wide bands are expected for valence orbitals meanwhile for the core ones they are expected to be narrow and separated by greater energy gaps. Furthermore, because of the relatively low differences in energy between metal valence orbitals their valence bands often overlap. That occurs for instance between the s and p band in alkali and alkaline-earth metals. Analogously, the $(n-1)d$, ns and np in many transition metals are expected to overlap. These features are described schematically in Fig. 1.27.

Alkali metal atoms in the crystal have half as many electrons as can be housed in the delocalized molecular orbital band. The presence of partially filled bands accounts for bonding and electrical conduction in metals. The application of an small potential difference between two regions of the metal is enough for

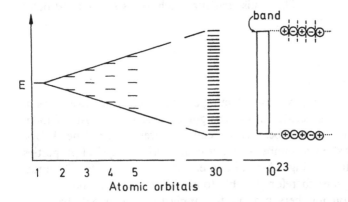

Fig. 1.26. Formation of an energy band by combination of atomic orbitals s in a linear atom chain

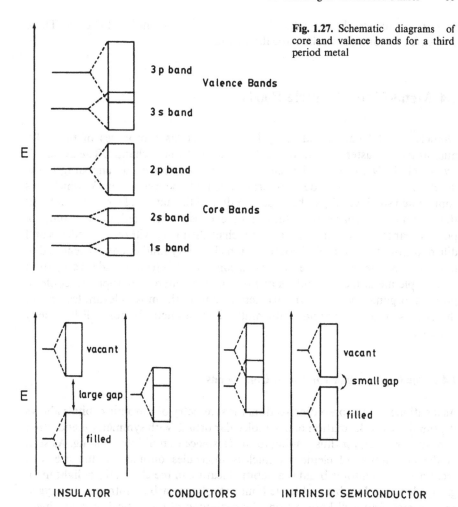

Fig. 1.27. Schematic diagrams of core and valence bands for a third period metal

Fig. 1.28. Band description of insulators, conductors and intrinsic semiconductors

exiting an electron to an unfilled level and sending it moving through the metal. A partially filled band is said to be a *conduction band*. Although in the case of the alkaline earth metals there are enough electrons to fill their valence band, metallic conductivity is observed because of the *s–p* band overlapping mentioned above. Conductivity in transition metals normally results from their unfilled *d*-bands.

In the case of systems in which the filled valence band and the next unoccupied band are separated by a considerable energy gap the electric potential required for the transition and hence to produce conductivity is large. Such material behaves as *insulators*. The intermediate case where the energy gap exist but is relatively small in comparison with thermal energies correspond to an *intrinsic semiconductor*. Silicon and germanium belong to this class. These concepts are compared schematically in Fig. 1.28.

1.4 Metal-Metal Multiple Bonds

Dinuclear metal compounds may be considered as prototypes of the poly-nuclear metal clusters which will be discussed in the next chapter. The availabil-ity of relatively simple molecular dinuclear compounds has encouraged the development of a great deal experimental and theoretical work which has contributed significantly to the understanding of the nature of metal-metal bond thus forming the basis of a branch of chemistry centered in the study of polynuclear metal compounds i.e. a branch of chemistry without precedents and different from the traditional coordination chemistry fundamentally centered in mononuclear chemical species. In such a context the detection and investigation of multiple metal-metal bonds is one of the more interesting topics in contem-porary inorganic chemistry. In this section some of the most relevant features of the chemistry of compounds with multiple metal-metal bonds will be briefly outlined.

1.4.1 Bonding in Dinuclear Metal Compounds

As mentioned in the preceding section, the symmetry of p atomic orbitals allows the simultaneous formation of one molecular orbital with symmetry σ and other two with symmetry π thus leading to the existence of multiple bonding. As seen in the description of element dinuclear molecules outlined in the previous section, such a multiple bond is a habitual feature in the chemistry of light main group elements. Double and triple bonds involving carbon, nitrogen or oxygen are common and well known from the beginning of theoretical organic chem-istry in the first third of the century. The knowledge about multiple bonding between metal atoms in which we are interested has only been developed in the last 30 years.

The possibility of forming high order bonds is indeed not privative of main group elements with valence orbital s and p but transition metals can also lead to multiple metal-metal bonding. In the orbital schemes illustrated in Fig. 1.29 the overlapping possibilities of d atomic orbitals can be appreciated. They can form indeed three types of molecular orbitals: orbitals type σ and π, i.e. with the same symmetry than those obtained from p atomic orbitals, and orbitals of the new type δ whose symmetry is defined by two nodal surfaces. In a given compound however, not all of these orbitals are always available for multiple metal-metal bonding. The number and symmetry of the substituents in the metal atoms should also be considered. Thus for instance in the species $[Re_2X_4]^{2-}$, the metal valence orbitals s, p and d_{x2-y2} are chiefly involved in metal-ligand

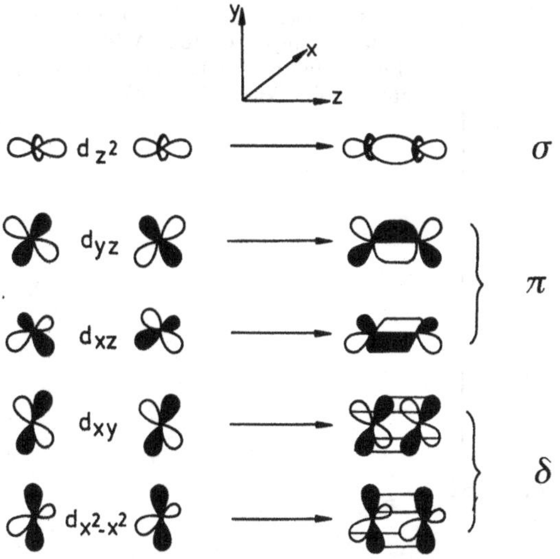

Fig. 1.29. Non-zero d–d overlaps between metal atoms

bonding. In this dinuclear dianion there are 8 valence electrons available for metal-metal bonding. The d-orbital overlapping of two adjacent metal atoms illustrated in the Figure above can be therefore an acceptable description of the metal-metal bond in this species.

According to the magnitudes of the overlapping integrals as well as to the nodal planes that characterize these molecular orbitals, their energies are expected to be in the order

$$\sigma \ll \pi < \delta < \delta^* < \pi^* \ll \sigma^*.$$

The electron configuration associated to the formation of this quadruple bond is therefore $\sigma^2 \pi^4 \delta^2$.

With 8 electrons the electron configuration with maximal bond order is reached. Thus species with bond orders 3.5, 3, 2.5, 2, 1.5 or 1 may in principle be obtained by either reduction or oxidation of dimetallic compounds with quadruple bonds.

1.4.2 Bonding Multiplicity and Internal Rotation

Before proceeding to the discussion of multiple metal-metal bonding it is convenient to analyze briefly the relationship existing between bond order and internal molecular rotation.

The rotation about a σ-bond is free because the atomic orbital overlapping originating this molecular bond does not change with rotation around the bond axis. In the case of species such as ethylene, where beside the σ-bond there is a π one, any rotation about the molecular axis will induce a diminution in the

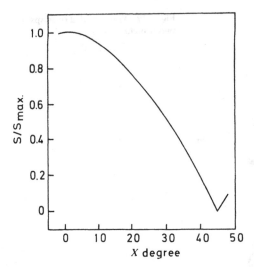

Fig. 1.30. Dependence of the δ overlap on the torsion angle. [Cotton FA, Walton RA (1982) Multiple bonds between metal atoms, John Wiley, New York]

overlapping of the p-orbitals forming the double bond and affecting thus the energy of the molecule. Systems with double bonds will tend therefore to acquire planar configurations. Nonetheless it is convenient to remember here that actual molecular configuration is determined by the totality of factors affecting the molecular energy among which steric hindrances often play an important role.

In a triple bond with an electron configuration $\sigma^2\pi^4$, the set of molecular orbitals has a cylindrical symmetry again so the system can rotate internally in a free way.

The energy of a molecular orbital with symmetry δ depend upon internal torsion angle (χ). The orbital overlapping degree changes with the torsion angle as described in Fig. 1.30. Accordingly the highest bond order in species $[Re_2X_8]^{2-}$ is achieved for an eclipsed ligand configuration ($\chi = 0$). However, the shape of the curve in Fig. 1.30 permits us to infer that small deviations from this configuration will do not affect significantly the overlapping integral. Consequently steric effects, which are maximal for an eclipsed and minimal for a staggered configuration, will be doubtless determinant for the actual configuration of the species.

1.4.3 Criteria for Establishing Multiple Metal-Metal Bonding

Before considering with some detail selected examples of compounds containing multiple metal-metal bonds, it is convenient to analyze the criteria usually used for determining the existence of multiple bonds. Most of these criteria are direct or indirectly related with structural data. That is the reason why this chemistry could be developed only when relatively easy access to reliable structural data became possible.

Intermetalic distances, exceptionally short in relation to the sum of covalent radii, are in general considered a reliable indication of a high order chemical bonds, specially when no bridging groups or atoms are present. Quantitative conclusions about bond order cannot be obtained from internuclear distances. However sometimes the differences between experimental intermetallic distances and those calculated for single linkages from covalent radii are great enough to constitute credible evidence for the presence of multiple bonds. Selected examples of metal-metal internuclear distances observed in compounds with quadruple metal-metal bonds are shown in Table 1.7.

Stereochemical evidence may be also useful in determining the existence of multiple metal-metal bond. In a dinuclear derivative $L_n M-ML_n$ the most stable configuration of the ligands, from the point of view of the interelectronic repulsions, is a staggered configuration. The presence of a multiple bond however can change the energetic balance making possible the existence of an eclipsed configuration (Fig. 1.31). Therefore in the absence of other factors, e.g.

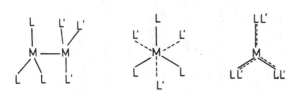

Fig. 1.31. Schematic representation of eclipsed and staggered ligand configurations in dinuclear metal complexes

Table 1.7. Selected internuclear bond distances in compounds with multiple metal-metal bonds

Compound	r (M–M) (Å)	Ref	Remarks
$Li_2Re_2Me_8 \cdot 2Et_2O$	2.178	1	r (Re–Re) in metal = 2.74Å
$Re_2Cl_6(PEt_3)_2$	2.222	2	
$Re_2(O_2CMe)_2Cl_4$	2.111	3	
$Re_2(O_2CCMe_3)_4Br_2$	2.234	4	
$K_4Mo_2Cl_8 \cdot 2H_2O$	2.139	5	r (Mo–Mo) in metal = 2.73Å
$Mo_2(O_2CMe)_4$	2.093	6	
$Mo_2(C_3H_5)_4$	2.183	7	
$Cr_2(O_2CMe_3)_4$	2.388	8	r (Cr–Cr) in metal = 2.50Å
$Cr_2(DMP)_4$	1.8747	9	DMP = dimethoxyphenyl anion
$Cr_2(TMP)_4$	1.849	10	DMT = trimethoxyphenyl anion

References

1 Cotton FA, Gage LD, Mertis K, Shive LW, Wilkinson G (1976) J. Am. Chem. soc. 98:6922
2 Cotton FA, Foxman BM (1968) Inorg. Chem. 7:2135
3 Kozmin PA, Suraz'hokaya MD, Larina TB (1979) Sov. J. Coord. Chem. 5:1201
4 Collins DM, Cotton FA, Gage LD (1979) Inorg. Chem. 18:1712
5 Brencic JV, Cotton FA (1969) Inorg. Chem. 8:7
6 Cotton FA, Mester ZC, Webb TR (1974) Acta Crystallogr. B30:2768
7 Cotton FA, Niswander RH, Sekutowski JC (1979) Inorg. Chem. 18:1149
8 Cotton FA, Extine MW, Rice GW (1978) Inorg. Chem. 17:176
9 Cotton FA, Koch SA, Millar M (1978) Inorg. Chem. 17:2087
10 Cotton FA, Millar M (1977) Inorg. Chim. Acta 25:L105

Table 1.8. Examples of binuclear complexes with multiple metal-metal bonding

Compound	Bond order	Multiple bond electron configurations	Remarks	Ref.
$Cr_2(O_2CR)_4L_2$	4	$\sigma^2\pi^4\delta^2$	$R = CH_3$; $L =$ donors e.g. H_2O, Py, etc. Bridging acetate groups. L are axial ligands.	1-4
$M'_4[Cr_2(CO_3)_4]$	4	$\sigma^2\pi^2\delta^2$	$M =$ alkalines, NH_4, Mg. Structure of dihydrated species are analogous to those of $Cr_2(O_2CR)_4L_2$	5-7, 4
$Cr_2(DMP)_4$	4	$\sigma^2\pi^2\delta^2$	DMP = 2,6-Dimethoxyphenyl anion. Very short metal-metal distance. $d(Cr-Cr) = 1.847$ Å.	8-10
$Li_4[Cr_2(CH_3)_8]\cdot 4THF$	4	$\sigma^2\pi^2\delta^2$	Compound without bridging ligands and a very short metal-metal distance (1.980 Å)	11, 12
$Cr_2(CO)_4Cp_2$	3	—	Low symmetry non-linear Cp-Cr-Cr-Cp axis and asymmetric CO bridges; short Cr-Cr distance (2.20 Å).	13-15
$Mo_2(C_2CR)_4$	4	$\sigma^2\pi^2\delta^2$	$R =$ H, Me, CMe_3, CF_3, C_6H_5. Structures similar to that of the acetate (Fig. 1.35). $d(Mo-Mo) = 2.09–2.10$ Å.	16, 17
$Mo_2(O_2CR)L_n$	4	$\sigma^2\pi^2\delta^2$	$n = 1$; $R = CF_3$, $L = PR_3$, $OPMe_3$, $n = 2$; $R = CHCl_2$, $L =$ Py, DMSO. Formation of adducts lows $\nu(Mo-Mo)$ in about 30 cm^{-1}	18-21
$[Et_4N]_n[Mo_2(O_2CCF_3)_4X_n]$	4	$\sigma^2\pi^2\delta^2$	$n = 1$; $X =$ Cl, Br, I, CF_3CO_2, $SnCl_3$. $n = 2$; $X =$ Br, I. Adducts with anions are in general more stable than those with neutral donors.	22
$K_4[Mo_2X_8]\cdot 2H_2O$	4	$\sigma^2\pi^2\delta^2$	$X =$ Cl, Br, SO_4. Anion $[Mo_2X_8]^{4-}$ are isostructural with $[Re_2Cl_8]^{2-}$. $d(Mo-Mo) = 2.138$ Å ($X =$ Cl).	23, 24
$Mo_2X_4(LL)_2$	3.5	—	$X =$ Cl, $L =$ dppe, arphos, dpae. IR spectra and structural determination ($X =$ Br, $L =$ arphos) show deviations of about 30° from eclipsed ligand configuration.	25-27
$Li_4Mo_2(CH_3)_8\cdot 4Et_2O$	4	$\sigma^2\pi^4\delta^2$	Centro-symmetric structure with eclipsed ligand configuration. $d(Mo-Mo) = 2.147$ Å	28
$Mo_2H_4(PMe_3)_6$	4	$\sigma^2\pi^4\delta^2$	Structure with $Mo_2(\mu-H)_2$ unit: $(Me_3P)_3HMo\cdot(\mu-H)_2MoH(PM_3)_3$. Very short Mo-Mo distance (2.19 Å)	29
$Mo_2(LL)_4$ Solv.	4	$\sigma^2\pi^4\delta^2$	LL = mhp (6-methyl-2-hydroxypyridine); dmhp (2,4-dimethyl-6-hydroxypyridine); map (2-amino-6-methylpyridin); $PyNHC(O)CH_3$ (N-2-pyridylacetamide); dmmp (4,6-dimethyl-2-mercapto-pyridin). Geometry around each Mo atom is trans-MoX_2Y_2, similar to Cr and W complexes. This class of compounds shows one of the shortest of the known Mo-Mo distances (2.03-2.07 Å).	30, 34

Compound	Bond order	Configuration	Description	Ref.
Mo$_2$R$_6$	3	$\sigma^2\pi^4$	R = CH$_2$SiMe$_3$. Mo$_2$C$_6$ skeleton with staggered ligand configuration (D$_{3d}$). d(Mo–Mo) = 2.167 Å.	35, 36
Mo$_2$(NR$_6$)$_6$	3	$\sigma^2\pi^4$	R = Me, d(Mo–Mo) = 2.211 and 2.217 Å.	37, 38
Mo$_2$(OR)$_6$	3	$\sigma^2\pi^4$	R = Bu^i, PhMe$_2$C, Me$_2$HC, Me$_3$CCH$_2$, Me$_3$Si, Et$_3$Si. Staggered ligand configuration. d(Mo–Mo) = 2.222 Å (R = Me$_3$CCH$_2$).	37, 39
Mo$_2$(CO)$_4$(Cp*)$_2$	3	—	Linear Cp*–Mo–Mo–Cp axis. d(Mo–Mo) = 2.448 Å.	40–43
Mo$_2$(OR)$_8$	2	$\sigma^2\pi^2$	R = Pr^i, Bu^i. d(Mo–Mo) = 2.523 Å.	44, 45
Mo$_2$[F$_2$PN(CH$_3$)PF$_2$]$_4$Cl$_2$	3	$\sigma^2\pi^4\delta^2\delta^{*2}$	Mo$_2$P$_8$ unit is twisted 21° from eclipsed configuration. d(Mo–Mo) = 2.457 Å, one of the longest known Mo–Mo triple bond.	46
W$_2$(O$_2$CCF$_3$)$_4$	4	$\sigma^2\pi^4\delta^2$	Very unstable species. Structure similar to Mo$_2$(O$_2$CCH$_3$)$_4$. d(W–W) = 2.221 and 2.207 Å (2 independent W$_2$ units).	47
Li$_4$[W$_2$(CH$_3$)$_8$]·4L	4	$\sigma^2\pi^4\delta^2$	L = ether, THF. Eclipsed ligand configuration. d(W–W) = 2.26 Å.	48–50
W$_2$(CH$_2$–SiMe$_3$)$_6$	3	$\sigma^2\pi^4$	M$_2$C$_2$ skeleton has staggered D$_{3n}$ configuration. d(W–W) = 2.255 Å.	51
W$_2$(NR)$_6$	2	$\sigma^2\pi^2$	d(W–W) = 2.479 Å, compatible with bond order 2.	52, 53
M$_2^+$[Re$_2$X$_8$]	4	$\sigma^2\pi^4\delta^2$	Eclipsed ligand configuration. d(Re–Re) = 2.222 Å (X = Cl, M = Bu$_4$N$^+$).	54–57
Re$_2$X$_6$L$_2$	4	$\sigma^2\pi^4\delta^2$	X = Cl, Br; L = PPh$_3$, PEtPh$_2$, PEt$_2$Ph, PEt$_3$, 2,5-dithiahexane, tetramethyl-thiourea. Re$_2$Cl$_6$(PEt$_3$) as a centrosymmetric eclipsed structure. d(Re–Re) = 2.222 Å.	58–61
Re$_2$(SO$_4$)(H$_2$O)$_2$	4	$\sigma^2\pi^4\delta^2$	Structure with sulphate bridges (Fig. 1.37). d(Re–Re) = 2.214 Å.	62
Re$_2$(O$_2$CR)$_4$X$_2$	4	$\sigma^2\pi^4\delta^2$	X = Cl, Br; R = alkyl, aryl. Structure with carboxilate bridges similar to that of the sulphate derivative. d(Re–Re) ca. 2.35 Å. Relatively weak axial-halide bonds.	63, 64
Re$_2$(O$_2$Cr)$_2$X$_4$	4	$\sigma^2\pi^4\delta^2$	X = Cl, Br; R = Me, CMe$_3$. cis-arrangement of bridging corboxylate groups. There are differences between halides in equatorial and axial positions. d(Re–X$_{ax}$) = 2.7 Å; d(Re–X$_{eq}$) = 2.31 Å (R = CH$_3$, X = Cl).	60, 65
Re$_2$(O$_2$CR)$_3$X$_3$	4	$\sigma^2\pi^4\delta^2$	X = Cl, R = CMe$_3$. Intermediate structure between those of Re$_2$(O$_2$CR)$_2$Cl$_4$ and Re$_2$(O$_2$CR)$_4$Cl$_2$. d(Re–Re) = 2.229 R = CMe$_3$, X = Cl).	66
Li$_2$Re$_2$(CH$_3$)$_8$·2Et$_2$O	4	$\sigma^2\pi^4\delta^2$	Eclipsed ligand configuration. Short intermuclear distance. d(Re–Re) = 2.17 Å.	67
Re$_2$(O$_2$CCH$_3$)$_4$(CH$_3$)$_2$	4	$\sigma^2\pi^4\delta^2$	Structure similar to Re(O$_2$CR)$_4$Cl with two bridging acetate groups and one acetate and one methyl group on each Re atom.	68
Re$_2$X$_4$(PR$_3$)$_4$	3	$\sigma^2\pi^4\delta^2\delta^{*2}$	X = Cl, Br, I; PR$_3$ = PMe$_3$, PEt$_3$, PPr$_3$, PBu$_3$, PPhMe, PPhEt. For X = Cl, R = Et: non-centrosymmetric eclipsed structure. d(Re–Re) = 2.232 Å.	58, 69, 70
Re$_2$Cl$_4$(dppe)$_2$	3	$\sigma^2\pi^4\delta^2\delta^{*2}$	dppe = 1,2 bis-(diphenilphosphino) ethane. Structure analogous to R$_2$Cl$_4$(PEt$_3$)$_4$ but with staggered ligand configuration.	71, 72
Re$_2$Cl$_5$(dth)$_2$	3	$\sigma^2\pi^4$	dth = dithiahexane. Very unsymmetrical, paramagnetic species [(dth)$_2$ClRe ≡ ReCl$_4$] with staggered ligand configuration. d(Re–Re) = 2.293 Å.	60, 73, 74

Table 1.8. (continued)

Compound	Bond order	Multiple bond electron configurations	Remarks	Ref.
$Re_2X_5(PR_3)_3$	3.5	$\sigma^2\pi^4\delta^2\delta^*$	X = Cl, Br; R = Me, Et, dppm [bis(diphenylphosphino)methane]. Structure with bridging dppm groups. Paramagnetic species. d(Re–Re) = 2.263 Å.	75
$Re_2Cl_6(dppm)_2$	2	$\sigma^2\pi^2\delta^2\delta^*$	di-(μ_2-Cl) bridged structure with Re–Re double bond. Re–Re intenuclear distance (2.616 Å) and the results of a theoretical treatment point to this electron configuration.	76, 77
$[Re_2X_4(LL)_2]PF_6$	3	$\sigma^2\pi^4\delta^2\delta^{*2}$	LL = dppe [bis(diphenylphosphino)ethane], or arphos [1-diphenylphosphino-2-diphenylarsinoethane]; X = Cl, Br, I, LL bridges the two metal center thereby conferring a staggered rotational geometry.	78
$Ru_2(O_2CC_3H_7)_4Cl$	2.5	$\sigma^2\pi^4\delta^2\pi^*\delta^*$	Paramagnetic species with structurally equivalent Ru atoms. d(Ru–Ru) = 2.281 Å.	79, 80
$[Ru_2L_2]Cl$	2.5	$\sigma^2\pi^4\delta^2\delta^{*2}\pi^*$	Paramagnetic species with 2 unpaired electrons. d(Ru–Ru) = 2.379 Å. With 1 unpaired electron d(Ru–Ru) = 2.267 Å	81-83
Ru_2L_2	2	$\sigma^2\pi^4\delta^2\delta^{*2}\pi^{*2}$	L = $[C_{22}H_{22}N_4]^{2-}$ = dibenzo tetraaza [14] annulene	81-83
$Os_2(hp)_4Cl_2 \cdot 2CH_3CN$	3	$\sigma^2\pi^4\delta^2\delta^{*2}$	hp = 2-hydrypyrine. Short Os–Os distance (2.357 Å).	84

References

1 Ocone LR, Block BP (1966) Inorganic Syntheses, vol 8. McGraw-Hill, New York, p 125
2 Herzog S, Kalies W (1967) Z. Anorg. Allg. Chem. 351:237
3 Cotton FA, Extine MW, Rice GW (1978) Inorg. Chem. 17:176
4 Cotton FA, Rice GW (1978) Inorg. Chem. 17:2004
5 Ouahes R, Amiel J, Suquet H (1970) Rev. Chim. Min. 7:789
6 Ouahes R, Pezerat H, Gayoso J (1970) Rev. Chim. Min. 7:849
7 Ouahes R, Devallez B, Amiel J (1970) Rev. Chim. Min. 7:855
8 Cotton FA, Koch S, Millar M (1977) J. Am. Chem. Soc. 99:7372
9 Cotton FA, Koch SA, Millar M (1978) Inorg. Chem. 17:2087
10 Cotton FA, Millar M (1977) Inorg. Chim. Acta 25:L105
11 Kurras E, Otto J (1965) J. Organomet. Chem. 4:114
12 Krausse J, Marx G, Schödl G (1970) J. Organomet. Chem. 21:159
13 King RB, Efraty A (1971) J. Am. Chem. Soc. 93:4951
14 King RB, Efraty A (1972) J. Am. Chem. Soc. 94:3773

15 Curtis MD, Butler WM (1978) J. Organomet. Chem. 155:131

16 Stephenson TA, Bannister E, Wilkinson G (1964) J. Chem. Soc. 2538

17 Cotton FA, Mester ZC, Webb TR (1974) Acta Crystallogr. B30:2768

18 Holste G, (1978) Z. Anorg. Allg. Chem. 438:125

19 Girolami GS, Mainz VV, Andersen RA (1980) Inorg. Chem. 19:805

20 Ketteringham AP, Oldham C (1973) J. Chem. Soc., Dalton Trans., 1067

21 Cotton FA, Norman Jr JG (1972) J. Am. Chem. Soc. 94:5967

22 Garner CD, Senior RG (1975) J. Chem. Soc., Dalton Trans., 1171

23 Brencic JV, Cotton FA (1969) Inorg. Chem. 8:7

24 Brencic JV, Leban I, Segedin P (1976) Z. Anorg. Allg. Chem. 427:85

25 Glicksman HD, Walton RA (1978) Inorg. Chem. 17:3197

26 Best SA, Smith TJ, Walton RA (1978) Inorg. Chem. 17:99

27 Sharp PR, Schrock RR (1980) J. Am. Chem. Soc. 102:1430

28 Cotton FA, Troup JM, Webb TR, Williamson DH, Wilkinson G (1974) J. Am. Chem. Soc. 96:3824

29 Jones RA, Chiu KW, Wilkinson G, Galas AMR, Hurthouse HB (1980) J. Chem. Soc., Chem. Commun., 408

30 Coton FA, Fanwick PE, Niswander RH, Sekutowski JC (1978) J. Am. Chem. Soc. 100:4725

31 Coton FA, Fanwick PE, Sekutowski JC (1978) Inorg. Chem. 17:3541

32 Coton FA, Fanwick PE, Sekutowski JC (1979) Inorg. Chem. 18:1152

33 Coton FA, Fanwick PE, Sekutowski JC (1979) Inorg. Chem. 18:1149

34 Cotton FA, Ilsley WH, Kaim W (1979) Inorg. Chem. 18:2717

35 Huq F, Mowat W, Shortland A, Skapski AC, Wilkinson G (1971) J. Chem. Soc., Chem. Commun, 1079

36 Mowat W, Shortland A, Yagupski G, Hill NJ, Yagupsky M, Wilkinson G (1972) J. Chec. Soc., Dalton Trans., 533

37 Chisholm MH, Reichert W (1974) J. Am. Chem. Soc. 96:1249

38 Chisholm MH, Cotton FA, Frenz BA, Reichert WW, Shive LW, Stults BR (1976) J. Am. Chem. Soc. 98:4469

39 Chisholm MH, Cotton FA, Murillo CA, Reichert WW (1977) Inorg. Chem. 16:1801

40 King RB, Bisnette MB (1967) J. Organomet. Chem. 8:287

41 Ginley DS, Bock CR, Wrighton MSA (1977) Inorg. Chim. Acta 23:85

42 Ginley DS, Wrighton MS (1975) J. Am. Chem. Soc. 97:3533

43 Klingler RJ, Butler WM, Curtis MD (1978) J. Am. Chem. Soc. 100:5034

44 Chisholm MH, Kirkpatrick CC, Huffman JC (1981) Inorg. Chem. 20:871

45 Chisholm MH, Cotton FA, Extine MW, Reichert WW (1978) Inorg. Chem. 17:2944

46 Cotton FA, Ilsley WH, Kaim W (1980) J. Am. Chem. Soc. 102:1918

47 Sattelberger AP, McLaughlin KW, Huffman JC (1981) J. Am. Chem. Soc. 103:2880

48 Cotton FA, Koch S, Mertis K, Millar M, Wilkinson G (1977) J. Am. Chem. Soc. 99:4989

49 Collins DM, Cotton FA, Koch S, Millar M, Murillo CA (1977) J. Am. Chem. Soc. 99:1259

50 Collins DM, Cotton FA, Koch S, Millar M, Murillo CA (1978) Inorg. Chem. 17:2017

51 Chisholm MH, Cotton FA, Extine M, Stults BR (1976) Inorg. Chem. 15:2252

52 Reagan WJ, Brubaker CH (1970) Inorg. Chem. 9:827

Table 1.8. (continued)

53 Anderson LB, Cotton FA, DeMarco D, Fang A, Ilsley WH, Kolthammer BWS, Walton RA (1981) J. Am. Chem. Soc. 103:5078
54 Cotton FA, Curtis NF, Johnson BFG, Robinson WR (1965) Inorg. Chem. 54:326
55 Cotton FA, Curtis NF, Robinson WR (1965) Inorg. Chem. 4:1696
56 Cotton FA, Harris CB (1965) Inorg. Chem. 4:330
57 Cotton FA, DeBoer BG, Jeremic M (1970) Inorg. Chem. 9:2143
58 San Filippo Jr J (1972) Inorg. Chem. 11:3140
59 Ebner JR, Walton RA (1975) Inorg. Chem. 14:1987
60 Cotton FA, Oldham C, Walton RA (1967) Inorg. Chem. 6:214
61 Cotton FA, Foxman BM (1968) Inorg. Chem. 7:2135
62 Cotton FA, Frenz BA, Shive LW (1975) Inorg. Chem. 14:649
63 Bennett MJ, Bratton WK, Cotton FA, Robinson WR (1968) Inorg. Chem. 7:1570
64 Collins DM, Cotton FA, Gage LD (1979) Inorg. Chem. 18:1712
65 Brant P, Salmon DJ, Walton RA (1978) J. Am. Chem., Soc. 100:4424
66 Cotton FA, Gage LD, Rice CE (1979) Inorg. Chem. 18:1138
67 Cotton FA, Gage LD, Mertis K, Shive LW, Wilkinson G (1976) J. Am. Chem. Soc. 98:6922
68 Hursthouse MB, Abdul Malik KM (1979) J. Chem. Soc, Dalton Trans, 409
69 Cotton FA, Frenz BA, Ebner JR, Walton RA (1974) J. Chem. Soc., Chem. Commun., 4
70 Cotton FA, Frenz BA, Ebner JR, Walton RA (1976) Inorg. Chem. 15:1630
71 Ebner JR, Tyler DR, Walton RA (1976) Inorg. Chem. 15:833
72 Cotton FA, Stanley GG, Walton RA (1978) Inorg. Chem. 17:2099
73 Bennett MJ, Cotton FA, Walton RA (1966) J. Am. Chem. Soc. 88:3866
74 Bennett MJ, Cotton FA, Walton RA (1968) Proc. R. Soc. A303:175
75 Bennett MJ, Brencic JV, Cotton FA (1969) Inorg. Chem. 8:1060
76 Barder TJ, Cotton FA, Lewis D, Schwotzer W, Tetrick SM, Walton RA (1984): J. Am. Chem. Soc. 106:2882
77 Shaik S, Hoffmann R, Fisel RC, Summerville RH (1980): J. Am. Chem. Soc. 102:4555
78 Cotton FA, Walton RA (1985) Struct. Bonding 62:1
79 Stephenson TA, Wilkinson G (1966) J. Inorg. Nucl. Chem. 28:2285
80 Bennett MJ, Caulton KG, Cotton FA (1969) Inorg. Chem. 8:1
81 Warren LF, Goedken VL (1978) J. Chem. Soc. Chem. Commun., 909
82 Anderson RA, Jones RA, Wilkinson G, Husthouse MB, Abndul KM (1977), J. Chem. Soc., Chem., Commun., 283
83 Cotton FA, Thompson JL (1980) Inorg. Chim. Acta 44:L247
84 Cotton FA, Thompson JL (1980) J. Am. Chem. Soc. 102:6437

steric hindrances, an eclipsed configuration will be indicative of specially strong metal-metal interactions.

Theoretical considerations are not only useful but must be considered for establishing the existence of metal-metal multiple bonds. Atomic orbitals must have the symmetry and energy adequate for permitting the orbital overlapping required by multiple bonding. Simple, qualitative considerations are often adequate for such analysis. However sometimes calculations leading to the quantitative determination of electronic energy levels are also possible permitting us to outline theoretical models to be compared with experimental features.

Although each of the criteria outlined above may be individually not enough for determining multiple bond existence, the concordance between them and with other physical and chemical properties of the compounds can lead to reliable conclusions about bond multiplicity.

The presence of metal-metal multiple bonds in general competes with cluster formation where bonds between more than two metal atoms should be present. Analogously as occurring for main group elements, valence saturation though multiple interactions prevents the formation of homonuclear metal skeletons.

Selected examples of compounds in which the existence of multiple metal-metal bonding has been actually detected are described in Table 1.8. Most compounds with high bond orders are formed by rhenium and the group 6 elements, chromium, molybdenum and tungsten. The ligands in such compounds are in general relatively hard Lewis bases such as halides, carboxylic acids, amines etc.. Nevertheless in some cases π-acceptor ligands as carbonyl, phosphine, nitriles etc. are also found. The examples of compounds with metal-metal multiple bonds containing multidentate ligands of recognized coordination ability in traditional monuclear chemistry such as ethylenediamine are in general rare. However bidentate ligands with three bridging atoms as the carboxylic acids appear to have a favorable geometry for supporting metal-metal multiple bond formation (Fig. 1.32). Organometallic compounds with high metal-metal bond orders are scarce. However they are often found among those with metal-metal double bond.

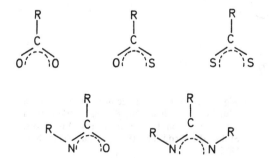

Fig. 1.32. Examples of triatomic bridging ligands frequently found in species with multiple metal metal bonding

1.4.4 Compounds with Bond Order Four

Because of orbital availability, bonding with order four is possible only between transition metals. Orbitals with angular quantum momentum number 2 or higher are necessary.

An acceptable description of multiple metal-metal bond can be obtained indeed by considering the overlap between d-orbitals of two adjacent metal atoms mentioned above. As described in Fig. 1.29, d atomic orbital overlapping occurring between two metal atoms at bond distances lead to molecular orbitals of type σ, π, and δ.

The first chemical species for which the existence of quadruple bond was determined is the anion $[Re_2Cl_8]^{2-}$. In Fig. 1.33 the structure of the *tert*-butylamonium salt of this anion is reproduced. In this structure, as well as in data in Table 1.7 it can be appreciated that the Re-Re distance is considerable shorter than that in metallic rhenium (2.75 Å). As observed in the structure in Fig. 1.33, chloride ligands do show an eclipsed configuration in spite of possible Cl–Cl interactions.

The molybdenum species $[Mo_2Cl_8]^{4-}$ has a structure similar to that of the rhenium derivative and interestingly, in both cases, there are only terminal ligands, so the short metal-metal distances cannot be attributed to ligand bridging effects. Quantitative theoretical descriptions of this kind of compounds have been achieved by a number of rigorous calculations. In all cases, results are qualitatively the same to those based on the symmetry properties of the metal atomic orbitals commented above. That can be clearly seen by observing the shapes of the molecular orbitals σ, π, and δ in the tetraanion $[Mo_2Cl_8]^{4-}$ whose contours are reproduced in the diagrams in Fig. 1.34. For somewhat more complex systems as the carboxylate derivatives $Mo_2(O_2CR)_4$ similar results are

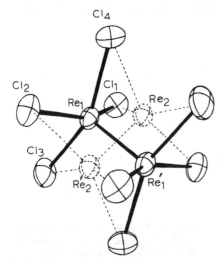

Fig. 1.33. Molecular structure of the anion $(Re_2Cl_8)^{2-}$ in $(t\text{-}Bu_4N)_2(Re_2Cl_8)$. Reproduced with permission from Cotton FA, Frenz BA, Stults BR, Wedd TR (1976) J. Am. Chem. Soc. 98:2768

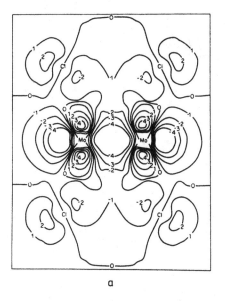

a

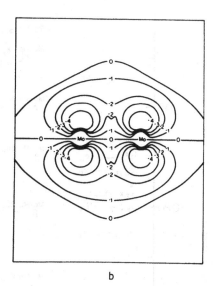

b

c

Fig. 1.34. Contour diagrams of (**a**) σ, (**b**) π, and (**c**) δ bonding orbitals in the anion $[Mo_2Cl_8]^{2-}$. Reproduced with permission from Norman Jr. JG, Kolari HJ (1975) J. Am. Chem. Soc. 97:33

obtained. In the molecular orbital diagram of the compound $Mo_2(O_2CH)_4$ reproduced in Fig. 1.35 it can be observed that HOMOs correspond to the orbitals δ, π, and σ, i.e. they have metallic character.

Spectroscopic properties of species with quadruple metal-metal bonds agree very well with theoretical descriptions. Low energy absorptions in UV-spectra normally assigned to δ–δ^* transitions as well as other transitions in the spectra accord well, at least qualitatively, with calculated MO schemes and confirm

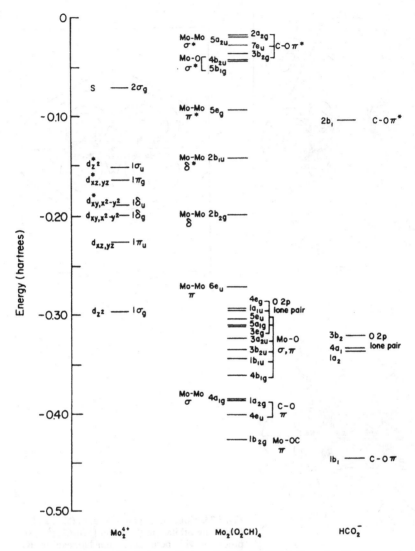

Fig. 1.35. Molecular orbital energy diagram for $Mo_2(O_2CH_4)$. Reproduced with permission from Norman Jr. JG, Kolari HJ, Gray HB, Trogler WC (1977) Inorg. Chem. 16:987

wholly assumed electron configurations. Direct measurements of orbital energy can be obtained from UV-photoelectron spectra. For compounds of type $Mo_2(O_2CR)_4$ the peak at lowest energy is normally assigned to photoelectrons arising from molecular orbitals with symmetry δ. Moreover, the differences in energy between δ and π photoelectrons accord with the corresponding ones deduced from theoretical calculations.

Although metal atoms in the octahalide derivatives appear to be coordinatively saturated, the tetracarboxylates $M_2(O_2CR)_4$ often show a certain

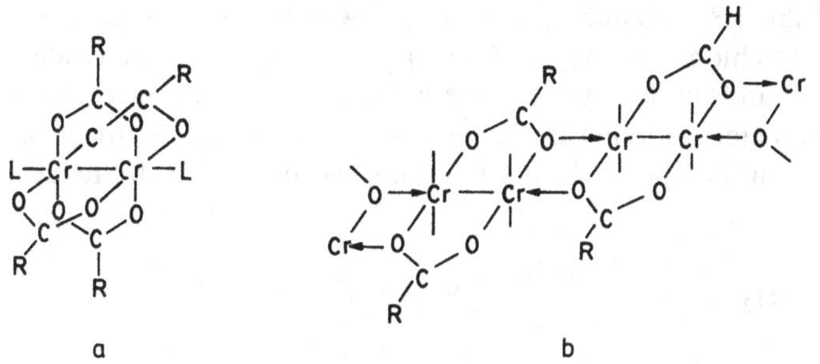

Fig. 1.36. Axial coordination of tetracarboxilates $Cr_2(O_2CR)_4$ leading (a) to discrete molecules, and (b) to polymeric aggregates

tendency to axial coordination bearing thus as shown in Fig. 1.36 to molecular species of type $M_2(O_2CR)_4L_2$ or to aggregates or polymerization by auto-association by oxygen bridges. Although for molybdenum derivative the bond M–M remains practically unaltered by effect of L, in the chromium derivatives the influence of L is noticeable. Table 1.9 shows selected structural data for adducts of $Cr_2(O_2CCH_3)_4$ which illustrate such influences permitting us to correlate the M–M internuclear distances with those of the corresponding M–L bonds. Interestingly, it can be observed that both M–M and M–L bonds appear to compete. Thus the metal-metal bond strength decreases with increasing M–L interaction. Theoretical calculations on both chromium and molybdenum carboxilates show that the influence of axial ligands L on Cr–Cr bonding arises from partial charge transfer from the L to antibonding MO δ^* in the complex. The effect is not observed in the molybdenum derivative where the HOMO-LUMO difference in energy is much greater than in the chromium species.

Table 1.9. Structural data of dichromium tetraacetate derivatives with axial ligands, $Cr_2(O_2CH_3)_4L_2$

L	d (Cr–Cr) (Å)	d (Cr–L) (Å)	Ref.
—	2.288 (2)	2.237 (4)	1
H_2O	2.362 (1)	2.272 (3)	2
CH_3CO_2H	2.300 (1)	2.306 (3)	3
Piperidine	2.342 (2)	2.338 (7)	3
Pyridine	2.369 (2)	2.335 (5)	4

References

1 Cotton FA, Rice CE, Rice GW (1977) J. Am. Chem. Soc. 99:4704
2 Cotton FA, DeBoer BG, La Prade MD, Pipal JR, Ucko DA (1971) Acta Crystallogr. B27:1664
3 Cotton FA, Rice GW (1978) Inorg. Chem. 17:2204
4 Cotton FA, Felthouse TR (1980) Inorg. Chem. 19:328

As observed in Table 1.8, all binuclear compounds with metal-metal quadruple bonds have the stoichiometry $[M_2L_8]^n$ which depending upon the oxidation state of the metal and on the nature of the ligands are cationic, neutral or anionic species. Since metal atoms in this class of compounds are in intermediate oxidation states they can be obtained from mononuclear derivatives by reductive condensations,

$$2[ReO_4]^- + 4H_2 + 8HCl \xrightarrow[380-330°C]{30 \text{ atm}} [Re_2Cl_8]^{2-} + 8H_2O$$

but also by oxidation of monuclear organometallic compounds.

$$2Mo(CO)_6 + 4HO_2CR \xrightarrow[\text{diglyme}]{} Mo_2(O_2CR)_4 + 12CO + 2H_2$$

$$2Cr(aq)^{2+} + 4(CH_3CO_2)^- + H_2O \xrightarrow{H_2O} Cr_2(O_2CCH_3)_4(H_2O)_2(s)$$

However, if a compound with metal-metal quadruple bonds is available, other species can be obtained by substitution processes.

$$[Re_2Cl_8]^{2-} + 4(CH_3CO_2)^- \rightarrow [Re_2(O_2CR)_4]^{2+} + 4HCl + 4Cl^-$$

The carboxylate derivatives may be used for preparing a number of other derivatives. In the chemistry of dichromium and dimolybdenum the following reactions have proved to be very useful.

$$[Mo_2(O_2CR)_4] + 8HX \xrightarrow{\text{conc. HX}} [Mo_2X_8]^{4-} + 4RCOOH + 4H^+$$

The addition of hard bases often permits ligand exchanges that, as mentioned above, can be a useful method of synthesis. The addition of soft bases however often leads to mononuclear compounds.

$$Mo_2X_4L_4 \xrightarrow[CH_2Cl_2]{NO} 2Mo(NO)_2X_2L_2$$

$$M_2(O_2CR)_4] \xrightarrow[\text{MeOH}]{CNPh} M(CNPh)_6 \qquad M = Mo, Cr.$$

$$Re_2(O_2CR)_4Cl_2 \xrightarrow[\text{MeOH, PF}_6^-]{CNR, \text{ reflux}} [Re(CNR)_6]PF_6 \qquad R = t\text{-Bu}$$

Probably, the reactions with soft Lewis bases leads to the formation of adducts which by subsequent restructuration yield the final degradation products:

This type of ligand addition in axial position is indeed relatively common in the chemistry of this class of compounds. The structures of the species illustrated in Figs. 1.37 and 1.38 exemplify such kind of adducts.

Quadruple bonded direhnium compounds possess a rich redox chemistry. As shown in the scheme in Fig. 1.39 the combination of chemical and electrochemical reactions leads to a variety of dimetal compounds.

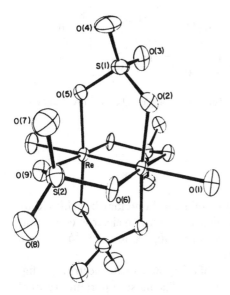

Fig. 1.37. Molecular Structure of the anion $[Re_2(SO_4)_4(H_2O)]^{2-}$ in the salt $Na_2Re_2(SO_4)_4 \cdot 8H_2O$. Reproduced with permission from Cotton FA, Frenz BA, Shive LW (1975) Inorg. Chem. 14:649

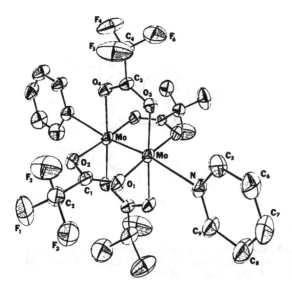

Fig. 1.38. Molecular structure of the compound $Mo_2(O_2CCF_3)_2(Py)_2$. Reproduced with permission from Cotton FA, Norman Jr. JG (1972) J. Am. Chem. Soc. 94:5697

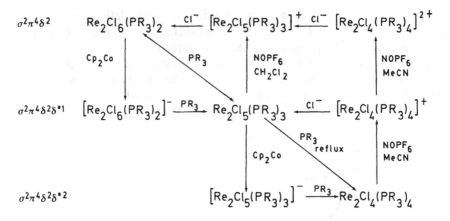

Fig. 1.39. Chemical oxidation-reduction processes of dirheniun complexes [Cotton FA, Walton RA (1985) Struct. Bonding 62:1]

1.4.5 Compounds with Bond Order Three

As mentioned above, species with a metal-metal triple bond may be achieved by both addition or substraction of electrons in species with an electron configuration $\sigma^2\pi^4\delta^2$. In both bond order three electron configurations $\sigma^2\pi^4\delta^2\delta^{*2}$ and $\sigma^2\pi^4$ the net δ component does not exist.

As observed in Table 1.8 most examples of compounds with electron configuration $\sigma^2\pi^4\delta^2\delta^{*2}$ arise from rhenium chemistry. The most important group of compounds with this configuration are those of stoichiometry $Re_2X_4(PR_3)_4$.

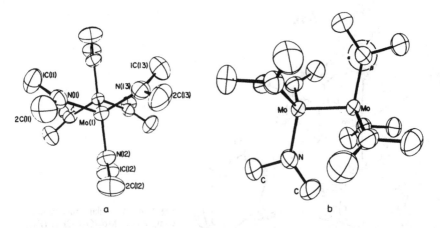

Fig. 1.40. Molecular Structure of $Mo_2(NMe_2)_6$ (a) front view emphasizing staggered ligand configuration and (b) side view. Reproduced with permission from Chisholm MH, Cotton FA, Frenz BA, Reichert WW, Shive LW Stults BR (1976) J. Am. Chem. Soc. 98:4469

Since this kind of electronic configuration does not restrict the rotation about the M–M bond, the spacial configuration is mainly determined by steric interaction of the ligands. Although in the compound $Re_2X_4(PEt_3)_4$ the ligands show an eclipsed configuration – probably due to interaction between phosphine groups – the interatomic distances as well as the spectroscopic properties clearly indicate a triple bond for the metal-metal linkage. Relatively sophisticated calculations as the application of relativistic Xα-SW calculations on the hypothetical molecule $Re_2Cl_4(PH_3)_4$ show that the HOMO actually corresponds to a M–M δ^* orbital. The redox behavior of rhenium compounds mentioned before (Fig. 1.39) also accords with the electron configuration assigned above.

Compounds of type M_2L_6 (M = Mo or W; L = N_2, OR) with structures as those illustrated in Fig. 1.40 possess a triple metal-metal bond with a $\sigma^2\pi^4$

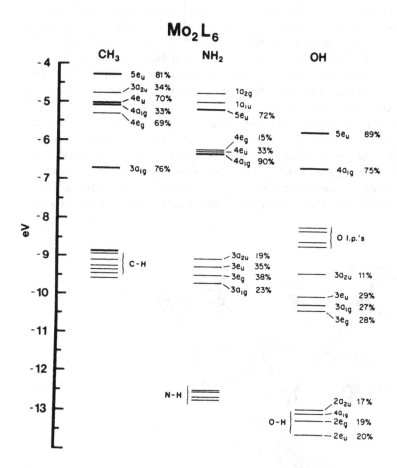

Fig. 1.41. Molecular orbital energy diagrams from SCF-Xα-SW calculations for $Mo_2(CH_3)_6$, $Mo_2(OH)_6$ and $Mo_2(NH_2)_6$ (only higher orbitals containing bonding and lone-pair electrons are shown). Reproduced with permission from Bursten BE, Cotton FA, Green JC, Seddon EA, Stanley GG (1980) J. Am. Chem. Soc. 102:4579

electron configuration. As observed in the Figure, ligands are found in these structures with a staggered configuration. This feature as well as the intermetalic distances agree with the assigned electron configuration. Results of theoretical calculations as those shown by the MO-schemes in Fig. 1.41 for three Mo_2L_6 molecules also accord with proposed electron configuration. In the schemes it can be observed that in the three cases the HOMO has a high metallic character which in turn is formed mainly by metal $d\pi$ contribution. For instance for the hydroxyl derivative the HOMO $5e_u$ is 89% metallic in character with a great contribution, 81%, from orbital $d\pi$.

1.4.6 Compounds with Bond Order Two

Most of the compounds with metal-metal double bonding do not arise directly from the analogue of higher bond order discussed above. Thus for instance in the chemistry of rhenium compounds the best known and also most interesting species are the Re(III) trinuclear clusters that will be discussed in the next chapter. In this sense the chemistry of compounds with metal-metal double bonds appears as a transition from the chemistry of dinuclear metal compounds to that of higher metal aggregates.

Group 5 elements in an oxidation state $+3$ form a series of dinuclear compounds which, according to their metal-metal internuclear distances and paramagnetic behavior, correspond to species with a double metal-metal bond and the electron configuration $\sigma^2\pi^2$. These π-orbitals are not always degenerate. Thus for instance, the compound $Ta_2Cl_6(THF)_3$ (Fig. 1.42) possesses a Ta-Ta internuclear distance corresponding to a double bond but it shows however

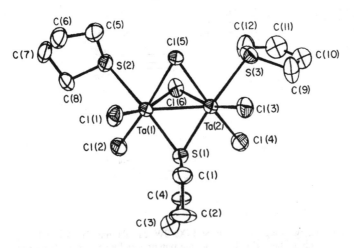

Fig. 1.42. Molecular structure of $Ta_2Cl_6(THT)_3$ (THT = tetrahydrothiophene). Reproduced with permission from Cotton FA, Najjar RC (1981) Inorg. Chem. 20:2716

diamagnetic behavior. Possibly the low system symmetry leads, in this case, to remove the π orbital degeneration.

The compound of ruthenium(II) with azulene Ru_2L_2 constitutes an example of electron-rich species with a metal-metal double bond. Thus for this compound, which has the same structure as its oxidized derivatives $(Ru_2L_2)BF_4$ and $(Ru_2L_2)(BF_4)_2$, an interatomic distance of 2.379 Å agreeing with a double metal-metal bond and an electron configuration $\sigma^2\pi^4\delta^2\delta^{*2}\pi^4$ is observed.

The number of organometallic compounds with a double metal-metal bond is notoriously higher than those with higher bond orders. This feature agrees with the relatively lower oxidation states of the metal in these compounds and it also points out to the idea of a transition toward the behavior observed in cluster compounds.

Chapter 2

Transition Metal Cluster Chemistry

The word *cluster* evokes the idea of a group, bunch, swelling, i.e. the idea of a set of equivalent individuals that through strong interactions among themselves can behave as a new entity clearly differentiated from the surroundings. In the chemistry, cluster are fundamentally a group of atoms – in the case of metal clusters, of metal atoms – directly linked to each other leading to a polyatomic metallic nucleus that can be found either isolated or associated with a given number of ligands. Some examples of this kind of compound are illustrated in Fig. 2.1.

Microcrystalline metal particles are species that agree well with the definition of isolated clusters. Investigations in this field have grown significantly in the last few years promoted to a great extent by the importance of such kinds of microcrystals for understanding and improving catalytic processes. However they have been studied from a physical rather than from a chemical point of view. Hence, the chemical knowledge about these systems is still incipient.

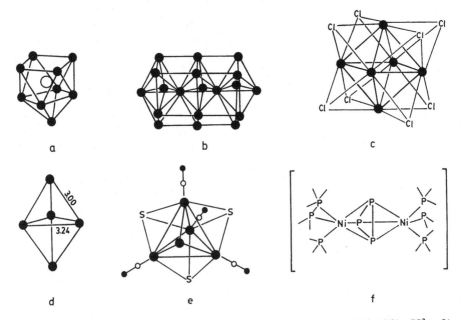

Fig. 2.1. Examples of different types of molecular clusters. Cluster of (**a**) $[Rh_9(CO)_{21}P]^{3-}$, (**b**) $[Pt_{19}(CO)_{22}]^{4-}$, (**c**) $[Mo_6Cl_8]^{4+}$, (**d**) Pb_5^{2-}, (**e**) $Fe_4S_4(NO)_4$, (**f**) $Ni_2 (triphos)_2(\mu - \eta^3\text{-}P_3)]^{2+}$. CO groups have been removed for clarity

For the benefit of clarity, this Chapter has been restricted fundamentally to the discussion of the chemistry of molecular transition metal clusters; no dinuclear compounds, which were analyzed to some extent in Chapter 1, nor microcrystalline metal particles are considered. For the same reason the main emphasis is given to homonuclear metal compounds. However, heteronuclear species with different transition metals or containing main group atoms are taken into account whenever they are useful for a better understanding of cluster chemistry.

The chemistry of cluster containing main group elements is discussed in Chapters 3 and 4. Moreover, selected aspects of the chemistry of some iron-sulfur clusters of great biological interest are analyzed in Chapter 5.

2.1 Classification of Clusters

Molecular clusters may be classified by considering the presence and nature of the ligands they exhibit. There is first a non-ligand class of isolated molecular cluster species known as *naked clusters* as those exemplified by the anions Pb_5^{2-} or Sn_5^{2-} in Fig. 2.1d. Strictly, naked clusters can be refered to only when these species are in a total inert medium or matrix. However the idea of naked can also be extended to species that existing in determined media, in solution or in the solid state, their interactions with the environment are considerably smaller that those among the constituents of the cluster. The naked cluster ions of post-transition metals, which under determined conditions can be stabilized in solution or in solid compounds, can be included in this class of molecular clusters. As mentioned above, the chemistry of this kind of compounds is described in Chapter 4. Nonetheless, most of the known molecular metal clusters are not bare metal conglomerates but follow a pattern similar to that found in classical metal complexes, i.e. a nucleus, consisting of, in this case, a metal network surrounded by a given number of ligands. These clusters can be classified by the type of ligands they contain.

There are two large classes of molecular clusters in which the ligands are all of the same nature: The halide cluster class and the carbonyl or, in general, organometallic cluster class. According to the formal oxidation numbers acquired by metal atoms in these two classes of clusters, they are often classified as high and low-valence clusters respectively.

In the high-valence cluster species, metal atoms forming the metal network appear to have positive intermediate oxidation states. However, metal oxidation states in cluster are always lower than those characteristic for the same metals in classical mononuclear complexes. The ligands associated to this class of cluster are normally good σ-donors that according to the classification of Pearson would have intermediate hardness. Among these compounds the most frequent are the halides, specially chlorides and bromides. An example of this cluster class is the molybdenum species $[Mo_6X_8]^{4+}$ shown in Fig. 2.1c.

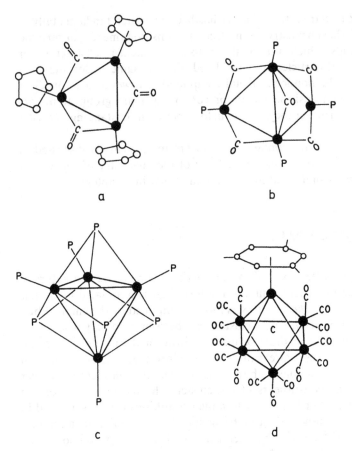

Fig. 2.2. Example of different types of low-valence molecular clusters. (a) $Rh_3(CO)_3(\pi\text{-}Cp)_3$, (b) $Pt_4(CO)_5[PPh(Me_2)]_4$, (c) $Co_4(PPh_3)_4(PPh)_4$, (d) $Ru_6C(CO)_{14}(C_6H_3Me_3)$

Low-valence clusters are species in which the metal atoms have an oxidation state zero or negative. They are always associated with ligands able to behave as π-acceptors. Some examples of cluster belonging this class are shown in Fig. 2.2. Carbon monoxide is by far the most representative of these ligands but there are also many examples of compounds containing other typical soft ligands such as phosphine, olefines, acetylenes, cyclopentadiene, etc.

2.1.1 Types of Ligands

In the next section, particular clusters with different nuclearity and shape will be analyzed. However, firstly, it is convenient to examine briefly the different types of ligands as well as the modes in which they are bonded to metal clusters. As will be seen in the numerous examples of metal clusters discussed in the next sections, the multitude of binding possibilities offered by organic ligands are in part responsible for the variety of species and structures observed for organometallic molecular clusters.

Carbon monoxide. As mentioned previously, the most important organic ligand in low-valence metal clusters is by far carbon monoxide. It is practically the only one that is able to form binary cluster compounds, $M_m L_n$. Moreover, almost all metal clusters containing other organic ligands are also metal carbonyls. Until now, there have been very few other binary organometal clusters. Most of them are isocyanide derivatives, a ligand very similar to carbon monoxide.

Structural determinations show that the CO-ligand in metal clusters can be bonded to the metal conglomerates fundamentally in the three modes illustrated schematically in Fig. 2.3:

1. Two-center or terminal carbonyl (μ_1-CO). The CO-group is bonded to one vertex of metal polyhedron. Bond M–C–O may be linear or bent.
2. Three-center or edge-bridging carbonyl (μ_2-CO). The CO-group is bonded simultaneously to two neighbour atoms in a metal polyhedron. The detection of bridging carbonyl groups often indicates the presence of metal-metal bonds.
3. Four-centered or triangular face-bridging carbonyl (μ_3-CO). The CO-group is bonded to three metal atoms sharing a face of the polyhedron.

A fourth mode of CO-coordination is also known, namely that found in the anion $[Fe_4(CO)_{13}H]^-$ in which, as can be observed in Fig. 2.4, one carbon monoxide ligand is coordinated by the carbon atom to all four Fe-atoms which are arranged in a butterfly configuration. As also indicated in Fig. 2.4, the CO-oxygen atom appears to be also bonded to the metal polyhedron.

While terminal and edge-bridged CO-ligands are normally found in monuclear and binuclear carbonyl complexes respectively, the face-bridging mode is characteristic for cluster complexes and, the bigger the nuclearity of the cluster, the more frequently μ_3-CO ligands are found.

Bridged CO ligands are commonly observed for the first and second-row transition metal complexes but they are however rare in neutral third-row metal

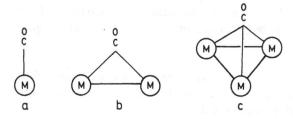

Fig. 2.3. Modes of bonding of carbon monoxide in metal clusters. (a) Terminal, (b) edge-bridging, (c) face-bridging

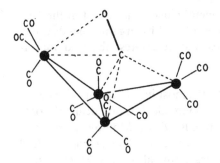

Fig. 2.4. Molecular structure of $[HFe_4(CO)_{13}]^-$ (Manassero M, Longoni E, Sansoni M (1976) J. Chem. Soc. Chem. Comm. 1976:919]

clusters. Indeed no bridging CO groups have been observed in binary neutral carbonyl cluster $M_m(CO)_n$ of the third-row metals. Only when the electron density on the metal atoms is augmented through substitution of the CO by better donor ligands or by development of formal negative charge, bridging CO interactions are generated. Such an effect is exemplified by the structures of the cluster species illustrated in Fig. 2.5. As observed, the dissipation of electronic charge into the ligands appears to play an important role in the distribution of the carbonyl ligands between terminal and bridging modes. Bridging carbonyl ligands are indeed better π-acids and more effective in removing electron density from the metal cluster than those bonded terminally.

The distinct bonding modes of CO-ligands appear to be associated with different π-acceptor capacities. That, similarly to mononuclear carbonyl complexes, is reflected by the strength of the CO bond. Thus, as the metal-carbon bonding changes from terminal to edge-bridging or to face-bridging an appreciable reduction of CO bond order is observed. Indeed carbon-oxygen bond distances are in general longer for both types of bridging species – about 20 pm for edge-bridging and 40 pm for face-bridging carbonyls – than the distances found for an equivalent terminal carbonyl. Frequency ranges for the stretching band $v(CO)$ in clusters are shown in Table 2.1.

Isocyanide. Isocyanide (RNC) are as ligand in transition metal complexes electronically similar to carbon monoxide. Nevertheless they are in general somewhat better donors than the carbonyl group. Terminal as well as three-center, edge-bridging bonding modes are known for isocyanides. Terminally isocyanides show practically linear RNC arrays; however in the bridging mode, bent RNC bonds are observed. Analogously to carbonyls, a reduction of the RNC bond order is produced on going from a terminal to a multicenter bridging bond.

Hydrocarbons. A variety of unsaturated hydrocarbons such as monoolefins, linear and cyclic dienes, cyclic tetraenes etc. have been found and structurally characterized as ligands bonded to metal clusters.

The scheme in Fig. 2.6 shows the modes of bonding more frequently observed for the coordination of ethylene or other simple monoalkenes to metal

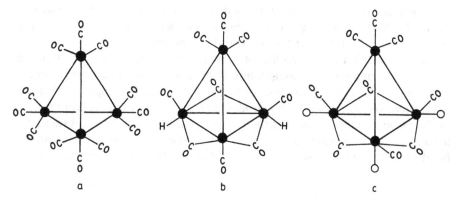

Fig. 2.5. Examples of the influence of cluster electron density on the bonding mode of CO in metal clusters. (a) $Ir_4(CO)_{12}$; (b) $[H_2Ir_4(CO)_{10}]^{2-}$; (c) $Ir_4(CO)_{10}(PPh_3)_2$

Table 2.1. Ranges for internuclear distances and stretching frequencies of carbon monoxide in metal cluster carbonyls

C–O Group	C–O Distance (Å)	ν(C–O) Frequency (cm^{-1})
terminal	1.12–1.19	2150–1950 (1850)[a]
edge-bridging	1.165–1.20	1900–1750 (1650)
face-bridging	1.19–1.22	1800–1700 (1600)

[a] In the presence of electron donors or negative charge lower frequency values are observed.

Reference

Chini P (1968) Inorg. Chim. Acta Rev. 2:31

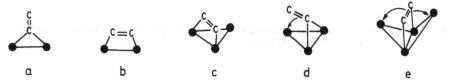

a   b   c   d   e

Fig. 2.6. Modes of bonding of alkenes in metal clusters

clusters. The simple conventional $p\pi$-$d\pi$ bonding well known for mononuclear complexes is practically not observed in metal cluster complexes. The coordination of olefines to clusters normally occurs with formation of metal-carbon bonds at the expense of olefine carbon-hydrogen bonds. Compounds with structures of type (a) and (b) in Fig. 2.6 are normally termed 1:1 and 1:2 complexes respectively, denoting thus the the number of carbon atoms involved in the σ-bonding to cluster. Cyclic monoene compounds bear only species with bonding mode (b) because mode (a) would imply a carbon-carbon bond rupture. Figure 2.7 illustrates the mode of bonding in other alkene systems. The bonding patterns are equivalent to modes (a) and (b) mentioned above.

Similar to alkenes, the modes of bonding of alkynes in clusters is more complex than in mononuclear compounds. In metal clusters, alkyne carbon atoms are normally bonded to two, three or four metal atoms. It is apparent that the interaction with the metal atoms produces an extensive orbital rearrangement on the carbon centers. Carbon-carbon distances are in this process substantially lengthened; in some cases distances near to those of a single carbon-carbon bond are observed. The scheme in Fig. 2.8 shows basic structures for alkyne

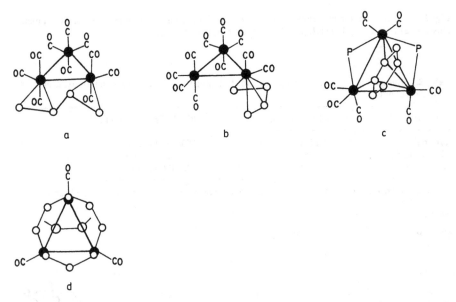

Fig. 2.7. Modes of bonding of some special alkenes in metal clusters. (**a**) *trans* and (**b**) *cis* isomers of $Os_3(CO)_{10}(C_4H_6)$; (**c**) $Os_3(CO)_7(PPh_2)_2(C_6H_4)$; (**d**) $Ni_3(CO)_3(CF_3CCCF_3)(C_8H_8)$

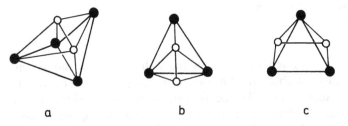

Fig. 2.8. Basic structures of some alkyne cluster complexes. (**a**) Bonded to a butterfly array of metal atoms bearing to an octahedral M_4C_2 framework; (**b**) and (**c**) isomeric forms of interaction with a M_3-cluster triangular face

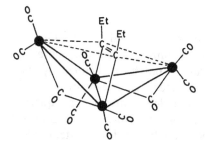

Fig. 2.9. Schematic representation of the structure of $Co_4(CO)_{10}(C_2H_5C = CC_2H_5)$. [Dahl LF, Smith DL, (1962) J. Am. Chem. Soc. 84:2450]

cluster complexes. Mode (a) involves an interaction of the ligand with four metal atoms leading to a butterfly array, and mode (b) or (c) with three metal atoms in either a triangular cluster or a M_3 triangular face in a larger cluster. As can be seen in the structure of the complex $Co_4(CO)_{10}(C_2H_5C\equiv CC_2H_5)$ shown in Fig. 2.9 the alkyne complex is practically a six-atom dicarbo-metal cluster.

2.2 Structural Characteristics of Molecular Metal Clusters

2.2.1 Structural Techniques

Among the techniques more frequently used for elucidating cluster structures are crystallographic studies by both single-crystal X-ray and neutron-diffraction techniques, multinuclear high resolution nuclear magnetic resonance (NMR), and infrared spectroscopy.

A great deal of the existing knowledge about structures, stereochemistry, and precise structural parameters of molecular clusters have been obtained by single crystal X-ray diffraction studies. To obtain adequate single crystals is the major limiting factor for the application of this technique. Precise location of the hydrogen atoms in hydride clusters often requires complementary neutron diffraction studies. Although the potential precision of crystallographic determinations is very high, it is not always accomplished because of disorder problems, small or poor quality of the crystals, or limited data sets.

High resolution nuclear magnetic resonance is the most effective structural technique for studies in solution. Although in some ideal cases, such studies can define structure and stereochemistry, it is not possible to obtain structural parameters by this method. Nuclear magnetic resonance techniques have proved to be an excellent method for studying dynamic processes in clusters.

Infrared analysis is employed largely as a diagnostic technique which is very useful, for instance, in determining the coordination mode of the ligands. Structural information as the molecular symmetry can be obtained by this method only occasionally when the clusters have a relative high symmetry and the use of isotope-substitution techniques is possible.

2.2.2 Metal Cluster Geometries

High symmetry is one of the principal characteristics of metal clusters. Generally, metal atoms are arranged in the cluster defining regular polyhedra as triangles, tetrahedra, octahedra etc.. Furthermore, polyhedra defined by the metal positions in metal clusters are largely deltahedra, i.e. polyhedra with all triangular faces. However, arrangements other than deltahedric such as square-planar or nearly square-planar clusters, trigonal prismatic, or C_{4v}-capped square-antiprism clusters are also represented though to a lower extent. The majority of metal atom arrangements M_m in clusters may be considered as representing fragments of any close-packed array of metal atoms, hexagonal close packing (hcp) or face centered cubic packed (ccp). Thus, relatively common cluster geometries such as triangle, tetrahedron, square-based pyramid, trigonal bipyramid, octahedron, etc. are formally fragments of a hcp array. Figure 2.10 illustrate some geometries found in molecular cluster structures and their relationship to close-packed arrays found in bulk metal.

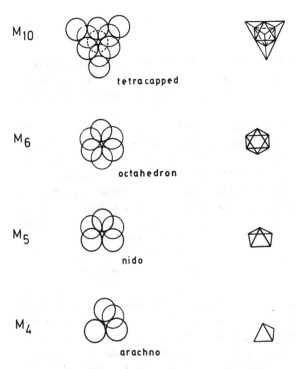

M_{10} tetracapped

M_6 octahedron

M_5 nido

M_4 arachno

Fig. 2.10. Some types of geometries frequently found in molecular cluster structures and their relationship to the close-packed arrays found in bulk metals (*arachno, nido* and tetracapped denominations refer to fundamental octahedral geometry)

2.2.3 Trinuclear Metal Clusters

The number of cluster species with a triangular array of metal atoms which have been characterized by X-ray diffraction analysis is rather large. Low as well as high-valence triangular clusters are known. The former are numerous and mostly carbonyl species; the later are practically restricted to rhenium halides. Selected examples of triangular trinuclear metal clusters are described in Table 2.2. In most cases the metal atoms in the cluster are equidistant, but there are also many examples in which the internuclear distances are not equivalent.

Table 2.2. Selected examples of trinuclear transition metal clusters

Compound	Remarks	Ref.
Nb_3Cl_8	Three-dimensional infinite structure formed by cross linking of Nb_3 triangles.	1
$[Nb_3(\mu\text{-Cl})_6(Cp^*)_3]^+$	Counterion chloride: Nb_3 triangle symmetrically bridged by μ_2-Cl atoms. d(Nb–Nb) 3.334 Å. With tetracyano-p-quixo-dimethane (tcnq) as counterion, a dication with similar molecular geometry is stabilized.	2, 3
$Nb_3Cl_3(\mu\text{-Cl})_3(\mu_3\text{-O})(\mu_3\text{-X})(Cp^*)_3$	X = OH, Cl. Triangular Nb_3 array in which one edge has one short (2.92 Å) and two long Nb–Nb distances (3.28 and 3.29 Å).	4
$Nb_3(CO)_6(\mu_3\text{-CO})(Cp)_3$	Distorted triangular Nb_3 unit. d (Nb–Nb) = 3.044–3.320 Å. Bridging CO ligand is really a $\eta^2\text{-}(\mu_3\text{-C}, \mu_2\text{-O})$ unit. d(CO) = 1.303 Å.	5
$[M_3(CO)_9(\mu\text{-CO})_3(\mu_3\text{-S})]^{2-}$	M = Cr, Mo and W (in Mo_2W). Cluster core is a tetrahedral M_3S unit with μ-CO ligands lying below M_3 plane.	6
$Zn_2Mo_3O_8$	Triangular Mo_3 units. d(Mo–Mo) = 2.53 Å.	7
$[Mo_3(\mu_3\text{-CMe})_2(\mu\text{-O}_2CMe)_6(H_2O)_3]^{2+}$	Triangular Mo_3 units d(Mo–Mo) = 2.814.	8
$Mo_3(\mu_3\text{-As})(\mu\text{-O})_3(Cp)_3$	Triangular array of Mo atoms d(Mo–Mo) = 2.666–2.809 Å.	9
$Mo_3(\mu_3\text{-S})(\mu\text{-S})_2(Cp)_3$	Triangular Mo_3 unit. d(Mo–Mo) = 2.812 Å.	10
$Mo_3(CO)_6(\mu_3\text{-CF})(Cp)_3$	Triangular array of Mo atoms.	11
$[W_3(\mu_3\text{-O})(\mu\text{-O})_2F_6]^{5-}$	Equilateral triangular W_3 unit. d(W–W) = 2.515 Å.	12
$W(R)_2(\mu_3\text{-O})(\mu\text{-OR}')_3(OR')_4$	R = CH_2Ph; R' = O-i-Pr. Triangular array of W atoms. Both hydrocarbyl ligands are equivalent.	13
$Mn_3(\mu_3\text{-NO})(\mu\text{-NO})_3(Cp)_3$	Triangular Mn_3-unit. d (Mn–Mn) = 2.506 Å.	15
$Mn_3(CO)_{12}(CH_3NN)$	Mn_3 bent configuration. μ_2 terminal-diazo-nitrogen. Second nitrogen coordinates the other $Mn(CO)_4$ fragment. d (Mn–Mn) = 2.826 and 2.807 Å. Angle Mn–Mn–Mn = 107.4 Å.	16
$[Mn_3(CO)_{14}]^-$	Linear Mn_3 unit (D_{4h}). d(Mn–Mn) = 2.895 Å.	17

Table 2.2. (continued)

Compound	Remarks	Ref.
Re_3Cl_9	Two dimensional polymeric sheets build by Re_3 units held together by chloride bridges (Fig. 2.13a). $d(Re-Re) = 2.49$ Å.	18
$[Re_3(\mu\text{-}Cl)_3Cl_9]^{3-}$	Equilateral triangular Re_3 unit. Cl-bridges in Re_3 plane plane (Fig. 2.13a). $d(Re-Re) = 2.47$ Å.	18
$Re_3(\mu_2\text{-}Cl)_3(CH_2SiMe_3)_6$	Re_3 units. $d(Re-Re) = 2.387$ Å.	19
$Re_3(\mu\text{-}Me)_3(CH_2Ph)_6$		20
$Fe_3(\mu\text{-}CO)_2(CO)_{10}$	Slightly distorted isosceles triangle (Fig. 2.11B) Bridged edge (2.558 Å) shorter than the mean of the other two (2.660 Å)	21
$Fe_3(CO)_9(\mu\text{-}CO)_2PPh_3$	Replacement of equatorial CO by PPh_3 in $Fe_3(CO)_{12}$. Asymmetry and $d(Fe-Fe)$ is retained.	22
$Fe_3(CO)_9(\mu_3\text{-}CO)(\mu_3\text{-}NSiMe_3)$	Triangular bicapped Fe_3 unit. $d(Fe-Fe) = 2.535$ Å.	23
$[Fe_3(CO)_9(\mu\text{-}CO)(\mu_3\text{-}CO)]^{2-}$	Triangular Fe_3 unit. $d(Fe-Fe) = 2.596$ Å.	24
$Fe_3(CO)_9(\mu_3\text{-}S)_2$	Open Fe_3 triangle. $d(Fe-Fe) = 2.58$ and 2.61 Å.	25
$Fe_3(CO)_9(\mu_3\text{-}SM(CO)_5(\mu_3\text{-}PBu^t)$	M = Cr, W. Fe atoms define open triangle capped by both, μ_3-PR and μ_3-$SM(CO)_5$ fragments. $d(Fe-Fe) = 2.65$–2.69 Å and 3.49 Å (non bonding).	26
$Fe_3(CO)_6(\mu_3\text{-}SC_6H_4NN)_2$	Three Fe atoms form a linear cluster. Each heteroatom, S and N, bridges two Fe-atoms but not the same pair. $d(Fe-Fe) = 2.494$ Å.	27
$Fe_3(\mu\text{-}CO)_3(\mu_3\text{-}NO)(Cp^*)_3$	Triangular array of Fe atoms. $d(Fe-Fe) = 2.533$ Å	28
$Ru_3(CO)_{12}$	All carbonyl groups are terminal (Fig. 2.11a). $d(Ru-Ru) = 2.854$ Å	29
$Ru_3(CO)_{12-n}(CNBu^t)_n$	n = 1,2. Triangular Ru_3 unit. $d(Ru-Ru) = 2.847$–2.866 Å.	30
$Ru_3(CO)_{10}(\mu\text{-}NO_2)_2$	Open triangle Ru_3. Di-bridged Ru-Ru is non bonding (3.15 Å). $d(Ru-Ru) = 2.866$ Å.	31
$Ru_3(CO)_6(\mu\text{-}CO)_3[MeSi(PBu_2)_3]$	Equilateral triangular Ru_3 unit. $d(Ru-Ru) = 2.917$. Silyl group lies above Ru_3 plane bonded with one P atom to each metal atom.	32
$Ru_3(CO)_8L_4$	L = PMe_2Ph and $P(OR)_3$ (R = Me, Et, Ph). Non-carbonyl ligands occupy equatorial positions in Ru_3 triangle.	33
$Ru_3(CO)_6(\mu\text{-}CO)(C_{10}H_8)$	$d(Ru-Ru) = 2.740$ (bridged) and 2.944 (unbridged) azulene coordinate to one Ru atom through a five-member ring, and to the other two through the seven-member ring.	34
$Ru_3(CO)_4(C_8H_8)_2$	Isosceles triangle of Ru atoms. $d(Ru-Ru) = 2.938$ and 2.782 Å each cyclo-octatetraene is bonded to a pair Ru atoms.	35
$[Ru_3(PMe_3)_8(\mu\text{-}CH_2)_4]^{2+}$	Linear arrangement of Ru atoms linked by bridging methylene groups and Ru-Ru bonds (2.637 Å)	36

Table 2.2. (continued)

Compound	Remarks	Ref.
$[Ru_3(\mu_3\text{-OMe})_2(Cp^*)_3]^-$	Triangular Ru_3 unit bridged symmetrically by two μ-OMe groups. d(Ru–Ru) = 2.958 Å.	37
$Os_3(CO)_{12}$	Isostructural with $Ru_3(CO)_{12}$ (Fig. 2.11a)	38
$Os_3(CO)_{11}\{P(C_6F_5)_3\}$	Os atoms form an approximate isosceles triangle. Phosphine ligand coordinated equatorially. d(Os–Os) = 2.867–2.920 Å.	
$Os_3(CO)_{11}(NCMe)$	Triangular array of Os atoms d(Os–Os) = 2.829–2.951 Å.	39
$Os_3(CO)_8(CS)(\mu_3\text{-S})_2$	Os_3 core form a wide "V". One edge of the Os_3 triangle is rather large (3.642 Å). μ_3-S ligands cap either side of the Os_3 triangle. d(Os–Os) = 2.780–2.830 Å.	40
$Os_3(CO)_{12}I_2$	Sraight chain of Os atoms. Equatorial Iodine atoms occupying trans positions to each other on terminal Os atoms. d(Os–Os) = 2.93 Å.	41
$[Co_3(CO)_6(\mu\text{-CO})_3(\mu_3\text{-CO})]^-$	Co atoms form a regular triangle. d(Co–Co) = 2.462–2.489 Å.	42
$Co_3(CO)_9(\mu_3\text{-PPh})$	Triangular Co_3 unit capped by the μ_3-PPh ligand. d(Co–Co) = 2.717 Å.	
$(\{Co_3(CO)_9(\mu_3\text{-C})\}_2SCO)$	Two $Co_3C(CO)_9$ units joined by a C(O)-S-chain d(Co–Co) = 2.480–2.472 Å.	43
$Co_3(CO)_6(\mu\text{-PCy}_2)_3$	Cy = Cyclohexyl. Co_3 triangle. The three phosphido groups lie above, in, and below the Co_3 plane. d (Co–Co) = 2.57 Å.	44
$Co_3(\mu\text{-CO})_2(\mu_3\text{-CO})(Cp)_3$	Triangular Co_3 unit. d (Co–Co) = 2.438–2.519 Å.	45
$[Co_3(\mu_3\text{-X})(\mu_3\text{-Y})(Cp)]^{n+}$	X = Co, Y = S n = 0; d(Co–Co) = 2.452 Å. X = Y = S, n = 0; av. d(Co–Co) = 2.687 Å. n = 1; d (Co–Co): two long (2.649) and one short (2.474 Å) distances.	46
$Co_3(\mu_3\text{-}\eta^1\text{-CS}) (\mu_3\text{-S}) (Cp)_3$	Triangular array of Co atoms with a triple bridging thiocarbonyl ligand. d(Co–Co) = 2.43 Å.	47
$Rh_3(CO)_6(PPh_2H)(\mu\text{-PPh}_2)_3$	Triangular array of Rh atoms. However internuclear distances $[d$(Rh–Rh) = 3.130–3.24 Å] imply negligible metal-metal bonding interactions. Metal and phosphor atoms form a near planar six-member Ph_3P_3 ring.	48
$[Rh_3(CO)_5(\mu_3\text{-PMe})(PMe_3)_4]^+$	Equilateral Rh_3 triangle capped, at most, symmetrically. This species is part of the compound $FeRh_8(CO)_{20}(PMe) (PMe_3)_5$ formed by this cation and the anion $[FeRh_5(CO)_{15}(PMe_3)]^-$.	49
$Rh_3(CO)_5(\mu\text{-PPh}_2)_3$	Triangular array of Rh atoms. d(Rh–Rh) = 2.698–2.806 Å.	50
$Rh_3(\mu\text{-CO})_3(Cp)_3$	Equilateral triangular unit. d(Rh–Rh) = 2.62 Å. There is other isomer which is isomorphous with $Co_3(\mu\text{-CO})_2(\mu_3\text{-CO}) (Cp)_3$.	51
$Ir_3(CO)_6(Ph)(\mu\text{-dppm}) (\mu_3\text{-PPh})$	Triangular array of Ir atoms. d(Ir–Ir) = 2.768–2.779 Å.	52

Table 2.2. (continued)

Compound	Remarks	Ref.
$Ir_3(CO)_3(Cp)_3$	Triangular Ir_3-unit. $d(Ir-Ir) = 2.6693-2.876$ Å.	53
$Ni_3(\mu_3-CO)_2(Cp)_3$	Triangular Ni_3 unit. $d(Ni-Ni) = 2.530$.	54
$Ni_3(\mu_3-CEt)(Cp)_3$	Triangular array of Ni atoms. $d(Ni-Ni) = 2.325-2.349$ Å.	55
$Ni_3(\mu_3-S)_2(Cp)_3$	Equilateral triangular Ni_3 unit. $d(Ni-Ni) = 2.80$ Å.	56
$Ni_3(\mu-NCBu^t)(Cp)_3$	Irregular Ni_3 triangle. $d(Ni-Ni)$ ranging from 2.334 to 2.386 Å.	57
$Pd_3(CO)_2(\mu-t-Bu_2P)_3Cl$	Triangle of Pd atoms. Two atoms bear a terminal CO group while the third is bonded to a Cl-ligand. $d(Pd-Pd) = 2.975$ Å.	58
$[Pd_3(PEt_3)_3(\mu-PPh_2)_2(\mu-Cl)]^+$	Isosceles Pd_3 unit. $d(Pd-Pd) = 2.93$ (twice), 2.89 Å. Cl-ligand is bridging shorter edge.	
$Pt_3(\mu-CO)_3[P(C_6H_{11})_3]_4$	Slightly distorted isosceles triangle. $d(Pt-Pt) = 2.675, 2736$ and 2.714 Å.	59
$Pt_3(ButNC)_6$	Equilateral triangular Pt_3 unit. $d(Pt-Pt) = 2.531$ Å.	60
$Pt_3(PEt_3)_4(PhC_2Ph)_2$	V-shaped arranged Pt-atoms. $d(Pt-Pt) = 2.905$ Å; angle $Pt-Pt-Pt = 144°$	61
$[Cu_3(\mu_3-\eta^1-C \equiv CPh)(\mu-dppm)_3]^{2+}$	Triangular array of Cu atoms. $d(Cu-Cu)$: two short (2.817 and 2.904 Å) and one larger (3.274 Å) distances suggest negligible metal–metal bonding interactions.	62
$[Au_3(\mu-(CH_2)_2PPh_2)_4]^+$	Chain of Au atoms in an spirocyclic arrangement of two diaurophospha cycles $[Au_2((CH_2)_2 PPh_2)_2]$ oriented at 90° relative to each other. $d(Au-Au) = 3.050$ Å.	63
$[Au_3(\mu-CPMe_3)(PPh_3)_3]^+$	Triangle of Au atoms which are about equidistant from common C atom. $d(Au-Au) = 3.18$ Å.	64

References

1 Schäfer H, Schnering HGV (1964) Angew. Chem. 76:833
2 Churchill MR, Chang SWY (1974) J. Chem. Soc., Chem. Commun., 248
3 Goldberg SZ, Spwack B, Stanley G, Eisenberg R, Braitsch RD, Miller JS, Abkowitz M (1977) J. Am. Chem. Soc. 99:110
4 Curtis MD, Real J (1988) Inorg. Chem. 27:3176
5 Herrmann WA, Biersack H, Ziegler ML, Weidenhammer K, Siegel R, Rehder D (1981) J. Am. Chem. Soc. 103:1692
6 Darensbourg DJ, Zalended DJ, Sanchez KM, Delord T (1988) Inorg. Chem. 27:821
7 McCarroll WH, Katz L, Ward R (1957) J. Am. Chem. Soc. 79:5410
8 Ardon M, Bino A, Cotton FA, Dori Z, Kaftory M, Kolthammer BWS, Kapon M, Reisner G (1981) Inorg. Chem. 20:4083
9 Neumann HP, Zeiglar ML, (1988) J. Chem. Soc., Chem. Commun., 498
10 Vergamini PJ, Vahrenkamp H, Dahl LF (1971) J. Am. Chem. Soc. 93:6327
11 Schulze M, Hartl H, Seppelt K (1987) J. Organomet. Chem. 319:77
12 Mennemann K, Mattes R (1976) Angew. Chem. Int. Edn. 15:118
13 Chisholm MH, Folting K, Eichhorn BW, Huffman JC (1987) J. Am. Chem. Soc. 109:3146
14 Hohman Von M, Krauth-Siegel L, Weidenhammer K, Schulze W, Ziegler ML (1981) Z. Anorg. Allg. Chem. 481:95
15 Elder RC (1974) Inorg. Chem. 13:1037

Table 2.2. (continued)

16 Herrman WA, Ziegler ML, Weidenhammer K (1977) Angew. Chem. Int. Edn. 15:368
17 Bau R, Kirtley SW, Sorrell TN, Winarko S (1974) J. Am. Chem. Soc. 96:998
18 Cotton FA, Mague JT (1964) Inorg. Chem. 3:1403 and 3:1094
19 Hursthouse MB, Malik KMA (1978) J. Chem. Soc., Dalton Trans., 1334
20 18/113 Wong ACC, Edwards PG, Wilkinson G, Reevallie M, Hurthouse MB (1988) J. Chem. Soc., Dalton Trans., 219
21 Cotton FA, Troup JM (1974) J. Am. Chem. Soc. 96:4155
22 Raper G, McDonald WS (1971) J. Chem. Soc. (A) 3430
23 Barnett BL, Krüger C (1971) Angew Chem. 83:969
24 Lo FYK, Longoni G, Chini P, Lower LD, Dahl LF (1980) J. Am. Chem. Soc. 102:7691
25 Wei CH, Dahl LF (1965) Inorg. Chem. 4:493
26 Winter A, Zsolnai L, Huttner G (1983) J. Organomet. Chem. 250:409
27 Pannell KH, Mayr AJ, van Derveer DU (1983) J. Am. Chem. Soc. 105:6186
28 Muller J, Sonn I, Akhnoukh T (1989) J. Organomet. Chem. 367:133
29 Churchill MR, Hollander FJ, Hutchinson JP (1977) Inorg. Chem. 16:2655
30 Bruce MI, Matisons JG, Wallis RC, Patrick JM, Skelton BW, White AH (1983) J. Chem. Soc., Dalton Trans., 2365
31 Norton J, Collman J, Dolcetti G, Robinson WT (1972) Inorg. Chem. 11:382
32 de Boer JJ, van Doom JA, Masters C (1978) J. Chem. Soc., Chem. Commun., 1005
33 Bruce MI, Liddell MJ, Bin Shawkataly O, Bytheway I, Skelton BW, White AH (1989) J. Organomet. Chem. 369:217
34 Churchill MR, Scholer FR, Wormald J (1971) J. Organomet. Chem. 28:C21
35 Bennett MJ, Cotton FA, Legzdins P (1968) J. Am. Chem. Soc. 90:6335
36 Jones RA, Wilkinson G, Galas AMR, Hursthouse MB, Malik KMA (1980) J. Chem. Soc., Dalton Trans., 171
37 Kolle U, Kossakowski J, Boese R (1989) J. Organomet. Chem. 378:449
38 Churchill MR, DeBoer BG (1977) Inorg. Chem. 16:878
39 Dawson PA, Johnson BFG, Lewis J, Puga J, Raithby PR, Rosales MJ 1982) J. Chem. Soc., Dalton Trans., 233
40 Broadhurst PV, Johnson BFG, Lewis J, Raithby PR (1980) J. Organomet. Chem. 194:C35
41 Cook N, Smart L, Woodward P (1977) J. Chem. Soc., Dalton Trans., 1744
42 Adams HN, Fachinetti G, Strähle J (1980) Angew. Chem., Int. Edn. 19:404
43 Gervasio G, Rossetti R, Stanghellini PL, Bor G (1983) J. Chem. Soc., Dalton Trans., 1613
44 Albright TA, Kang SK, Arif AM, Bard AJ, Jones RA, Leland JK, Sohwab ST (1988) Inorg. Chem. 27:1246
45 Bailey WI, Cotton FA, Jamerson JD, Kolthammer BWS (1982) Inorg. Chem. 21:3131
46 Frisch PD, Dahl LF (1972) J. Am. Chem. Soc. 94:5082
47 Werner H, Leonhard K, Kolb O, Röttinger E, Vahrenkamp H (1980) Chem. Ber. 113:1654
48 English RB, Haines RJ, Steen ND (1982) J. Organomet. Chem. 238:C34
49 Podlahova J, Podlaha J, Jegorov A, Hasek (1989) J. Organomet. Chem. 359:401
50 Haines RJ, Steen NDCT, English RB (1984) J. Chem. Soc., Dalton Trans., 515
51 Paulus EF, Fisher EO, Fritz HP, Schuster-Woldan H (1967) J. Organomet. Chem. 10:C3
52 Harding MM, Nicholls BS, Smith AK (1983) J. Chem. Soc., Dalton Trans., 1479
53 Shapley JR, Adair PC, Lawson RJ, Pierpont CG (1982) Inorg. Chem. 21:1701
54 Maj JJ, Rae AD, Dahl LF (1982) J. Am. Chem. Soc. 104:3054
55 Lehmkuhl H, Krüger C, Pasynkiewiez S, Poplawska J (1988) Organometallics 7:2038
56 Vahrenkamp H, Uchtman VA, Dahl LF (1968) J. Am. Chem. Soc. 90:3272
57 Kamijyo N, Watanabe TW (1974) Bull. Chem. Soc. Japan, 47:373
58 Arif AM, Heaton DE, Jones RA, Nunn CM (1987) Inorg. Chem. 26:4228
59 Abinati A, Carturan G, Musco A (1976) Inorg. Chim. Acta 16:L3
60 Green M, Howard JAK, Murry M, Spencer JL, Stone FGA (1977) J. Chem. Soc., Dalton Trans., 1509
61 Boag NM, Green M, Howard JAK, Spencer JL, Stansfield RFD, Stone FGA, Thomas MDO, Vicente J, Woodward P (1977) J. Chem. Soc., Chem. Commun., 930
62 Gamasa MP, Gimero JJ, Lastra E, Aquire A, Garcia-Granda S (1989) J. Organomet. Chem. 378:C11
63 Schmidbauer H, Hartmann C, Raber G, Muller G (1987) Angew. Chem. Int. Edn. 26:1146
64 Schmidbauer H, Scherbaum F, Huber B, Müller (1988) Angew. Chem. Int. Edn. 27:419

Carbonyl Metal Clusters. The equivalence of metal atoms in the clusters and thus the symmetry of the arrangement is determined by the distribution of the carbonyl ligands between the terminal and bridging modes discussed above. In the compounds $Os_3(CO)_{12}$ and $Ru_3(CO)_{12}$ (Fig. 2.11a) with all carbonyl groups as terminal ligands metal atoms form an equilateral triangle. However, in the equivalent iron compound, $Fe_3(CO)_{12}$ (Fig. 2.11b), there are both terminal and bridges carbonyl groups. The asymmetric distribution of carbonyl groups leads in this case to non-equivalent iron atoms so forming an isosceles triangle. There the Fe-Fe distances between CO-bridged iron atoms are shorter than those to the vertex with all terminal carbonyl groups.

There are numerous triangular metal complexes containing hydrogen. Structural features of hydride metal clusters will be separately described in Chapter 3. The study of cluster hydrides is of considerable interest since, as it will be discussed later in this Chapter (Sect. 2.6), they are intimately associated with important catalytic processes e.g. Fischer Troppch processes for ammonia and methane synthesis.

Many non-carbonyl ligands such as NO, S, PR, O or RC can be edge or face-bridging ligands in triangular metal clusters (Fig. 2.12). Phosphines PR_3 are typical donor terminal ligands in trinuclear triangular clusters.

High-Valence Trinuclear Rhenium Clusters. Trinuclear rhenium arrangements with triangular geometries are stabilized by π-donor ligands as being the halide ions. The X-ray diffraction analysis of rhenium(III) chloride shows that the compound is not a simple ionic compound but a polymer built up by $[Re_3Cl_9]$ units bonded by chloride bridges (Fig. 2.13a). The same type of structure with three edge-bridging chlorides and three terminal ligands has the substituted monomeric species $Re_3Cl_9L_3$ (Fig. 2.13b). As it can be observed from data in Table 2.2, the intermetallic distances in these rhenium clusters are about 2.48 Å, i.e. ca 0.27 Å shorter than in the metal (2.75 Å) and significantly shorter than those expected for simple Re-Re bonds. This feature is interpreted in terms of double bonds between the metal atoms and explained, as it will be seen in the next section, by considering the number of electrons disposable for building the cluster structure.

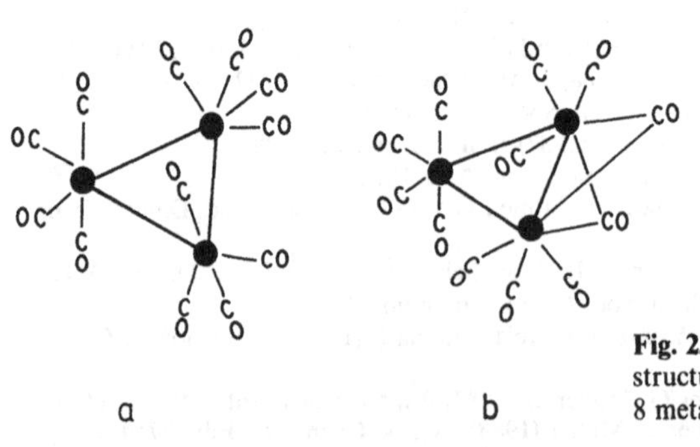

Fig. 2.11. Schematic representation of the structures of dodecarbonyls of group 8 metals (a) Ru, Os; (b) Fe

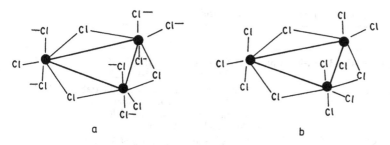

Fig. 2.12. Schematic structures of (a) Rhenium (III) chloride and (b) the anion $[Re_3Cl_{12}]^{3-}$

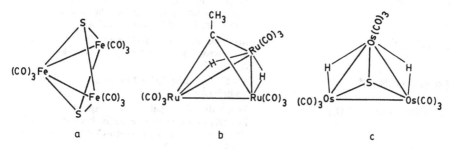

Fig. 2.13. Examples of structures of triangular metal clusters containing main-group bridging ligands

2.2.4 Tetranuclear Metal Clusters

Selected examples of tetranuclear clusters are described in Table 2.3.

Many of the known tetranuclear clusters show a tetrahedral arrangement of metal atoms with six metal-metal bonds. However, there are also other geometries. For instance, the type of structure observed for the compound $Ru_4(CO)_{11}$ (μ_4-C_8H_{10}) illustrated in Fig. 2.14a known as "butterfly" structures and which can be formally considered to result from the rupture of one of the tetrahedral metal-metal bonds. The interplanar dihedral angles in butterfly arrangements are not always small; there are indeed examples of practically planar butterfly structures as occurs for instance in the anion $[Re_4(CO)_{16}]^{2-}$ (Fig. 2.14b). Square-planar arrangement of the metal atoms, as that observed for the clusters $Mo_4(OP_r^i)_8Cl_4$ and $Fe_4(CO)_{11}(NEt)(ONEt)$ (Fig. 2.15), are also possible for tetranuclear clusters. Energy differences between butterfly and tetrahedral metal arrangements appear to be relatively small, thus dynamic rearrangement of the former through tetrahedral geometry is often possible.

Among tetrahedral metal clusters the carbonyl complexes are the more representative. Neutral carbonyl clusters of cobalt and rhodium have similar structures (Fig. 2.16) with nine terminal and three edge-bridging carbonyl

Table 2.3. Selected examples of tetranuclear metal clusters

Compounds	Remarks	Ref.
$Ti_4(\eta^1\text{-}\eta^5 C_5H_4)(\eta^5\text{-}\eta^5\text{-}C_{10}H_8)(\mu_3\text{-}N_2)(Cp)_5$	$[Ti_2Cp_2(\eta^5\text{-}\eta^5 C_{10}H_8](\mu_3\text{-}N_2)[Ti_2Cp_3(\eta^1\text{-}\eta^5\text{-}C_5H_4)]$; $v(N \equiv N) = 1282\ cm^{-1}$	1
$Ti_4(MeC_5H_4)(\mu_3\text{-}S)_4$	Cubane cluster: Interpenetrated tetrahedra of S and Ti atoms where the latter are bonded to terminal ligands. Two sets of d(Ti–Ti), 3.00 (four) and 2.93 Å (two).	2
$V_4(MeC_5H_4)(\mu_3\text{-}S)_4$	Interpenetrated tetrahedra of S and V atoms (cubane cluster). d(V–V) = 2.873 Å (av. two slightly different sets of distances).	2, 3
$Cr_4(\mu_3\text{-}S)_2(\mu_3\text{-}L)_2(Cp)_4$	Cubane-like structures; L = CO, S.	4
$Cr_4(\mu_5\text{-}O)_4(Cp)_4$	Cr atoms form an approximate tetrahedron capped by the Cp rings and by the O-atoms above each face. d(Cr–Cr) = 1.937 Å.	3, 5
$[Re_4(CO)_{16}]^{3-}$	Re-atoms define a parallelogram with Re–Re bond across the shortest diagonal. d(Re–Re) 2.596–2.96 Å.	6
$[Re_4(\mu_3\text{-}X)_4(CN)_{12}]^{4-}$	X = S or Se. d(Re–Re) = 2.755(S) or 2.805(Se) Å.	7
$[Fe_4(CO)_9(\mu\text{-}CO)_3(\mu_3\text{-}CO)]^{2-}$	Tetrahedral Fe_4 configuration. Apical $Fe(CO)_3$ group is symmetrically bonded (258 Å) to the basis $Fe_3(CO)_6(\mu\text{-}CO)_3(\mu_3\text{-}CO)$ [d(Fe–Fe) = 2.50 Å].	8
$Fe_4(CO)_{12}(\mu\text{-}CO)(\mu_4\text{-}C)$	Fe atoms in a butterfly configuration (dihedral angle 101°). d(Fe–Fe) = 2.545–2.642 Å.	9
$[Fe_4(CO)_{12}(\mu\text{-}C)]^{2-}$	Butterfly arrangement of Fe atoms (dihedral angle 102.4°) with C atom bridging the central Fe–Fe bond. d(Fe–Fe) = 2.533–2.637 Å.	10
$[Fe_4(CO)_{12}(\mu_4\text{-}CCO_2Me)]^-$	Fe_4 unit with butterfly configuration (dihedral angle 130°). d(Fe–Fe) = 2.430–2.489 Å.	11
$Fe_4(CO)_{12}(\mu_3\text{-}\eta^2\text{-}CS)(\mu_3\text{-}S)$	Three Fe atoms define an equilateral triangle; the fourth is terminally bonded to the triangle. CS-group is terminally bonded to Fe(4) via sulfur. The C atom is capping the Fe_3 triangle. d(Fe–Fe) = 2.564 (av. in the triangle) and 2.710 Å.	12
$Fe_4(CO)_{10}(\mu\text{-}CO)(\mu_4\text{-}PR)_2$	R = p-tolyl. Fe-atoms define a quadrilateral bicapped by the μ_4-PR groups. d(Fe–Fe): three ca. 2.69 Å and one shorter (2.42 Å) assigned to one double bond.	13
$Fe_4(CO)_{10}(\mu\text{-}CO)\{\mu_4\text{-}P(tol\text{-}p)\}_2$	Fe and P atoms define the vertices of a distorted octahedron with PR-groups in axial positions. One CO-group is bridging an equatorial Fe–Fe bond (2.44 Å). d(Fe–Fe) = 2.69 Å.	14
$Fe_4(CO)_8(py)_4$	Fe atoms define the vertices of a rhombus. Fe atoms on the shortest diagonal are each linked to 2 pyridine groups. d(Fe–Fe) = 2.55 Å.	15
$Ru_4(CO)_{12}(\mu\text{-}CO)(\mu\text{-}Cl)$	Butterfly arrangement of Ru atoms with a Cl atom bridging the two wing-type metal atoms. μ-CO group is bridging central Ru–Ru bond. d(Ru–Ru) = 2.832 Å.	16

Table 2.3. (continued)

Compounds	Remarks	Ref.
$Ru_4(CO)_{12}(C_2Ph_2)$	Butterfly Ru_4 configuration capped by the PhCCPh group. d(Ru–Ru) = 2.73–2.85 Å.	17
$Ru_4(CO)_{11}(\mu_4\text{-PPh})(\mu_4\text{-}\eta^4\text{-}C_6H_4)$	Square Ru_4 unit capped on one side by a μ_4-phosphinidene ligand and, on the other, by the benzyne (ortho-dimetallated benzene). d(Ru–Ru) = 2.791–2.943 Å.	18
$Ru_4(CO)_{11}(PhC_2Ph)(\mu\text{-NPh})$	Pentagonal bipyramide with equatorial plane containing two Ru atoms, two acetylene residues, and the nitrene atom. This plane is capped by two $Ru(CO)_2$ units with two edge-bridging CO groups. d(Ru–Ru) = 2.7–2.8 Å.	19
$Ru_4(CO)_{10}(\mu_2\text{-Cl})_2(\mu_3\text{-OEt})_2$	Ru-atoms adopt a planar geometry that can be described as two triangles sharing a common elongated edge. Unshared edges of one of the triangles are bridged by chlorine atoms. The same triangle is capped on either side by the oxygen atoms of the two ethoxy groups.	20
$Ru_4(CO)_5\{P(OMe)_3\}(\mu_3\text{-}C_5H_4)_2(Cp)_2$	Ru_4 chain which can be considered as two Ru_2 units (2.760 Å) jointed by the central, longer (2.887 Å) Ru–Ru bond. Terminal Ru atoms each with a Cp-ligand. The other C_5 rings are bridging three Ru atoms, η^5 to one and η^1 to each of the other two.	21
$Os_4(CO)_n(PMe_3)$	n = 13, 14, 15. n = 15: $Os(CO)_4(PMe_3)$ acts as a ligand to an $Os_3(CO)_{11}$ fragment. n = 14 planar Os_4 skeleton with two adjacent short Os–Os (2.779–2.784 Å) and two long (2.982–3.013 Å). n = 13 tetrahedral framework (d(Os–Os) = 2.765–2.869 Å).	22
$Os_4(CO)_{12}(\mu_3\text{-NMe})$	Os atoms define distorted tetrahedron capped by the N-atom of methylnitrene group on the face containing three longer Os–Os distances (2.834 Å) d(Os–Os) = 2.75–2.835 Å.	23
$Os_4(CO)_{12}(\mu\text{-}\eta^2\text{-}C_2H_2)$	Os atoms adopt a butterfly configurations. Three terminal CO-groups are bonded to each metal. The acetylenic ligand lies over the butterfly.	24
$Os_4(CO)_{12}(\mu_3\text{-S})_2$	Butterfly Os_4 unit with sulfido ligands bridging the two open triangular faces. d(Os–Os): two long (3.002–3.091 Å) and three short (2.914–2.940 Å) distances.	25
$Os_4(CO)_{12}(\mu_3\text{-HC}_2Ph)(\mu_3\text{-S})$	Butterfly Os_4 unit with triply bridging sulfido and alkyne ligands on the two open triangular faces. d(Os–Os) = 2.904–3.083 Å.	26
$Os_4(CO)_{11}L(\mu_3\text{-S})$	L = PMe_2Ph, CN-t-Bu. Butterfly array of Os atoms with μ_3-S groups on the two open triangular faces. L replaces a CO-group in a wing-type Os atom d(Os–Os) = 2.987 (L = PMe_2Ph).	27

Table 2.3. (continued)

Compounds	Remarks	Ref.
$Co_4(CO)_9(\mu\text{-}CO)_3$	Apical $Co(CO)_3$ group is bonded to basal $CO_3(CO)_6(\mu\text{-}CO)_3$ fragment. $d(Co–Co) = 2.457–2.527$ Å.	28
$[Co(CO)_8(\mu_2\text{-}CO)_3I]^-$	Tetrahedral Co_4 unit. μ-CO ligands define the basal triangle. Iodine is terminally bonded to apical metal atom. $d(Co–Co) = 2.493$ Å.	29
$Co_4(CO)_{10}(\mu_2\text{-}CO)(\mu_4\text{-}GeMe)_2$	Co and Ge atoms define a irregular square bipyramidal structure with Ge atom in apical positions. $d(Co–Co) = 2.58$ Å (bridged) and 2.692–2.721 Å.	30
$Co_4(CO)_{10}(EtC_2Et)$	Butterfly Co_4 configuration. Dihedral angle 118°. $d(Co–Co) = 2.55$ Å.	31
$Co_4(CO)_8(\mu_2\text{-}CO)_2(\mu_4\text{-}PPh_2)_2$	Rectangular Co_4 configuration; longer edges are unbridged. $d(Co–Co) = 2.519$ and 2.697 Å.	32
$Co_4(CO)_7(\mu\text{-}CO)_3\{P(OMe)_3\}_2$	Co atoms define a tetrahedron bridged on three edges by CO-groups (C_{3v}). $d(Co–Co) = 2.523–2.542$ (eq–ax) and 2.468–2.491 Å (eq–eq).	33
$Co_4(CO)_4(\mu\text{-}CO)(\mu_3\text{-}CO)_2(Cp)_2$	Co atoms are forming a distorted tetrahedron. $d(Co–Co) = 2.417–2.494$ Å.	34
$Co_4(NO)_4(\mu_3\text{-}NCMe_3)_4$	Distorted tetrahedral Co_4 configuration. Isocyanide groups are capping the four tetrahedron faces.	35
$Co_4(PPh)(B_2H_2)(Cp)_4$	Pentagonal bipyramid containing four Co, one P, and two B atoms in cluster core. Axial positions occupied by Co atoms. Cp ligands on metal atoms. $d(Co–Co) = 2.517–2.521$ Å.	36
$Rh_4(CO)_9(\mu\text{-}CO)_3$	Isostructural with $Co_4(CO)_{12}$ (vide supra). $d(Rh–Rh) = 2.70–2.80$ Å.	37
$Rh_4(CO)_4(\mu_2\text{-}CO)_4\{P(OEt)_3\}$ (tripod)	tripod = $HC[P(C_6H_5)_2]_3$. Distorted tetrahedral Rh_4 array with three P atoms of the tripod-ligand bonded in axial positions. Each edge of basal triangle and one axial-equatorial edge are bridged by CO groups.	38
$Rh_4(CO)_5(\mu_2\text{-}CO)_3[P(OPh)_3]_4$	Phosphorus ligand replaces terminal CO-groups in $Rh_4(CO)_{12}$. $d(Rh–Rh) = 2.72$ Å.	39
$[Rh_4(CO)_5(\mu\text{-}PPh_2)_5]^-$	Rh atoms define an *arachno* or butterfly structure that contains only PPh^- groups as bridging ligands.	40
$Rh_4(\mu\text{-}CO)_4\{P(OPh)_3\}_4(\mu\text{-}SO_2)_3$	Rh atoms arranged in a butterfly structure with a relatively small dihedral angle (82.4°). One SO_2 ligand is bridging the wing-type metal atoms. The other two are bridging the central metal–metal bond. $d(Rh–Rh) = 2.711–2.782$ Å.	41
$[Rh_4\{CN(CH_2)_3NC\}_4Cl_2]^{5+}$	Linear arrangement of Rh atoms mode up of two $[Rh_2(LL)_4]^{3+}$ units $[d(Ru–Ru) = 2.926$ Å$]$ linked by Rh–Rh bond (2.475 Å).	42
$Ir_4(CO)_{12}$	Tetrahedral Ir_4 unit. $d(Ir–Ir) = 2.293$ Å.	43

Table 2.3. (continued)

Compounds	Remarks	Ref.
$[Ir_4(CO)_8(\mu\text{-}CO)_3Br]^-$	Structurally similar to $Co_4(CO)_{12}$ (vide supra). One basal, terminal CO-group is replaced by bromine. $d(Ir–Ir) = 2.695–2.745$ (basal) and 2.696 Å (apical).	44
$Ir_4(CO)_9\{\mu_3\text{-}(PPh_2)_3CH\}$	Ir atoms define the vertices of a tetrahedron with the tripod-P-ligand capping a face. CO-groups are terminally bonded. $d(Ir–Ir) = 2.684–2.695$ Å.	45
$Ir_4(CO)_5(\mu\text{-}CO)_3(PMe_3)$	Metal tetrahedron displays a trigonal pyramidal distortion with average apical and basal edges of 2.747 and 2.894 Å respectively.	46
$Ir_4(CO)_5(dppm)_2$		47
$Ir_4(CO)_2(\mu\text{-}CO)_2(\mu\text{-}AsBu_2)_4$	Planar, rhomboidal Ir_4 unit. Short axis (2.592 Å) is consistent with an Ir–Ir double bond. $d(Ir–Ir)$ around the outer edges of Ir_4 parallelogram: 2.807–2.866 Å.	48
$Ni_4(N_2CO)_6\{P(CH_2CH_2CN)_3\}_4$	Tetrahedral Ni_4 unit. $d(Ni–Ni) = 2.51$ Å.	49
$Ni_4(CO)_4(CF_3C_2CF_3)_3$	Tetrahedral Ni_4 unit. $CF_3C_2CF_3$ groups are coordinated to three triangular faces. $d(Ni–Ni) = 2.378$ (basal) and 2.660 Å (apical).	50
$Pd_4(\mu\text{-}CO)_6(PBu_3)_4$	Tetrahedral arrangement of Pd atoms with CO ligands symmetrically bridged. $d(Pd–Pd) = 2.778–2.817$ Å.	51
$Pd_4(\mu_4\text{-}CO)_4(OAc)_4](AcOH)_2$	Distorted square planar Pd_4 unit. Two short (2.663 Å) and two long (2.909 Å) Pd–Pd distances. Pd–Pd–Pd angles differ from 90°.	52
$Pt_4(\mu\text{-}COOCH_3)_8$	Square planar Pt_4 geometry. $d(Pt–Pt) = 2.492–2.498$ Å.	53
$Pt_4(\mu_2\text{-}CO)_2(\mu_2\text{-}dppm)_3\{Ph_2PCH_2P(O)(Ph_2)\}$	Butterfly arrangement of Pt atoms (dihedral angle 83.7°). The five edges are bridged by CO and dppm ligands. $d(Pt–Pt) = 2.611–2.739$ Å.	54
$Cu_4(MeN_3Me)_4$	Cu atoms define parellelogram with angles 67° and 113°. $d(Cu–Cu) = 2.64$ Å.	55

References

1 Pez GP, Apgar P, Crissey RK (1982) J. Am. Chem. Soc. 104:428
2 Darkwa J, Lockemeyer JR, Boyd PDW, Rauchfuss TB, Rheingold AL (1988) J. Am. Chem. Soc. 110:141
3 Pasynskii, AA, Eremenko IL, Katugin AS, Gasanov G Sh, Turchanova EA, Ellert OG, Struchov Yu T, Shklover VE, Berberova NT, Sogomonova AG, Okhlobystin O. Yu (1988) J. Organomet. Chem. 344:195
4 Chen W, Goh LY, Mark TCW (1986) Organometallics 5:1997
5 Bottomley F, Praez DE, White PS (1981) J. Am. Chem. Soc. 103:5581
6 Churchill MR, Bau R (1968) Inorg. Chem. 7:2606
7 Laing M, Kieman PM, Griffith WP (1977) J. Chem. Soc., Chem. Commun., 221
8 Doedens RJ, Dahl LF (1966) J. Am. Chem. Soc. 88:4847
9 Bradley JS, Ansell GB, Leonowicz ME, Hill EW (1981) J. Am. Chem. Soc. 103:4968
10 Boehme RF, Coppens P (1981) Acta Crystallogr., Sect. B. 37:1914
11 Bradley JS, Ansell GB, Hill EW (1979) J. Am. Chem. Soc. 101:7417
12 Broadhurst PV, Johnson BFG, Lewis J, Raithby PR (1980) J. Chem. Soc., Chem. Commun. 812

Table 2.3. (continued)

13 Vahrenkamp H, Wolters D (1982) J. Organomet. Chem. 224:C17
14 Vahrenkamp H, Wucherer EJ, Wolters D (1983) Chem. Ber. 116:1219
15 Fachinetti G, Fochi G, Funaioli T, Zanazzi PF (1987) J. Chem. Soc., Chem. Commun. 89
16 Steinmetz GR, Harley D, Geoffroy GL (1980) Inorg. Chem. 19:2985
17 Johnson BFG, Lewis J, Reichert BE, Schorpp KT, Sheldrick GM (1977) J. Chem. Soc., Dalton Trans. 1417
18 Knox SAR, Lloyd BR, Orpen AG, Vinas JM, Weber M (1987) J. Chem. Soc., Chem. Commun. 1498
19 Rheingold AL, Staley DL, Han SH, Geoffrey GL (1988) Acta Cristallogr. C44:570
20 Johnson BFG, Lewis J, Mace JM, Raithby PR, Vargas MD (1987) J. Organomet. Chem. 321:409
21 Feasey ND, Forrow NJ, Hogarth G, Knox SAR, MacPherson KA, Morris MJ, Orpen AG (1984) J. Organomet. Chem. 267:C41
22 Martin IR, Einstein FWB, Pomeroy RK (1988) Organometallics 7:294
23 Lin YC, Knobler CB, Kaesz HD, (1981) J. Organomet. Chem. 213:C41
24 Jackson R, Johnson BFG, Lewis J, Raithby PR, Sankey SW (1980) J. Organomet. Chem. 193:C1
25 Adams RD, Yang LW (1982) J. Am. Chem. Soc. 104:4115
26 Adams RD, Wang S (1987) J. Am. Chem. Soc. 109:924
27 Adams RD, Horváth IT, Natarajan K, (1984) Organometallics 3:1540
28 Carve FH, Cotton FA, Frenz BA (1976) Inorg. Chem. 15:380
29 Albano VG, Braga D, Longoni G, Campanella S, Cerriotti A, Chini P (1980) J. Chem. Soc., Dalton Trans. 1820
30 Foster SP, MacKay KM, Nicholson BK (1982) J. Chem. Soc., Chem. Commun. 1156
31 Dahl LF, Smith DL (1962) J. Am. Chem. Soc. 84:2450
32 Ryan RC, Dahl LF (1975) J. Am. Chem. Soc. 97:6904
33 Darensbourg DJ, Incorvia MJ (1981) Inorg. Chem. 20:1911
34 Altbach MI, Fronczek FR, Butler IG (1987) Acta Crystallogr. C43:2283
35 Gall RS, Connelly NG, Dahl LF (1974) J. Am. Chem. Soc. 96:4017
36 Feilong J, Fehlner T, Rheingold AL (1987) J. Chem. Soc. Chem. Commun. 1393
37 Wei CH (1969) Inorg. Chem. 8:2384
38 Kennedy JR, Selz P, Rheingold AL, Trogler WC, Basolo F (1989) J. Am. Chem. Soc., 111:3615
39 Albano VG, Ciani G, Martinengo S, Chini P, Giovdano G (1975) J. Organomet. Chem. 88:381
40 Kreter PE, Meek DW, Christoph GG (1980) J. Organomet. Chem. 188:C27
41 Briant CE, Theobald BRC, Mingos DMP (1981) J. Chem. Soc., Chem. Commun. 963
42 Mann KR, DiPierro MJ, Gill TP (1980) J. Am. Chem. Soc., 102:3965
43 Churchill MR, Hutchinson JP (1978) Inorg. Chem. 17:3528
44 Chini P, Ciani G, Garlaschelli L, Manassero M, Martinengo S, Sironi A, Canziani F (1978) J. Organomet. Chem. 152:C35
45 Clucas JA, Harding MM, Nicholls BS, Smith AK (1984) J. Chem. Soc., Chem. Commun. 319
46 DeMartin F, Manassero M, Sansoni M, Garlaschelli L, Sartorelli U (1981) J. Organomet. Chem. 204:C10
47 Harding MM, Nicholls BS, Smith AK (1984) Acta Crystallogr. C40:790
48 Arif AM, Jones RA, Schwab ST, Whittlesey BR (1986) J. Am. Chem. Soc. 108:1703
49 Bennett MJ, Cotton FA, Winquist BHC (1967) J. Am. Chem. Soc. 89:5366
50 Davidson JL, Green M, Stone FGA, Welch AJ (1975) J. Am. Chem. Soc. 97:7490
51 Moiseev II, Stromnova TA, Vargafting MN, Mazo GJ, Kuzmina LG, Struchkov YT (1978) J. Chem. Soc., Chem. Commun. 27
52 Bogdanovic B, Goddard R, Rubach M (1989) Acta Crystallogr. C45:1511
53 Carrondo MAAF de CT, Skapski AC (1978) Acta Cryst. B34:1857
54 Frew AA, Hill RH, Manojlovic-Muir L, Muir KW, Puddephatt RJ (1982) J. Chem. Soc., Chem. Commun. 198
55 O'Connor JE, Janusonis GA, Corey ER (1968) J. Chem. Soc., Chem. Commun. 445

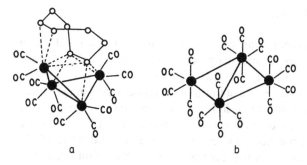

Fig. 2.14. Structures type "butterfly" (a) $Ru_4(CO)_{11}(\mu_4-C_8H_{10})$ (b) $[Re_4(CO)_{16}]^{2-}$

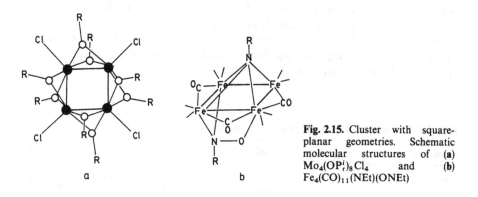

Fig. 2.15. Cluster with square-planar geometries. Schematic molecular structures of (a) $Mo_4(OP_r^i)_8Cl_4$ and (b) $Fe_4(CO)_{11}(NEt)(ONEt)$

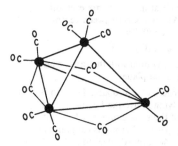

Fig. 2.16. Schematic molecular structures of the dodecacarbonyls of cobalt and rhodium

groups. The analogous compound of iridium shows indeed only terminal CO-ligands leading to a very symmetric structure. As discussed above, tetrairidium clusters with bridging CO-ligands are observed only in compounds as $[H_2Ir_4(CO)_{10}]^{2-}$ and $Ir_4(CO)1_{10}(PPh_3)_2$ (see Fig. 2.5) where a charge excess in metal is induced.

2.2.5 Pentanuclear Metal Clusters

The number of possible geometric arrays of the atoms in the metal skeleton increases with increasing cluster nuclearity. Thus, for pentanuclear clusters, a variety of structures may be found. Selected examples of clusters with nuclearity five are described in Table 2.4. Some of the geometries occurring in pentanuclear clusters are schematically illustrated in Fig. 2.17. Low symmetry arrays may be formally derived from the trigonal bipyramide. Structure b in Fig. 2.17 can be considered as arising from breaking one equatorial metal-metal bond of the trigonal bipyramide. The scission in the same polyhedron also of one bond but an axial one gives rise to the edge-bridged tetrahedral geometry c. From structures b and c the structures with bow-tie, d, and bi-diminished pentagonal bipyramid, e, geometries can be derived by rupture of two and one bonds respectively. These geometries are related, as it will be seen in the next section, by changes in the number of cluster electrons.

Solid angles and distortion degree in open structures depend on a number of factors as, for instance, the number of electrons, the nature of disposable orbitals, and the properties of the ligands.

Table 2.4. Selected examples of pentanuclear metal clusters

Compounds	Remarks	Ref.
$V_5(\mu_3\text{-S})_6(Cp^*)_5$	VCp* units form a trigonal bipyramid face-capped by S-atoms. d(V–V) = 3.062 (eq-ax) and 3.206 Å (eq-eq).	1
$[Cr_2(\mu\text{-SCMe}_3)(\mu_3\text{-S})(Cp)_2]_2Cr$	Pentanuclear "bow-tie" frame (see Fig. 2.17d) with a central Cr(II) atom. Angle between the triangle planes 90°. Sulfur atoms cap the Cr_3 triangle. Thiolate groups are bridging peripheral Cr atoms which are each other double-bonded (2.655 Å). d(Cr–Cr) (between central and peripheral atoms) = 2.889–2.933 Å.	2
$[Mo_5Cl_{13}]^{2-}$	Square pyramidal Mo_5 unit (Fig. 2.23).	3
$Fe_5(CO)_{15}(\mu_5\text{-C})$	Fe atoms define a square pyramid with the carbide just below the center of the basal plane.	4
$Fe_5(CO)_{13}(\mu_3\text{-S})_2(\mu_4\text{-CS})$	Fe atoms define the vertices of a 5 square. The fifth is terminal to the square lying above the Fe_4 plane and forming with an edge of the square an open Fe_3 unit capped on each side by μ_3-S. CS group is capping the Fe_4 plane. d(Fe–Fe) = 2.46 (capped edge) to 2.78 Å.	5
$Ru_5(CO)_{15}(\mu_4\text{-PR})$	R = C_6H_5, CH_3, CH_2CH_3, $CH_2C_6H_5$. Square pyramidal arrangement of Ru atoms with phosphor ligand capping the square face.	6
$Ru_5(CO)_5(MeCN)(\mu_5\text{-C})$	Four Ru-atoms form butterfly arrangement with the wing-type Ru atoms bridged by a fifth one (Fig. 2.17d). Carbido atom is located above the central Ru–Ru bond and is coordinated to all five Ru atoms.	7
$Ru_5(CO)_{13}(\mu_4\text{-PPh})(\mu_5\text{-}\eta^6\text{-}C_6H_4)$	Square Ru_4 unit bridged on one side by a $Ru(CO)_3L$ fragment (L = η^2-double bond of benzyne which is η^2-η^2-η^2-coordinated to the former triruthenium unit.	8

Table 2.4. (continued)

Compounds	Remarks	Ref.
$Ru_5(CO)_{15}(\mu_5\text{-C})$	Ru atoms define the vertices of a square pyramid. Carbon atom lies slightly below the square base of Ru_5. $d(Ru–Ru) = 2.83–2.85$ Å. $d(Ru–C) = 2.04$ Å.	9
$Os_5(CO)_{15}(\mu_5\text{-C})$	Isostructural with $Fe_5\text{-C}(CO)_{15}$. $d(Os–Os) = 2.85–2.88$ Å; $d(Os–C) = 2.06$ Å.	10
$[Os_5(CO)_{15}(\mu_5\text{-C})I]^-$	Bridged-butterfly open structure in which the metal atoms define five of the seven vertices of a pentagonal bipyramid (*arachno* geometry) (Fig. 2.17d). Carbido atom lies approximately at the centre of the polyhedron.	10
$Os_5(CO)_{19}$	Metal-atom skeleton build by two triangles sharing a vertex (Bowtie geometry Fig. 2.17d). $d(Os–Os)$ involving common vertex are rather longer (2.916–2.945 Å) than the other two (2.85 Å).	11
$Os_5(CO)_{16}\{P(OMe)_3\}_3$	Structurally similar to $Os_5(CO)_{19}$ in which three CO groups have been replaced by phosphite ligands.	11
$Os_5(CO)_{13}(\mu_5\text{-}\eta^2\text{-}C_2PPh_2)(\mu\text{-PPh}_2)$	Three edge-fused Os_3 triangles with a "swallowlike" arrangement. PPh_2 group bridges the non-fused edge of the central triangle. $d(Os–Os) = 2.742–2.934$ Å.	12
$Co_5(CO)_9(\mu\text{-CO})_2(\mu\text{-PMe}_2)_3$	Trigonal bipyramidal arrangement of Co atoms with one Co(ax)–Co(eq) bond ruptured (edge bridged tetrahedron, Fig. 2.17c). $d(Co–Co)$ in central $Co_3(CO)_6$ unit: 2.41–2.47 Å.	13
$[Rh_5(CO)_{10}(\mu\text{-CO})_5]^-$	Metal atoms form a trigonal bipyramid (C_2). $(\mu\text{-CO})$-bridges are spanning the vertices of equatorial plane and two basal-apical edges. $d(Rh–Rh) = 2.73$ Å (eq. plane) and 2.923–3.032 Å.	14
$Ir_5(CO)_8(\mu\text{-CO})_2(\mu_4\text{-Ph})(Cp^*)$	Square-pyramidal metal skeleton with the $(\mu_4\text{-PPh})$ group capping the square base $d(Ir–Ir) = 2.747–2.784$ Å.	15
$[Ni_5(CO)_9(\mu\text{-CO})_9(\mu\text{-CO})_3]^-$	$Ni_3(CO)_3(\mu\text{-CO})_3$ unit is capped by two $Ni(CO)_3$ groups. $d(Ni–Ni)$ 2.36 (eq) and 2.74–2.865 Å (ax).	
$Pt_5(CO)(\mu_2\text{-CO})_2(\mu\text{-SO}_2)_3(PPH_3)_4$	Pt atoms arranged as the corners of an edge-bridged-tetrahedron (bridged-butterfly geometry, Fig. 2–17c). $d(Pt–Pt) = 2.751–2.877$ Å.	16
$Cu_5(mes)_5$	mes = mesytil. Pentameric cyclic structure consisting in a ten-membered ring where Cu atoms bridged by Phenyl groups are forming a five-pointed star-shaped skeleton. $d(Cu–Cu) = 2.473–2.469$ A.	17
$Au_5(mes)_5$	Mes = mesityl. Ten-membered rings of alternate Au and C atoms similar to $Cu_5(mes)_5$. $d(Au–Au) = 2.692–2.710$ Å.	18

References

1 Eremenko II, Katugin AS, Pasynskii, AA, Struchkov Yu T, Shklover VE (1988) J. Organomet. Chem. 345:79
2 Pasynskii AA, Eremenko IL, Orazsakhatov B, Gasanov G Sh, Shklover VE, Struchkov Yu T (1984), J. Organomet. Chem. 269:147
3 Jodden K, Schnering HGV, Schäfer H (1975) Angew. Chem. Internat. Edn. 14:570
4 Braye EH, Dahl LF, Hübel W, Wampler DL (1962) J. Am. Chem. Soc. 84:4633

Table 2.4. (continued)

5 Broadhurst PV, Johnson BFG, Lewis J, Raithby PR (1981) J. Am. Chem. Soc. 103:3198
6 Natarajan K, Zsolnai L, Huttner G (1981) J. Organomet. Chem. 209:85
7 Johnson BFG, Lewis J, Nicholls JN, Oxton IA, Raithby PR, Rosales MJ (1982) J. Chem. Soc., Chem. Commun., 289
8 Knox SAR, Lloyd BR, Orpen AG, Vinas JM, Weber M (1987) J. Chem. Soc., Chem. Commun., 1498
9 Farrar DH, Jackson PF, Johnson BFG, Lewis J, Nicholls JN, McPartlin M (1981) J. Chem. Soc., Chem. Commun., 415
10 Jackson PF, Johnson BFG, Lewis J, Nicholls JN, McPartlin M, Nelson WJH (1980) J. Chem. Soc., Chem. Commun., 564
11 Farrar DH, Johnson BFG, Lewis J, Raithby PR, Rosales MJ (1982) J. Chem. Soc., Dalton Trans., 2051
12 Daran J-C, Cabrera E, Bruce MI, Williams ML (1987) J. Organomet. Chem. 319:239
13 Keller E, Vahrenkamp H (1977) Angew. Chem. Intern. Edn. 16:731
14 Fumagalli A, Koetzle TF, Takusagawa F. Chini P, Martinengo S, Heaton BT (1980) J. Am. Chem. Soc. 102:1740
15 Braga D, Vargas MD (1988) J. Chem. Soc., Chem. Commun., 1443
16 Briant CE, Evans DG, Mingos DMP (1986) J. Chem. Soc., Dalton Trans., 1535
17 Gambarotta S, Floriani C, Chiesi-Villa A, Guastini C (1983) J. Chem. Soc., Chem. Commun., 1156
18 Gambarotta S, Floriani C, Chiesi-Villa A, Guastini C (1983) J. Chem. Soc., Chem. Commun., 1304

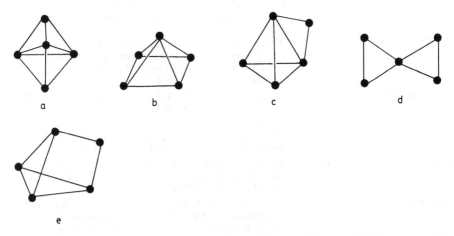

Fig. 2.17. Types of geometries in pentanuclear clusters. (**a**) Trigonal bipyramid; (**b**) Square pyramid; (**c**) Edge-bridged tetrahedron; (**d**) Bow-tie geometry; (**e**) Diminished pentagonal bipyramid or bridged-butterfly geometry

From the point of view of the stability of clusters, which appears to be preferred in the case of deltahedral geometries, the trigonal bipyramid should be the most favorable structure for pentanuclear clusters. Many of known clusters with nuclearity five indeed have this class of structure.

As it will be seen in next Section, metal cluster structures are closely related to the number of electrons and orbitals existing in the cluster core. In general,

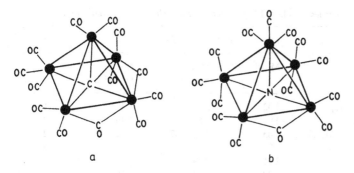

Fig. 2.18. Examples of pentanuclear clusters with square-base pyramidal geometries. (a) $[Fe_3C(CO)_{14}]^{2-}$; (b) $[Fe_5N(CO)_{14}]^{-}$

electron-rich species need more open structures than the electron-deficient ones. One special characteristic of open structures is the possibility of adding an atom or group capping the open face of the polyhedron. Square-based pyramidal structures are indeed very appropriate for the inclusion of heteroatoms. Carbon, nitrogen, sulfur, and pholpholidene groups RP are known to occupy the capping position on the square face of the pyramid leading thus to structures that can be considered as closed polyhedrons. Cluster structures illustrated in Fig. 2.18 may be thus considered as distorted octahedra. Examples of carbides in penta and hexanuclear metal clusters are described in Chapter 3.

The same type of stabilization by capping the open face of the polyhedron is observed for species with a bidiminished pentagonal bipyramidal geometry (structure e in Fig. 2.17) that, as noted above, has one metal-metal bond less than the square-based pyramid.

Some pentanuclear osmium cluster complexes show structures with edge-bridged tetrahedral geometries (c in Fig. 2.17). These structures have the same number of metal-metal bonds as the square-based pyramid but only triangular faces, so the stabilization of intersticial atoms is not possible there.

2.2.6 Hexanuclear Metal Clusters

There are an important number of clusters with nuclearity six. Organometallic as well as high-valence hexanuclear cluster species are indeed well represented.

High-Valence Metal Clusters. Structural properties of selected hexanuclear high-valence cluster complexes are described in Table 2.5. A series of molybdenum and tantalum derivatives of type $[(M_6Y_8)X_6]^{2-}$ and $[(M_6Y_8)L_8]^{4+}$ (M = Mo or W; X = halide or alkoxide; Y = halide or other monovalent anion; and L = neutral Lewis-base) are known. The structures of these species are like that of the anion $[(Mo_6Cl_8)Cl_6]^{2-}$ illustrated in Fig. 2.19 in which the molybdenum atoms are in the vertex of an octahedron. Metal-metal distances of about

Table 2.5. Structural properties of selected high-valence hexanuclear cluster compounds

Compound	Internuclear distances (Å)			Ref.
	M–M	M–X Bridged	M–X (Terminal)	
$(Me_4N)_2[Nb_6(\mu_2\text{-}Cl)_{12}Cl_6]$	3.02	2.43	2.48	1
$(PyH)_2[Nb_6(\mu_2\text{-}Br)_{12}Cl_6]$	3.07	2.56	2.46	2
$(PyH)_2[Nb_6(\mu_2\text{-}Cl)_6(\mu_2\text{-}Br)_6Cl_6]^3$	3.03	2.52, 2.57	2.62	2
$Nb_6(\mu_2\text{-}Cl)_{12}Cl_{6/2}$	2.92	2.41	2.58	3
$K_4[Nb_6(\mu_2\text{-}Cl)_{12}Cl_6]$	2.91	2.49	2.60	4
$Ta_6(\mu_2\text{-}Cl)_{12}Cl_{6/2}$	2.92	2.43	2.56	5
$H_2[Ta_6(\mu_2\text{-}Cl)_{12}]6H_2O$	2.96	2.41	2.51	6
$[Mo_6(\mu_3\text{-}Cl)_8Cl_2]Cl_{4/2}$	2.61	2.47	2.50	7
$Cs_2[Mo_6(\mu_3\text{-}Cl)_8Br_6]$	2.62	2.48	2.58	8
$Cs_2[W_6(\mu_3\text{-}Cl)_8Br_6]$	2.62	2.49	2.57	8
$[W_6(\mu_3\text{-}Br)_8Br_4]Br_4$	2.64	2.58	2.58	9

References

1 Koknat FW and McCarley RE (1972) Inorg. Chem. 11:812
2 Spreckelmeyer B, Schnering HG (1971) Z. Anorg. Chem. 386:27
3 Simon VA, Schnering HG, Wohrle H, Schafer H (1965) Z. Anorg. Allg. Chem. 339:155
4 Simon VA, Schnering HG (1968) Z. Anorg. Allg. Chem. 361:235
5 Bauer D, Schnering HG (1968) Z. Anorg. Chem. 361:259
6 Thaxton CB, Jacobson RA (1971) Inorg. Chem. 10:1460
7 Schäfer H, Schnering HG, Tillack J, Kuhnen T, Wöhrle H, Baumann H (1967) Z. Anorg. Chem. 353:281
8 Healy PC, Kepert DL, Taylor D, White AH, (1973) J. Chem. Soc. Dalton, 646
9 Siepmann R, Schnering HG (1968) Z. Anorg. Chem. 357:289

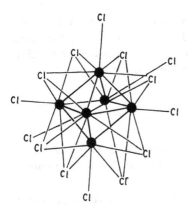

Fig. 2.19. Schematic structure of the anion $[(Mo_6Cl_8)Cl_6]^{2-}$

2.65 Å are shorter than those observed in the pure metals (2.725 Å). Eight chloride ions are coordinating the faces of the octahedron and six are linked terminally to the metal atoms. In these compounds, analogous to rhenium halides, the distances metal-terminal halides are somewhat shorter than those associated to bridging halides.

Niobium and tantalum also give rise to hexanuclear cluster species. Many ions with formulas $[(M_6Y_{12})X_6]^{4-}$ and $[(M_6Y_{12})L_6]^{2+}$ (M = Nb or Ta; Y = halides; X = halide or other monovalent anion; L = neutral Lewis base) are known. Figure 2.20 illustrates the type of structure found for this class of clusters. Metal atoms build octahedra which often undergo certain distortions along the tetragonal axis. The twelve halide ions bridge all the edges of the octahedron. Metal-metal distances in these Nb or Ta clusters are about 2.9 Å, i.e. a value near to that of the interatomic distances in the pure metal (ca. 2.86 Å). Similar compounds of scandium and zirconium are also known.

The versatility of halide ligands able to act as terminal, edge and face-bridging ligands often cause, similar to that which occurs for the $ReCl_3$, the clusters to be aggregated into more complex tridimentional extended structures.

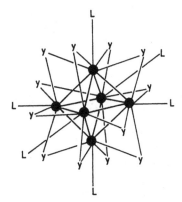

Fig. 2.20. Schematic structure of cations $[(M_6Y_{12})L_6]^{2+}$

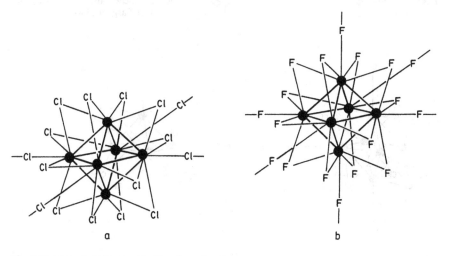

Fig. 2.21. Extended structures arising from the condensations of (a) Niobium chloride Nb_6Cl_{14} and (b) Niobium flouride Nb_6F_{15}

The replacement of neutral ligands by halides in $[Mo_6Cl_8L_8]^{4+}$ or $[Nb_6Cl_{12}L_8]^{2+}$ results indeed in the generation of extended structures as illustrated in Fig. 2.21 for two niobium halide derivatives.

Sulfur, selenium, tellurium, and antimony may also act as π-donor ligands to form metal clusters of some early transition elements. Although the compounds are often polymeric solids, structural determinations have shown that they are built up by condensed cluster units. Many of these compounds e.g. the Chevrel-Sergent compounds $[M_6S_8]\, M'$ ($M' = Pb^{2+}$, Cu^{2+}, and other divalent cations [Chevrel R, Gougeon P, Potel M, Sergent M (1985) Solid State Chem. 57:25]) (Fig. 2.22) show interesting magnetic and electrical properties.

Most of the known high-valence metal cluster species belong to the tri and hexanuclear clusters discussed above. However, there are also some examples of

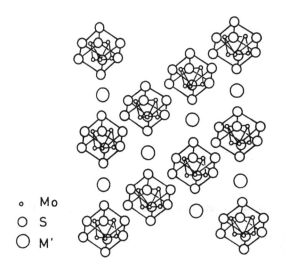

o Mo
O S
O M'

Fig. 2.22. Schematic structure of a Chevrel–Sergent phase

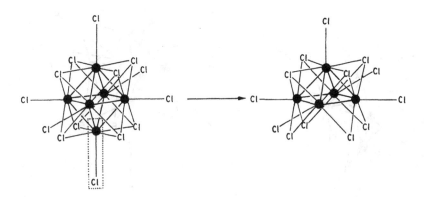

Fig. 2.23. Schematic structures of the anions $[M_5Cl_{13}]^{2-}$ M = Mo, W and their relationship with the hexanuclear species

this type of cluster with nuclearities of four and five. As shown schematically in Fig. 2.23, the pentanuclear cluster anion $[M_5Cl_{13}]^{2-}$ can be considered as being the result of a loss of one Mo-Cl unit from $[Mo_6Cl_{14}]^{2-}$. The structures observed for the tetranuclear compounds may be also rationalized as a fragment of the same type of octahedral compounds. The relationship between the structure of the hexanuclear species $[Mo_6X_8]^{2-}$ and those observed for Mo_4 compounds is analyzed in Fig. 2.24.

Hexanuclear Organometallic Clusters. Although the octahedral arrangement of metal atoms is the most representative one for hexanuclear cluster compounds, a number of other geometries have also been observed. The structural characteristics of a selected group of hexanuclear compounds are briefly described in Table 2.6. The different types of structures found for these clusters are illustrated schematically in Fig. 2.25.

Among hexanuclear cluster with an octahedral array of metal atoms and twelve metal-metal bonds, a number of inclusion compounds with interstitial atoms should also be counted, e.g. carbon and hydrogen atoms located inside the octahedron. Further examples of these types of compounds are described in Chapter 3.

An interesting feature being observed in Table 2.6 and in Fig. 2.25 is that from the types of geometries adopted by hexanuclear clusters all, but the octahedron and the trigonal prism, derive from other more simple arrangements by sharing additional atoms to edges or faces of polyhedra of lower nuclearity. Thus, the bicapped tetrahedron also with twelve metal-metal bonds is an alternative to the octahedron.

The capped square-based pyramid is an open structure with eleven metal-metal bonds. The open, square face is frequently shared by a μ_4-face-bridging

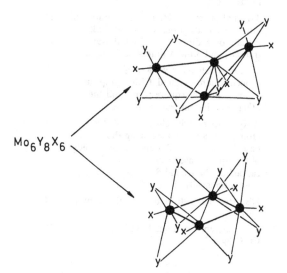

$Mo_6Y_8X_6$

Fig. 2.24. Relationship between Mo_4-compounds and the hexamolybdenum species $[Mo_6X_8]^{2-}$

Table 2.6. Selected examples of hexanuclear metal clusters

Cluster	Remarks	Ref.
$Ti_6(\mu_5\text{-}O)_8(Cp)_6$	Ti atoms define the vertices of a regular octahedron. d(Ti–Ti) = 2.891 [shorter than in metal (2.951 Å)].	1
$[M_6(\mu\text{-}X)_{12}X_6]^{n-}$	M = Nb, Ta; X = Cl, Br; n = 2, 4. d(M–M) = 2.91–3.07 Å. M_6 unit has regular octahedral geometry.	2–9
$[M_6(\mu_3\text{-}I)_8]^{n+}$	M = Nb, Ta; n = 2, 3.	10
$M_6(\mu_3\text{-}X)_8Y_4$	M = Mo, W; X, Y = Cl, Br. Metal atoms define the vertices of a little distorted octahedron. d(M–M) = 2.61–2.64 Å.	11–13
$Mo_6(\mu_3\text{-}Cl)_8(Cl)_2(Et)_2(PBu_3)_2$	Ethyl groups, two chlorine atoms, and two tributylphosphine groups are bounded to the Mo_6Cl_8 core in mutually trans positions. Cluster core dimensions are almost the same as that of the chlorine complex.	14
$[Fe_6(CO)_{13}(\mu\text{-}CO)(\mu_6\text{-}C)]^{2-}$	Carbide atom lies in the centre of a Fe_6 octahedron.	15
$Ru_6(CO)_{16}(\mu\text{-}CO)(\mu_6\text{-}C)$	Octahedral arrangement of Ru atoms with carbon atom in the centre. d(Ru–Ru) = 2.05 Å.	16
$[Ru_6(CO)_{20}(CN)_2]$	Two parallel Ru triangles linked by two axial CN-bridges. d(Ru–Ru) = 2.870–2.88 Å.	17
$Ru_6(CO)_{13}(\mu\text{-}CO)(\text{-}C_6H_3Me)(\mu_6\text{-}C)$	Carbide atom nearly in the centre of a Ru_6 octahedron. d(Ru–Ru) = 2.815–2.955 Å.	18
$Os_6(CO)_{18}$	Bicapped tetrahedron. d(Os–Os) = 2.754 Å in central tetrahedron and 2.731 Å in bonds with the other two Os atoms.	19
$Os_6(CO)_{18-n}(PPh_3)_n$	n = 1, 2. Substitution of CO groups by phosphine ligands on capped Os atoms in $Os_6(CO)_{18}$.	20
$[Os_6(CO)_{18}(\mu_6\text{-}P)]^-$	Trigonal prismatic Os_6 unit containing an interstitial P atom that coordinates all six metal atoms. d(Os–Os) = 2.930 (basal triangle) and 3.140 Å (inter-basal edges).	21
$Os_6(CO)_{16}(\mu_3\text{-}NCMe)(\mu_4\text{-}CMe)$	Os atoms define the vertices of a mono-capped square pyramid. CMe groups are capping, one the square plane, and the other, an adjacent triangular face.	22
$Os_6(CO)_{16}(\mu_3\text{-}S)(\mu_4\text{-}S)$	A square-pyramidal cluster of five Os atoms with a S atom spanning the square base. The sixth metal atom bridges an edge of the square base forming a triangle capped by the second S atom. Common edge is unusually short (2.686 Å).	23
$[Co_6(CO)_6(\mu_3\text{-}CO)_8]^{4-}$	Distorted octahedron. d(Co–Co) = 2.49 Å (av.)	24
$Co_6(CO)_{16}(\mu\text{-}\eta^3\text{-}C_2S_2)(\mu_3\text{-}S)$	Two different Co_3 units linked by the CCS_2 ligand.	25

Table 2.6. (continued)

Cluster	Remarks	Ref.
$[Co_6(CO)_9(\mu\text{-}CO)_4(\mu_6\text{-}N)]^-$	Octahedral array of Co atoms with interstitial N atom.	26
$Co_6(CO)_{12}(\mu_3\text{-}S)_2(\mu_6\text{-}C)$	Trigonal prism with the two ends capped by S atoms. d(CO–CO) = 2.437 and 2.669 Å in triangles and rectangular faces respectively.	27
$Rh_6(CO)_{12}(\mu_3\text{-}CO)_4$	$Rh_6(CO)_{12}$ octahedron. d(Rh–Rh) = 2.776 Å.	28
$[Rh_6(CO)_{11}(\mu_3\text{-}CO)_4I]^-$	Derived from $Rh_6(CO)_{16}$ by replacement of CO by iodine. d(Rh–Rh) = 2.746 Å.	29
$[Rh_6(CO)_7(\mu\text{-}CO)_6(\mu_6\text{-}C)]^{2-}$	Distorted octahedron of Rh atoms with the carbide in the centre. d(Rh–Rh) = 2.733–3.188 Å.	30
$[Rh_6(CO)_8(\mu_3\text{-}CO)_4\{P(OPh)_3\}_4]$	Rh atoms define the vertices of a slightly distorted octahedral cluster. d(Rh–Rh) = 2.789 Å. P(OPh)₃ ligands are bound to four equatorial coplanar Rh atoms.	31
$[Rh_6(CO)_{10}(\mu\text{-}CO)_4(\eta^3\text{-}C_3H_5)]^-$	Octahedral arrangement of Rh atoms. Allyl ligands are bonded to one of the Rh atoms. d(Rh–Rh) = 2.703–2.822 Å.	32
$[Rh_6(CO)_6(\mu\text{-}CO)_5(dppm)(\mu_6\text{-}C)]^{2-}$	Rh atoms define a distorted octahedron containing interstitial carbide. d(Rh–Rh) = 2.76 and 3.05 Å for the CO-bridged and unbridged edges respectively.	33
$[Ir_6(CO)_{12}(\mu\text{-}CO)_3]^{2-}$	Octahedral Ir₆ cluster with three symmetrical edge-bridging carbonyl groups. d(Ir–Ir) = 2.736 and 2.785 Å for bridged and unbridged edges respectively.	34
$Ni_6(Cp)_6$	Ni atoms are located at the corners of an octahedron with a Cp-ligand coordinated at each metal atom. d(Ni–Ni) = 2.441 Å.	35
$Pt_6(\mu\text{-}Cl)_{12}$	Pt₆ octahedron with a structure similar to Nb and Ta halides $[M_6X_{12}]^{2+}$ (vide supra). d(Pt–Pt) = 3.32–3.40 Å.	36
$Cu_6(\mu_3\text{-}2\text{-}MeNC_6H_4)_4(\mu\text{-}Br)_2$	Distorted octahedron. d(Cu–Cu) = 2.64–2.70 Å.	37
$[Au_6\{P(C_6H_7)_3\}_6]^{2+}$	Distorted octahedron. d(Au–Au) = 2.932–2.990 and 3.043–3.091 Å.	38

References

1 Huffman JC, Stone JG, Krusell WC, Caulton KG (1977) J. Am. Chem. Soc. 99:5830
2 Koknat FW, McCarley RE (1972) Inorg. Chem. 11:812
3 Field RA, Kepert DL, Robinson BW, White AH (1973) J. Chem. Soc., Dalton Trans., 1858
4 Spreckelmeyer B, Schnering HG (1971) Z. Anorg. Chem. 386:27
5 Simon VA, Schnering HG, Wohrle H, Schafer H (1965) Z. Anorg. Allg. Chem. 339:155
6 Simon VA, Schnering HG (1968) Z. Anorg. Allg. Chem. 361:235
7 Bauer D, Schnering HG (1968) Z. Anorg. Chem. 361:259
8 Thaxton CB, Jacobson RA (1971) Inorg. Chem. 10:1460
9 Burbank RD (1966) Inorg. Chem. 5:1491

Table 2.6. (continued)

10 Bateman LR, Blount JF, Dahl LF (1966) J. Am. Chem. Soc. 88:1082
11 Healy PC, Kepert DL, Taylor D, White AH (1973) J. Chem. Soc., Dalton, 646
12 Schäfer H, Schnering HG, Tillack J, Kuhnen T, Wöhrle H, Baumann H (1967) Z. Anorg. Chem. 353:281
13 Guggenberger LJ, Sleight AW (1969) Inorg. Chem. 8:2041
14 Saito T, Nishida M, Yamagate T, Yamagata Y, Yamaguchi Y (1986) Inorg. Chem. 25:1111
15 Churchill MR, Wormald J, Knight J, Mays MJ (1971) J. Am. Chem. Soc. 93:3073
16 Sirigu A, Bianchi M, Benedetti E (1969) J. Chem. Soc., Chem. Commun., 596
17 Lavigne G, Lugan N, Bonnet J-J (1987) J. Chem. Soc., Chem. Commun., 957
18 Farrugia LJ (1988) Acta Crystallogr. C44:997
19 Mason R, Thomas KM, Mingos DMP (1973) J. Am. Chem. Soc. 95:3802
20 Couture C, Farrar DH, Gomez-Sal MP, Johnson BFG, Kamarudin RA, Lewis J, Raithby PR (1986) Acta Crystallogr. Sect. C42:163
21 Colbran SB, Hay CM, Johnson BFG, Lahoz FJ, Lewis J, Raithby PR (1986) J. Chem. Soc., Chem. Commun., 1766
22 Eady CR, Fernandez JM, Johnson BFG, Lewis J, Raithby PR, Sheldrick GM (1978) J. Chem. Soc, Chem. Commun., 421
23 Adams RD, Horvath IT, Yang LW (1983) J. Am. Chem. Soc. 105:1533
24 Albano V, Bellon PL, Chini P, Scatturin V (1969) J. Organomet. Chem. 16:461
25 Stanghellini PL, Gervasio G, Rossetti R, Bor G (1980) J. Organomet. Chem. 187:C37
26 Ciani G, Mertinengo S (1986) J. Organomet. Chem. 306:C49
27 Bor G, Gervasio G, Rossetti R, Stanghellini PL (1978) J. Chem. Soc., Chem. Commun., 841
28 Paquette MS, Dahl LF, (1980) J. Am. Chem. Soc. 102:6621
29 Albano VG, Bellon PL, Sansoni M (1971) J. Chem. Soc. (A) 678
30 Albano VG, Braga D, Martinengo S (1981) J. Chem. Soc., Dalton Trans., 717
31 Ciani G, Manassero M, Albano VG (1981) J. Chem. Soc., Dalton Trans., 515
32 Ciani G, Sironi A, Chini P, Ceriotti A, Martinengo S (1980) J. Organomet. Chem. 192:C39
33 Bordini S, Heaton BT, Seregni C, Strona L, Goodfellow RJ, Hursthouse MB, Thornton-Pett M, Martinengo S (1988) J. Chem. Soc., Dalton Trans., 2103
34 Demartin F, Manassero M, Sansoni M, Garlaschelli L, Martinengo S, Canziani F (1980) J. Chem. Soc., Chem. Commun., 903
35 Paquette MS, Dahl LF (1980) J. Am. Chem. Soc. 102:6621
36 Brodersen K, Thiele G, Schnering HG (1965) Z. Anorg. Chem. 337:120
37 Guss JM, Mason R, Thomas KM, van Koten G, Noltes JG (1972) J. Organomet. Chem. 40:C79
38 Bellon P, Manassero M, Sansoni M (1973) J. Chem. Soc., Dalton Trans., 2423

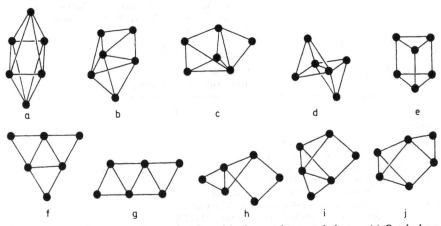

Fig. 2.25. Types of structures frequently adopted by hexanuclear metal clusters. (**a**) Octahedron (**b**) bicapped tetrahedron; (**c**) capped square pyramid; (**d**) edge-sharing bitetrahedron; (**e**) trigonal-prism; (**f**) six-atom-raft triangles; (**g**) six-atom-raft rhomb; (**h**), (**i**), and (**j**) edge-bridged bi-diminished bipyramidal or edge-bridged butterfly geometries

Table 2.7. Examples of clusters with seven and more metal atoms

Compound	Type of structure in fig. 2.26	Number of valence electrons	Ref.
Seven metal atoms			
$[Re_7C(CO)_{22}]^-$	a	98	1
$Os_7(CO)_{21}$	a	98	2
$[Rh_7(CO)_{16}]^{3-}$	a	98	3
$Ir_7(CO)_{12}(C_8H_{12})(C_8H_{11})(C_8H_{10})$	a	98	4
Eight metal atoms			
$Pd_8(CO)_8(PMe_3)_7$	b	110	5
$[Re_8C(CO)_{24}]^{2-}$	c	110	6
$[Os_8(CO)_{22}]^{2-}$	c	110	7
Nine metal atoms			
$[Co_9Si(CO)_{21}]^{2-}$	d	129	8
$[Rh_9(CO)_{19}]^{3-}$	e	122	9
$[Rh_9P(CO)_{21}]^{2-}$	d	130	10
$[Ge_9]^{2-}$	f	128	11
$[Ge_9]^{4-}$	d	130	11
$[Bi_9]^{5+}$	f	130	12
Ten metal atoms			
$[Os_{10}C(CO)_{24}]^{2-}$	g	134	13
$Os_{10}C(CO)_{23}(POMe_3)I_2$	g	134	14
$[Os_{10}C(CO)_{22}(NO)I]^{2-}$	g	136	14
$[Rh_{10}P(CO)_{22}]^-$	h	142	15
$[Rh_{10}S(CO)_{22}]^{2-}$	h	142	16
$Pd_{10}(CO)_{14}(PBu_3)_4$	i	136	17
Eleven metal atoms			
$[Os_{10}AuC(CO)_{24}(PPh_3)]^-$	j	146	18
$[Rh_{11}(CO)_{23}]^{3-}$	k	148	19
$[Rh_9Pt_2(CO)_{22}]^{3-}$	k	148	20
$Au_{11}(PPh_3)_7(SCN)_3$	l	138	21
$Au_{11}(PPh_3)_7I_3$	l	138	22
Twelve metal atoms			
$[Rh_{12}C_2(CO)_{24}]^{2-}$	m	166	23
$[Rh_{12}C_2(CO)_{23}]^{3-}$	m	167	24
$[Rh_{12}C_2(CO)_{23}]^{4-}$	m	168	25
$[Ni_9As_3(CO)_{15}Ph_3]^{2-}$	n	170	26
$[Ni_{10}As_2(CO)_{18}Me_2]^{2-}$	n	170	26
Thirteen metal atoms			
$[HRh_{13}(CO)_{24}]^{4-}$	o	170	27
$[H_2Rh_{13}(CO)_{24}]^{3-}$	o	170	28
$[Au_{13}(PMePh_2)_{10}]^{3+}$	p	162	29
$[Au_{13}(Ph_2PCH_2PPh_2)_6]^{4+}$	p	162	30
Fourteen metal atoms			
$[Rh_{14}(CO)_{26}]^{2-}$	q	180	31
$[Rh_{14}(CO)_{25}]^{4-}$	q	180	32
Fifteen metal atoms			
$[Pt_{15}(CO)_{30}]^{2-}$	r	212	33
Sixteen metal atoms			
$[Ni_{16}(CO)_{23}(C_2)_2]^{4-}$	s	216	34
Seventeen metal atoms			
$[Rh_{17}(CO)_{30}]^{3-}$	t	216	35

Table 2.7. (continued)

References

1 Beringhelli T, D'Alfonso G, De Angelis M, Ciani G, Sironi A (1987) J. Organomet. Chem. 322:C2
2 Eady CR, Johnson BFG, Lewis J, Mason R, Hitchcock PB, Thomas KM (1977) J. Chem. Soc., Chem. Commun., 385
3 Albano VG, Bellon PL, Ciani G (1969) J. Chem. Soc., Chem. Commun., 1024
4 Pierpont CG (1979) Inorg. Chem. 18:2972
5 Bachmann M, Hawkins I, Hursthouse MB, Short RL (1987) Polyhedron 6:1987
6 Ciani G, D'Alfonso G, Freni M, Romiti P, Sironi A (1982) J. Chem. Soc., Chem. Commun., 705
7 Jackson PF, Johnson BFG, Lewis J, Raithby PR (1980) J. Chem. Soc., Chem. Commun., 60
8 Mackay VM, Nicholson BK, Robinson WT, Sims AW (1984) J. Chem. Soc., Chem. Commun., 1276
9 Martinego S, Fumagalli A, Bonfichi R, Ciani G, Sironi A (1982) J. Chem. Soc., Chem. Commun., 825
10 Vidal JL, Walker WE, Pruett RL, Schoening RC (1979) Inorg. Chem. 18:129
11 Belin CHE, Corbett JD, Cisar A (1977) J. Am. Chem. Soc. 99:7163
12 Friedman RM, Corbett JD (1973) Inorg. Chem. 12:1134
13 Jackson PF, Johnson BFG, Lewis J, Nelson WJH, McPartlin M (1982) J. Chem. Soc., Dalton Trans., 2099
14 Goudsmit RJ, Johnson BFG, Lewis J, Nelson WJH, Vargas MD, Braga D, McPartlin M, Sironi A (1985) J. Chem. Soc., Dalton Trans., 1795
15 Vidal JL, Walker WE, Schoening RC (1981) Inorg. Chem. 20:238
16 Ciani G, Garlaschelli L, Sironi A, Martinengo S (1981) J. Chem. Soc., Chem. Commun., 536
17 Mednikov EG, Eremenko NK, Slovokhotov YL, Struchkov YT, Gubin SP (1983) J. Organomet. Chem. 258:247
18 Braga D, Henrick K, Johnson BFG, Lewis J, McPartlin M, Nelson WJH, Vargas MD (1986) J. Chem. Soc., Chem. Commun., 975
19 Fumagalli A, Martinengo S, Ciani G, Sironi A (1983) J. Chem. Soc., Chem. Commun., 453
20 Fumagalli A, Martinengo S, Ciani G (1984) J. Organomet. Chem. 273:C46
21 McPartlin M, Mason R, Malatesta L (1969) J. Chem. Soc., Chem. Commun. 334
22 Smits JMM, Beurskens PT, Van der Velden JWA, Bour JJ (1983) J. Cryst. Spectrosc. 13:373
23 Albano VG, Braga D, Chini P, Strumolo D, Martinengo S (1983) J. Chem. Soc., Dalton Trans., 249
24 Albano VG, Braga D, Martinengo S, Seregni C, Strumolo D (1983) J. Organomet. Chem. 252:C93
25 Albano VG, Braga D, Strumolo D, Seregni C, Martinengo S (1985) J. Chem. Soc., Dalton Trans., 1309
26 Rieck DF, Montag RA, McKechnie TS, Dahl LF (1986) J. Am. Chem. Soc. 108:1330
27 Ciani G, Sironi A, Matinengo S (1981) J. Chem. Soc., Dalton Trans., 519
28 Albano VG, Ciani G, Martinengo S, Sironi A (1979) J. Chem. Soc., Dalton Trans., 978
29 Briant CE, Theobald BRC, White JW, Bell LK, Mingos DMP (1981) J. Chem. Soc., Chem. Commun., 201
30 Van der Velden JWA, Vollenbroek FA, Bour JJ, Beurskens PT, Smits JMM, Bosman WP, (1981) Rec. J. R. Neth. Chem. Soc. 100:148
31 Martinengo S, Ciani G, Sironi A (1980) J. Chem. Soc., Chem. Commun., 1140
32 Ciani G, Sironi A, Martinengo S (1982) J. Chem. Soc., Dalton Trans., 1099
33 Calabrese JC, Dahl LF, Chini P, Longoni G, Martinengo S (1974) J. Am. Chem. Soc. 96:2614
34 Ceriotti A, Longoni G, Manassero M, Masciocchi N, Piro G, Resconi L, Sansoni M (1984) J. Chem. Soc., Chem. Commun. 1402
35 Ciani G, Magni A, Sironi A, Matinengo S (1981) J. Chem. Soc., Chem. Commun., 1280

ligand as occurs for instance in $Os_6(CO)_{17}S$. There are however compounds in which the square face remains open as found in $H_2Os_6(CO)_{18}$.

Trigonal prismatic structures are not found very often in simple hexanuclear clusters. Most compounds with this metal atom arrangement are carbides and nitrides in which the heteroatom occupies the center of the prism.

2.2.7 Clusters with Seven or More Metal Atoms

In Table 2.7, a select group of organometallic clusters with different nuclearities are described. Geometries adopted by some of the structures of large clusters in this table are shown schematically in Fig. 2.26.

The tendency towards the formation of structures formally produced by the condensation of structures of lower nuclearity by sharing edges and faces observed for hexanuclear clusters is reinforced in larger clusters.

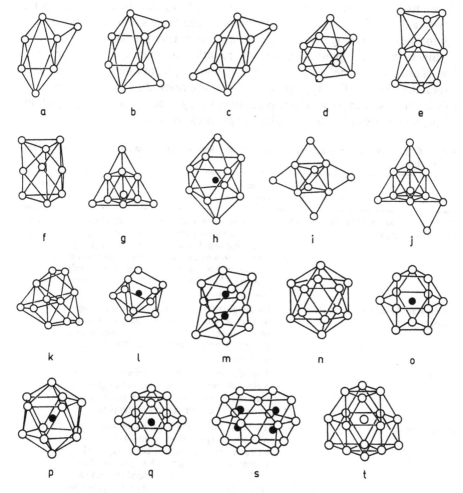

Fig. 2.26. Examples of types of geometric arrangements of metal atoms found in large clusters

2.3 Bonding in Metal Clusters

As discussed in the last section, the metal network in clusters may be considered, especially for high nuclearity clusters, as a finite portion of a compact metal structure stabilized by external ligands. The feature that the number of atoms is relatively small does not make the description of the system simpler. On the contrary, for pure metals as well as for traditional mononuclear compounds, there are theories which allow us to produce descriptions of their properties which, however, do not fit for clusters. In the very short history of this new class of compounds many efforts have been made to obtain some generalizations that permit us to rationalize experimental features as being structures and reactivity, as well as to perform quantitative theoretical calculations. The most important approaches to bonding in metal cluster are summarized in Table 2.8.

Since cluster compounds in general are rather complex species the application of quantitative methods for describing bonding is not only difficult but also impractical. Qualitative approaches and empiric rules often play an important role in this case. Simple methods permitting the systematization and projection of known features are very valuable for practical chemists.

Table 2.8. Relevant concepts and approaches to chemical bonding in main group and transition metal clusters

Approach	Remarks Description	Ref.
The Effective Atomic Number (EAN) or Inert-Gas Rule	The total number of cluster valence electrons (i.e. metal valence electrons plus those apported by the ligands) is such that each main group atom as well as each transition metal atom in the cluster has 8 and 18 electrons respectively. Clusters which follow this rule are known as "electron precise" species.	1
Polyhedral Skeletal Electron Pair Theory	Bonding pattern in metal clusters is analogous to that in the boranes (electron defficient species) where each atom contributes with 3 orbitals and 2 electrons to the skeletal bonding.	2–9
Dever Principle	The addition of secuensive pairs of electrons to a *closo*-cluster opens its structure originating *nido* and *arachno* structures, i.e. producing the remotion of cluster vertices (decapping).	4, 10, 11
Isobal Analogy	There are definite similarities between the number energies, shapes, and nodal characteristics of both main group and transition metal carbonyl fragments. This analogy has proved to be useful in accounting structural similarities between main group and transition metal clusters.	12–16
Capping Principle	Capping a face by a ML_3 fragments or bridging an edge by a C_{2v}-ML_4 fragment on a given polyhedron leads to no change in the number of skeletal bonding MOs.	8, 9 17, 18

Table 2.8. (continued)

Approach	Remarks Description	Ref.
Polyhedral Fusion	Condensed clusters may be regarded as the product of two or more polyhedra sharing a vertex, an edge, or a face.	19, 20
Polyhedral Inclusion	High nuclearity cluster may be formally divided into an incapsulated, internal polyhedron and other surface, external polyhedron.	21, 22
Free electron cluster bonding models	Skeletal bonding is described in terms of a free electron bonding model similar to those used for bulk metals.	23
Tensor Surface Harmonic theory	Cluster atoms are constrained to lie on the surface of a single sphere cluster; molecular orbitals are thus derived from a spherical potential model.	24–28
Graph-Theory derived Approach	Ideas derived from topology and graph-theory are used to model the skeleton chemical bonding in clusters.	29–33

References

1 Sidgwick NV, Powell HE (1940) Proc. Roy. Soc. A176:153
2 Teo BK (1984) Inorg. Chem. 23:1251
3 Teo BK, Longoni G, Chung FRK (1984) Inorg. Chem. 23:1258
4 Wade K (1971) J. Chem. Soc., Chem. Comm., 792
5 Wade K (1976) Adv. Inorg. Chem. Radiochem. 18:1
6 Wade K (1972) Inorg. Nucl. Chem. Lett. 8:559, 563, 823
7 Wade K (1972) Nature Phys. Sci. 240:71
8 Mingos DMP (1972) Nature Phys. Sci. 236:99
9 Mason R, Thomas KM, Mingos DMP (1973) J. Amer. Chem. Soc. 95:3802
10 Rudolph RW, Pretzer WR (1972) Inorg. Chem. 11:1974
11 Rudolph RW (1976) Acc. Chem. Res. 9:446
12 Elian M, Hoffmann R (1975) Inorg. Chem. 14:1058
13 Elian M, Chen MML, Mingos DMP, Hoffmann R (1976) Inorg. Chem. 15:1148
14 Hertler WR, Klanberg F, Muetterties EL (1967) Inorg. Chem. 6:1696
15 Hoffmann R (1981) Science 211:995
16 Hoffmann R (1982) Angew. Chem. Int. Edn. 21:711
17 Mingos DMP, Forsyth MI (1977) J. Chem. Soc., Dalton Trans., 610
18 Evans DG, Mingos DMP (1983) Organometallics 2:435
19 Mingos DMP (1983) J. Chem. Soc., Chem. Comm., 706
20 Mingos DMP (1984) Acc. Chem. Res. 17:311
21 Mingos DMP (1985) J. Chem. Soc., Chem. Comm., 1352
22 Mingos DMP (1986) Chem. Soc Rev. 15:31
23 DeKock RL, Gray HB (1980) Chemical Structure and Bonding. Benjamin/Cummings, Menjo Park (California), p. 419
24 Stone AJ (1980) Mol. Phys. 41:1339
25 Stone AJ (1981) Inorg. Chem. 20:563
26 Stone AJ, Alderton MJ (1982) Inorg. Chem. 21:2297
27 Stone AJ (1984) Polyhedron 3:2051
28 Fowler PW, Porterfield WW (1985) Inorg. Chem. 24:3511
29 King RB, Rouvray DH (1977) J. Amer. Chem. Soc. 99:7834
30 King RB (1981) Inorg. Chim. Acta 49:237
31 King RB (1982) Polyhedron 1:132
32 King RB (1986) Inorg. Chim. Acta 116:99, 109, 119, 125
33 King RB (1990) Isr. J. Chem. 30:315

2.3.1 The Inert-Gas Shell Configuration Approach

Mononuclear Metal Complexes. A large proportion of mononuclear transition metal complexes can be considered to posses 18 electrons around the metal atoms, i.e. an inert-gas shell configuration. The central metal atom utilizes its nine metal valence orbital – one s, three p and five d orbitals – to accommodate both metal valence electrons and ligand electron pairs. These features constitute the well-known *Effective Atomic Number* (E.A.N.) or *18-electron rule*. The large majority of low oxidation states, diamagnetic complexes specially binary carbonyls and other compounds with π-acceptor ligands, obey this rule. Indeed, the rule has been exceptionally successful as a means of both predicting and rationalizing the structures of low-oxidation state, transition metal complexes. Exceptions do occur, particularly with d^8 metal ions, where many examples of 16-electrons (square-planar) complexes are known. In these complexes the p_z orbital has a relative high energy and it is found to be non-bonding and empty. Exceptions to the 18-electron rule are also found for the group 11. Au(I) forms primarily two-coordinate, 14 electron complexes. Such deviations may be considered to be due to the relatively large s (or d) -p promotion energies found for the free atoms. As discussed in Chapter 1, as the atomic number increases across a transition metal series, the energy of the s and d orbitals drops more rapidly than that of the p orbitals thus increasing the s (or d) -p promotion energies.

The 18 electron-rule normally does not apply to transition metal complexes with π-donor ligands and metal atoms in normal oxidation states. In this case the metal orbitals with π-symmetry lie at relatively high energy and they stay empty or are only partially filled. In this way the formation of paramagnetic species can be possible.

Metal Clusters. The 18-electron rule as an approach to the bonding in metal cluster supposes the skeletal atoms are held together by a network of localized bonds and that each individual cluster atom utilizes its nine atomic orbitals to form metal-metal bonds as well as to accommodate both metal valence electrons and ligand electron pairs. Consequently, in the application of the 18-electron rule to clusters, the following three assumptions should be taken into account:

1. Metal-metal bonds are localized on the polyhedron edges.
2. All metal-metal bonds are two-center/two-electron bonds.
3. Ligands are in general one electron pair donors irrespective of their charges. Cyclopentadiene (π-Cp) and arenes are 5 and 6 electron donors respectively.

The number of metal-metal bonds according to this rule should be the exact one to permit the metal atoms to fill up all their valence orbitals (18 electrons per metal atom).

Although this rule can be applied to individual atoms in a cluster, it is more satisfactory to consider the cluster as a whole. The total number of electrons in the system corresponds to the sum of the metal valence electrons plus the electrons donated by the ligands. Thus the number of electrons available for metal-metal bonds is given by following relationship.

$$\begin{array}{c} \text{Number of electrons} \\ \text{in Metal-metal bonds} \end{array} = \begin{array}{c} 2 \times (\text{Number of} \\ \text{metal orbitals}) \end{array} + \begin{array}{c} \text{Total number} \\ \text{of electrons} \end{array}$$

As it will be seen below, the counting of the electrons in the cluster can provide information not only about the number of the metal-metal bonds and the presence of multiple bonds but also about the geometry of the metal polyhedron.

A number of selected examples of the application of the 18-electron rule to a number of clusters with different nuclearities are described in Table 2.9. Predicted structural features are also compared there with the experimental evidences.

It is easy to see in Table 2.9 that successfully application of the rule to cluster carbonyl systems is in general restricted to the smaller cluster containing five or fewer metal atoms. However, also among the low nuclearity clusters there are

Table 2.9. Application of the effective atomic numbers (EAN) or 18-electron rule to metal clusters

Compound	Number of valence electrons			Number of electrons after inert-gas rule	Number of metal-metal bonds	Geometry
	Metal atoms	Ligands	Total available			
$[Re_3Cl_{12}]^{3-a}$	12	30	42	54	6	Triangle
$Fe_3(CO)_{12}$	24	24	48	54	3	Triangle
$Os_3(CO)_{10}H_2^b$	24	22	46	54	4	Triangle
$[Mn_3(CO)_{14}]^-$	22	28	50	54	2	Chain
$Ru_3(CO)_8(Cp)_2^c$	24	26	50	54	2	Chain
$[Re_4(CO)_{16}]^{2-}$	28	34	62	72	5	Butterfly geometry
$Re_4(CO)_{12}H_4$	28	28	56	72	8	Tetrahedron
$[Fe_4(CO)_{13}]^{2-}$	34	26	60	72	6	Tetrahedron
$Co_4Cp_4H_4^{b,c}$	36	24	60	72	6	Tetrahedron
$Pt_4(CH_3CO_2)_8$	32	32	64	72	4	Square planar
$Fe_5C(CO)_{15}$	40	34	74	90	8	Square pyramid
$Os_5(CO)_{16}$	40	32	72	90	9	Trigonal bipyramid
$Os_5S(CO)_{15}$	40	34	74	90	8	Square pyramid
$[Ni_5(CO)_{12}]^{2-}$	52	24	76	90	7	Trigonal pyramid
$Ru_6C(CO)_{17}$	48	38	86	108	11	Octahedron
$Os_6(CO)_{18}$	48	36	84	108	12	Bicapped tetrahedron
$[Co_6(CO)_{15}]^{2-}$	56	30	86	108	11	Octahedron
$[Rh_6(CO)_{15}I]^-$	54	32	86	108	11	Octahedron
$Ir_6(CO)_{16}$	54	32	86	108	11	Octahedron

[a] Contribution of halides to electron counting: μ_1-X = 2 electrons, μ_2-X = 4 electrons and μ_3-X = 6 electrons.
[b] H is considered to contribute with 1 electron.
[c] Cp = η^5-C_5H_5 is considered to contribute with 5 electrons.

Table 2.10. Relationship between structure and number of cluster valence electrons according to the 18-electron rule

No. of Cluster valence electrons	Metal cluster structure
48	Triangle
50	Open triangle
60	Tetrahedron
62	"Butterfly"
74	Square pyramid
76	Trigonal bipyramid
86	Octahedron

many exceptions to the rule. Thus for instance, the existence of paramagnetic trinuclear species as $Co_3(CO)_9S$ and $Ni_3(CO)_2(\pi\text{-}Cp)_3$ indicates the presence of partially empty metal valence orbitals. The compound $Co_3(CO)_4(SEt)_5$ in which the group SEt comports as 3-electron donor does not fulfil the requirements of the 18-electron rule either, because it has 50 electrons, more than the 48 electrons required for the nucleus Co_3.

For hexanuclear clusters, deviations of the 18-electron rule are almost systematic since these clusters normally have 86 electrons, i.e. two electrons more than the required for an inert-gas shell configuration.

In spite of the apparent restrictions to the application of the 18-electron rule it is certainly interesting to discuss some of its consequences or corollaries:

1. There is a determined correspondence between the type of polyhedron built by metal-metal simple bonds and the number of electrons in the system. There is a kind of magic numbers for building polyhedra. These numbers for some common polyhedra are given in Table 2.10. Those clusters with an electron number higher or lower than the corresponding magic number are seen as electron-rich or electron-poor cluster respectively.

2. In the case of electron-poor systems the electron deficit must be compensated by the formation of multiple bonds. Thus, the compound $H_2Os_3(CO)_{10}$ that actually has only 46 electrons should have an Os–Os double bond leading to an isosceles triangular array of osmium atoms. Such a prediction agrees with structural data (Table 2.2) which show indeed one Os-Os distance of 2.60 Å, shorter than the other two (2.8 Å).

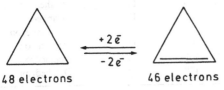

many many many many many many many many many many many many many many many many many many many many many many many many many many many many many many many many many many many many many many

48 electrons   46 electrons

Analogously, the Rhenium hydride $H_4Re_4(CO)_{12}$ with 56 electrons is also an unsaturated species. Two double bonds delocalized in a tetrahedral structure as shown in following resonance structures should agree with the relatively symmetric structure found for that compound (see Table 2.3).

etc...

The rhenium halide anion $[Re_3X_{12}]^{2-}$ with 42 electrons is also an electron-poor system. The principal characteristic of a series of tri and tetrarhenium clusters are shown in Table 2.11. The peculiarly short Re-Re distances observed for the halide derivatives agree with the three Re-Re double bonds predicted by the 18-electron rule.

3. The electron excess regarding the magic numbers for a given structure implies the rupture of bonds and cluster opening, so bonding orbitals are converted in non-bonding ones for housing the electron excess.

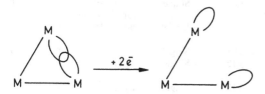

As occurring in the clusters $Re_4(CO)_{14}$ (tetrahedron, 60 electrons), $[Re_4(CO)_{16}]^{2-}$ (butterfly, 62 electrons), and $Pt_4(CH_3CO_2)_8$ or $Os_4S(CO)_{12}$-(CCHPh) (planar 64 electrons) (see Table 2.3 and Fig. 2.14), close and open structures are related by differences of two electrons.

2.3.2 Polyhedral Skeletal Electron Pair Theory

The polyhedral skeletal electron pair approach retains the condition that each metal atom uses its nine valence atomic orbitals. However it does not consider all the electron in the system but only those involved in the metal network. This approach is fundamentally grounded on the analogy existing between metal clusters and boron hydrides and carboranes.

Bonding in Boranes and Carboranes. Although in Chapter 4 some aspects of the boron cluster chemistry will be considered, it is convenient to discuss here some structure and bonding features characteristic for boron clusters.

Boron hydrides and carboranes adopt structures in which their skeletal boron and/or carbon atoms form complete or nearly complete deltahedra, i.e. regular polyhedra with triangular faces.

Table 2.11. Correlation between metal polyhedron geometry and cluster valence electron counting. Structural properties of selected rhenium clusters

Compound	Polyhedron geometry	Distances Re–Re (Å)	Type of Re–Re bonding	Ligands bridging Re–Re bonds	Number of valence electrons	Ref.
Metal	—	2.75	—	—	—	
Re_3Cl_9	Equilateral triangle	2.49	Double	μ_2-Cl	42	1
$[Re_3Cl_{12}]^{3-}$	Equilateral triangle	2.47	Double	μ_2-Cl	42	2, 3
$Re_3Cl_3(CH_2SiMe_3)_6$	Equilateral triangle	2.39	Double	μ_2-Cl	42	4
$[H_4Re_3(CO)_{10}]^-$	Isosceles triangle	3.18	Single	μ_2-H	46	5
		2.82	Double	bis (μ_2-H)		
$[H_3Re_3(CO)_{10}]^{2-}$	Isosceles triangle	3.03	Single	one (μ_2-H) for two bonds	46	6
		2.80	Double	bis (μ_2-H)		
$[HRe_3(CO)_{12}]^{2-}$	Isosceles triangle	3.01	Single	—	48	7
		3.12	Single	μ_2-H		
$[H_2Re_3(CO)_{12}]^-$	Isosceles triangle	3.18	Single	μ_2-H	48	8
		3.04	Single	—		
$HRe_3(CO)_{14}$	Bent. Angle Re–Re–Re = 90°	3.29	Single	μ_2-H	50	9
$H_4Re_4(CO)_{12}$	Tetrahedron	2.90–2.95	Partially double	μ_3-H	56	10
$[H_6Re_4(CO)_{12}]^{2-}$	Tetrahedron	3.14–3.17	Single	μ_2-H	60	11
$[Re_4(CO)_{16}]^{2-}$	Two fused coplanar triangles	2.96–3.02	Single	—	62	12

References

1 Cotton FA, Mague JT (1964) Inorg. Chem. 3:1094
2 Bertrand JA, Cotton FA, Dollase WA (1963) Inorg. Chem. 2:1166
3 Robinson WT, Fergusson JE, Penfold BR (1963) Proc. Chem. Soc., 116
4 Hursthouse MB, Malik KMA (1978) J. Chem. Soc., Dalton Trans, 1334
5 Ciani G, D'Alfonso G, Freni M, Romiti P, Sironi A, Albinati A (1977) J. Organomet. Chem. 136:C49
6 Bertolucci G, Freni M, Romiti P, Ciani G, Sironi A, Albano VG (1976) J. Organomet. Chem. 113:C61
7 Ciani G, D'Alfonso G, Freni M, Romiti P, Sironi A (1978) J. Organomet. Chem. 157:199
8 Churchill MR, Bird PH, Kaesz HD, Bau R, Fontal B (1968) J. Am. Chem. Soc. 90:7135
9 Teller RG, Bau R (1982) Struct. Bonding 44:1
10 Bau R, Kirtley SW, Sorrell TN, Winarko S (1974) J. Am. Chem. Soc. 96:998
11 Ciani G, Sironi A, Albano VG (1977) J. Organomet. Chem. 136:339
12 Churchill MR, Bau R (1968) Inorg. Chem. 7:2606

The structure of most known boranes and carboranes are based on the regular deltahedra illustrated in Fig. 2.27. Three types of structures are known for this class of compounds: *closo* (closed), *nido* (nest-like) and *arachno*-structures ("cobweb" structures). The relationship between these three types of

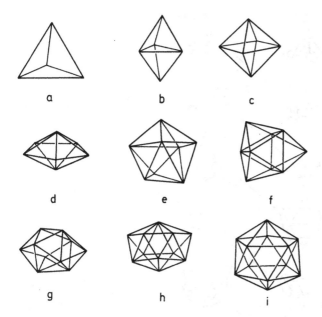

Fig. 2.27. Regular deltahedra often found in molecular structures of boranes and carboranes. (**a**) Tetrahedron; (**b**) trigonal bipyramid; (**c**) octahedron; (**d**) pentagonal bipyramid; (**e**) dodecahedron; (**f**) tricapped trigonal prism; (**g**) bicapped archimedean antiprism; (**h**) octahedron (**i**) icosahedron

structures can be appreciated in Fig. 2.28. In *closo*-structures, skeletal boron or carbon atoms occupy all the vertexes of the polyhedron. In the case of *nido*- and *arachno*-structures one and two of the vertexes of the appropriate polyhedron remain unoccupied respectively.

On each skeleton boron or carbon atom there is always a hydrogen atom or some other simple terminal ligand which emerges radially away from the center of the polyhedron and that is linked to the cluster by a single two-electron bond. Therefore, the skeleton distracts two electrons per atom for such interactions. The remaining valence-shell electrons are regarded as skeletal bonding electrons.

Each carbon or boron atom contributes with four orbitals. Due to the pseudo-spherical surface of these atom polyhedra, a convenient hybridation for the atomic orbitals of skeletal atoms should be, for instance, two radially oriented hybrid *sp*-orbitals and two pure *p* orbitals as shown schematically in Fig. 2.29. If the outward pointing *sp* hybrid orbitals are assigned to the linkages with the radial ligands, three atomic orbitals on each boron or carbon atom would remain available for the skeletal bonding. One of these orbitals is an inward pointing, radially oriented *sp* hybrid orbital. The other two are *p* orbitals oriented tangentially at the cluster surface.

In a polyhedron of n vertices there are then 3n atomic orbitals available for the boron or boron-carbon skeleton. These orbitals may be then linearly

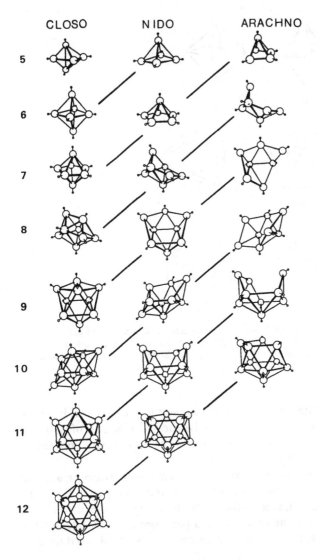

Fig. 2.28. *Closo, nido* and *arachno* boranes. Reproduced with permission from Rudolph RW, Pretzer WR (1972) Inorg. Chem. 11:1974

combined to a set of 3n molecular orbitals arranged according to the symmetry properties of the cluster. Of these 3n skeleton molecular orbitals, (n + 1) are bonding orbitals. Figure 2.30 describes schematically the linear combinations generating the skeleton bonding molecular orbitals in the octahedral cluster anion $[B_6H_6]^{2-}$. There, the six BH groups are linearly combined to give seven bonding molecular orbitals. An unique orbital of symmetry a_{1g} arises from the combination in phase of the six radial *sp* atomic orbitals. The interaction of the

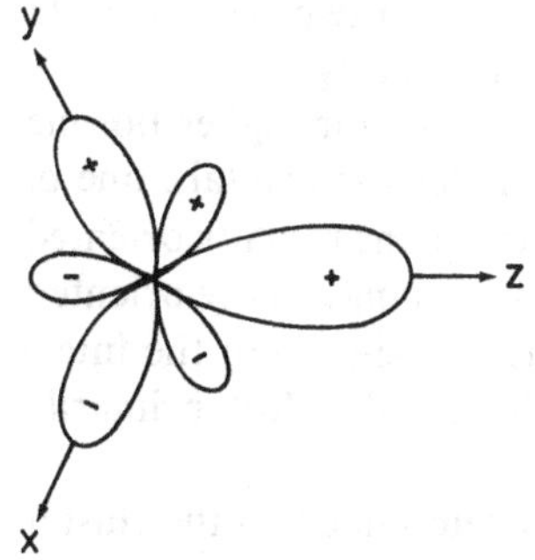

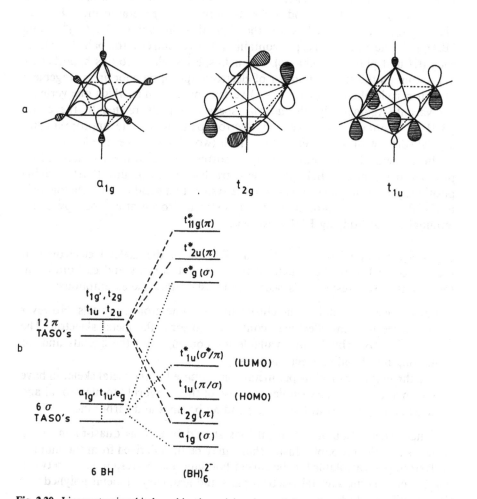

Fig. 2.29. Hybrid orbitals of boron useful in bonding description of boranes and carboranes

Fig. 2.30. Linear atomic orbital combinations giving rise to skeleton bonding molecular orbitals in the anion $[B_6H_6]^{2-}$. (a) Orbital sketches; (b) qualitative Molecular Orbital energy diagram

12 tangentially oriented p orbitals lead on the other side to two sets of triple degenerated bonding molecular orbitals of symmetry t_{2g} and t_{2u}.

The same molecular orbital distribution pattern persists for higher borane anions $[B_nH_n]^{2-}$. There are always $(n + 1)$ bonding molecular orbitals, one of a_1 or a_g symmetry representing the in phase combination of the radially oriented sp hybrid atomic orbitals but without contribution of the tangentially oriented p orbitals. The remaining n bonding molecular orbitals arising from the interaction of the 2n tangentially oriented p atomic orbitals are therefore primarily polyhedron surface orbitals.

Since the number of bonding molecular orbitals is determined by the cluster symmetry, the same polyhedron can serve as the basis for the structures of an ample range of isoelectronic species as neutral carboranes with formula $C_2B_{n-2}H_n$ or the naked clusters as Ge_9^{2-} that will be dealt with in Chapter 4.

Formally, neutral *nido* and *arachno*-boranes with general formula B_nH_{n+4} and B_nH_{n+6} are related with the hypothetical anions $[B_nH_n]^{4-}$ and $[B_nH_n]^{6-}$. The anion $[B_nH_n]^{4-}$ contains n + 2 skeletal electron pairs (one from each BH unit plus anion charge). The *nido*-species related to this hypothetical anion contain therefore a number of electrons appropriate for an arrangement of their n skeletal atoms in an $(n + 1)$-vertex polyhedron in which one vertex is left vacant. Analogously, the *arachno*-species formally related to the anion $[B_nH_n]^{6-}$ contain n + 3 skeleton electron pairs which are appropriate for a polyhedron with n + 2 vertices in which two of the vertices are left vacant.

In the neutral *nido*-boranes B_nH_{n+4}, corresponding to the addition of four protons to the anion $[B_nH_n]^{4-}$, the extra hydrogens occupy BHB bridging positions preserving the symmetry of the system to a great extent. Analogously in the B_nH_{n+6} *arachno*-boranes the six extra hydrogen atoms occupy endo-terminal BH or bridging BHB positions.

Transition Metal Clusters. For the application of the skeletal electron pair theory deduced from the structural properties of boranes and carboranes to transitions metal clusters, it is necessary to establish some assumptions:

1. Each transition metal in the cluster uses its nine atomic orbitals. However only three of them effectively contribute to form the metal skeleton. The remaining six orbitals are available for bonding with the ligands and for housing non-bonding electron pairs.
2. The three atomic orbitals per metal compromised in the metal skeleton have a form and a symmetry similar to those of the sp hybrid atomic orbital and tangential p orbitals in the corresponding boranes and carboranes.

Under such assumptions metal ions should behave as quasi-main group elements and the probable cluster shape may be then derived from the number of electron pairs available for the cluster bonding. The correspondence between number of bonding skeletal electron pairs and geometry of metal polyhedra is shown in Table 2.12.

Table 2.12. Correlation between metal polyhedron geometry and the number of skeletal electron pairs

Number of skeletal electron pairs	Fundamental metal polyhedron
6	Trigonal bipyramid
7	Octahedron
8	Pentagonal bipyramid
9	Dodecahedron
10	Tricapped trigonal prism
11	Bicapped square antiprism
12	Octadecahedron
13	Icosahedron

A convenient procedure for electron counting in metal clusters is achieved by making a balance between the number of electrons in the system and the atomic orbitals to which they can be assigned.

The total number of electrons associated with the valence shell of cluster metal atoms corresponds to the sum of the metal valence electrons, v, plus those provided by the ligands, L.

As mentioned already, from the nine orbitals provided by each transition metal atom, three are left to form the metal skeleton. The remaining six orbitals are to be used in the metal-ligand bonds as well as for housing non-bonding metal electrons. In the case of species with π-acceptor ligands all these six metal orbitals are stabilized by metal-ligand interactions. Some of them accepting electron density of the σ-donor ligands and the others by back-donation towards empty π-acceptor orbitals in the ligands. Accordingly, in the distribution of the cluster valence-electrons, these six orbitals will be filled preferentially.

Thus, each transition metal atom in the cluster will always spend twelve electrons for accommodating ligands and non-bonding electrons furnishing $(v + 2L - 12)$ electrons for skeletal bonding.

The same procedure is also valid for mixed clusters containing both transition metals and main group elements. Each main group element will leave three atomic orbitals available for cluster bonding and one remaining valence-shell atomic orbital either to accommodate a lone pair or to bond a suitable ligand. The number of electrons supplied by one atom E or group EL for cluster is $(v + y - 2)$ where v is the number of valence shell of the cluster atom E and y the number of electrons furnished by the ligand L. Table 2.13 summarizes the electron contribution of a some main group cluster units.

A number of applications of these electron counting procedure are illustrated in Table 2.14.

An alternative procedure for counting electrons according to the Skeleton Electron Pair Theory is to regard the cluster as being made up of a group of fragments. Thus $Os_5(CO)_{16}$ may be considered as a combination of five $Os(CO)_3$ fragments and an additional CO-ligand. By the same arguments

Table 2.13. The number of skeletal bonding electrons contributed by some main group cluster fragments[a]

v	Main group element	E $(X = 0)$	EH $(X = 1)$	EH$_2$ or EL $(X = 2)$
3	B, Al, Ga, In, Tl	1	2	3
4	C, Si, Ge, Sn, Pb	2	3	4
5	N, P, As, Sb, Bi	3	4	5
6	O, S, Si, Te	4	5	6
7	F, Cl, Br, I	5	[6]	[7]

[a] No. of skeletal electrons $= v + X - 2$ where v is the number of valence shell electrons on the main group element E and X is the number of electrons from ligand(s) on E.

Reference

Wade K (1975) Chem. Br. 11:177

Table 2.14. Skeletal electron counting. Examples and applications

Compound	Cluster valence electrons	Ligand and non-bonding electrons	Skeleton electron pairs	Cluster structure
Os$_5$(CO)$_{16}$	72	60	6	Trigonal bipyramid
[B$_5$H$_5$]$^{2-}$	22	10	6	Trigonal bipyramid
Fe$_3$Bi$_2$(CO)$_{12}$	52	40	6	Trigonal bipyramid
Pb$_5^{2-}$	22	10	6	Trigonal bipyramid
[Mo$_6$Cl$_{14}$]$^{2-}$	86	72	7	Octahedron
Co$_6$(CO)$_{16}$	86	72	7	Octahedron
C$_2$B$_4$H$_6$	26	12	7	Octahedron
Co$_3$B$_3$H$_5$(Cp)$_3$	56	42	7	Octahedron
FeB$_4$C$_2$H$_6$(CO)$_3$	40	24	8	Pentagonal bipyramid
[B$_7$H$_7$]$^{2-}$	30	14	8	Pentagonal bipyramid
B$_6$C$_2$H$_8$	34	16	9	D$_{2d}$-Dodecahedron
[Ni$_9$(CO)$_{18}$]$^{2-}$	128	108	10	Stacked trigonal prism
B$_7$C$_2$H$_9$	38	18	10	Tricapped trigonal prism
[MnB$_6$C$_2$H$_8$(CO)$_3$]$^-$	48	28	10	Tricapped trigonal prism
Pb$_9^{4-}$	38	16	11	Capped square antiprism
B$_9$H$_9$S	40	18	11	Capped square antiprism
Os$_5$(CO)$_{16}$	72	60	6	Trigonal bipyramid
Ni$_3$B$_5$CH$_6$(Cp)$_3$	70	48	11	Capped square antiprism
B$_9$C$_2$H$_{11}$	46	22	12	Octadecahedron
CoB$_8$C$_2$H$_{10}$Cp	56	32	12	Octadecahedron
Co$_2$B$_8$C$_2$H$_{10}$(Cp)$_2$	70	44	13	Icosahedron

discussed above the number of electrons contributed by each M(CO)$_x$ fragment for skeletal bonding in the cluster is $(v + 2x - 12)$ where v is the number of valence electrons of the metal M.

Since each unit in Os$_5$(CO)$_{16}$ contributes with two electrons $[8 + (2 \times 3) - 12]$, the five units plus the additional two electrons of the co-group give a total of 12 electrons for the skeletal bonding. Thus there are six

Table 2.15. The number of skeletal bonding electrons contributed by selected organo transition metal fragments

V	Cr	Mn	Fe	Co	Ni
			$Fe(CO)$ -2	$Co(CO)$ -1	$Ni(CO)$ 0
			$Fe(CO)_2$ 0	$Co(CO)_2$ 1	$Ni(CO)_2$ 2
	$Cr(CO)_3$ 0	$Mn(CO)_3$ 1	$Fe(CO)_3$ 2	$Co(CO)_3$ 3	$Ni(CO)_3$ 4
$V(CO)_4$ 1	$Cr(CO)_4$ 2	$Mn(CO)_4$ 3	$Fe(CO)_4$ 4	$Co(CO)_4$ 5	
		$Mn(\pi\text{-}Cp)$ 0	$Fe(\pi\text{-}Cp)$ 1	$Co(\pi\text{-}Cp)$ 2	$Ni(\pi\text{-}Cp)$ 3

Reference

Mingos DMP (1972) Nature Phys. Sci. 236:99

skeletal electron pairs for the m + 1 (m = 5) available skeletal orbitals in a trigonal bipyramide metal arrangement.

Table 2.15 summarizes the number of electrons contributed by a variety of common organometallic fragments.

2.3.3 Isolobal Relationships

The basis of the Skeleton Electron Pair Theory described above rest on the assumption that the constituent fragments of main group and transition metal clusters contribute both the same number of electrons and orbitals for the cluster bonding and that these orbitals have similar nodal properties.

Semiempirical molecular orbital analyses of the frontier molecular orbitals of metal carbonyl fragments confirm that there are definite similarities between the numbers, energies, shapes and nodal characteristics of these orbitals and those in main group fragments. Fig. 2.31 illustrates qualitative orbital descriptions of three pair of fragments. Similarity between the fragments is apparent. Although they are neither isostructural nor isoelectronic each one of them has a frontier orbital which is approximately equivalent. When this kind of situation occurs – in which the number, symmetry properties, approximate energy and shape of the frontier orbitals, and the number of electrons are similar – they are said to be isolobal fragments and are symbolized with a double arrow as indicated at the bottom of the schemes in Fig. 2.31. In Table 2.16 is summarized a series of isolobal relationships between a variety of main group and transition organometallic fragments.

Some more rigorous, non-parametric calculations also confirm the validity of the isolobal principle. Fenske-Hall LCAO-MO-SCF calculations of the structure of B_5H_9 and some of its ferroborane derivatives, i.e. 1-$Fe(CO)_3B_4H_8$, 2-$Fe(CO)_3B_4H_8$ and 1,2-$[Fe(CO)_3]_2B_3H_3$ have been completed. A summary of a qualitative description of the properties of frontier molecular orbitals in

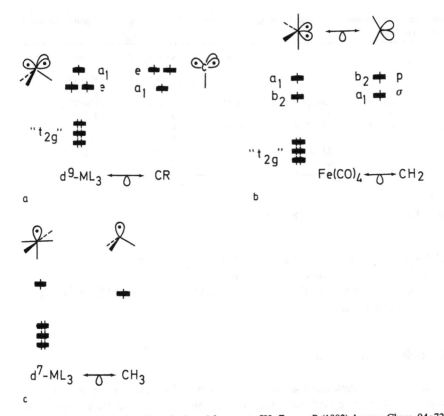

Fig. 2.31. Isolobal relationships of selected fragments [Hoffmann R (1982) Angew. Chem. 94:725]

C_5H_9 and 1-$Fe(CO)_3B_4H_8$ can be observed in Fig. 2.32. The HOMO-1 and HOMO-2 of B_5H_9 are an e and an a_1 orbital respectively. Both of these orbitals are involved in cluster bonding between the apical boron atom and the basal B_4H_8 group. The a_1 orbital also contains significant (BH)-apical character. The electronic structure of 1-$Fe(CO)_3B_4H_8$ is very similar in terms of the cluster e and a_1 orbitals. In addition there are three non-bonding orbitals that are predominantly localized on $Fe(CO)_3$ ("d^6").

The shape and nodal characteristics of the e and a_1 orbitals in both complexes observed in the contour diagrams reproduced in Fig. 2.33 provide a more quantitative assessment of the isolobal principle for these compounds.

The isolobal principle is widely appreciated as one of the most valuable generalizations in modern inorganic chemistry. Among numerous applications of the isolobal arguments to clusters is the rationalization of the transition metal cluster geometries by identifying isolobal fragments. These arguments are also considerably useful as a tool for designing synthetic pathways for cluster species.

The isolobal principle wholly arises from considerations about the electronic structure of molecular fragments at a microscopic level. Kinetic and thermodyn-

Table 2.16. Examples of isolobal fragment relationships[a]

CH_3	CH_2	CH
H	μ_2-S	EL_3(E = P, As)
$Mn(CO)_5$	$Fe(CO)_4$	$Co(CO)_3$
$Fe(CO)_2(Cp)$	$Rh(CO)(Cp^*)$	$Ni(Cp)$
$Mo(CO)_3(Cp)$	$Re(CO)_2(Cp)$	$W(CO)_2(Cp)$
	$Cr(CO)(NO)(Cp)$	$[Mn(CO)_2(Cp)]^+$
$Co(CO)_4$	$Cr(CO)_5$	$Re(CO)_4$
$Zn(Cp)$	$Cu(Cp^*)$	$Rh(\eta^6\text{-}C_6H_6)$
$Au(PPh_3)$	$IrCl(CO)_2$	$Re(CO)_3Br_2$
$Rh(PPh_3)_2(\eta^5\text{-}C_2B_9H_{11})$	$Fe(CO)_2(\eta^5\text{-}C_2B_9H_{11})$	$Mn(CO)(h^5\text{-}C_2B_9H_{11})$
	$TaMe(Cp)_2$	$TaCl(PMe_3)_2(Cp)$

CH_3^+	CH_2^+	CH^+
$Cr(CO)_5$	$Mn(CO)_4$	$Fe(CO)_3$
$Mn(CO)_2(Cp)$	$Fe(CO)(Cp)$	$Rh(Cp)$
BH_3	BH_2	BH

[a] Isolobal correlations are not one-to-one relationships, so a given fragment may be isolobal with more than one other fragment.

References

Stone FGA (1984) Angew. Chem. 96:85
Mingos DMP, Johnston RL (1987) Struct. Bonding 68:31

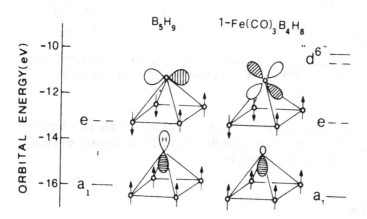

Fig. 2.32. Qualitative description of upper MOs in B_5H_9 and 1-$Fe(CO)_3B_4H_8$. Reproduced with permission from DeKock RL, Fehlner TP (1982) Polyhedron 1:521

amic aspects of the problem can lead to deviations in the predictions based on the isolobal principle.

Finally, it should be remembered that the orbitals of isolobal fragments are frequently non identical. There are differences, specially between fragments

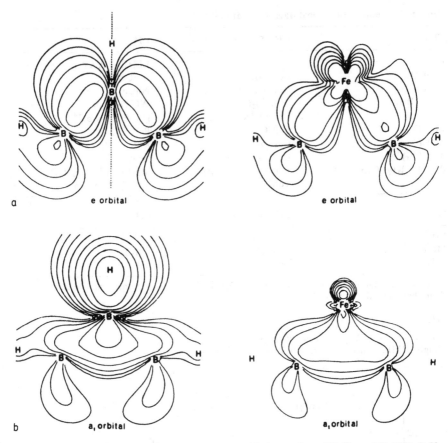

Fig. 2.33. Orbital contour diagrams for molecular orbitals e and a_1 of B_5H_9 and $Fe(CO)_3B_4H_8$. Reproduced with permission from DeKock RL and Fehlner TP (1982) Polyhedron 1:521

constituted by main group elements and by transition metals in which the d-electrons always play an important role that may have chemical significance affecting the applicability of the principle.

2.3.4 Capping Principle

The polyhedral Skeletal Electron Pair Theory as described above is particularly effective for rationalizing the structures of the species which adopt spherical deltahedral geometries or fragments derived from them as being the *nido* and *arachno* geometries. However, only a limited number of large clusters adopt as their structure polyhedra of this type and they are often structurally better described in terms of condensed polyhedra derived from triangular, tetrahedral, trigonal prismatic, and octahedral components (see Figs. 2.25 and 2.26).

The Capping Principle has been derived for evaluating the electron count in condensed polyhedra from those of its component polyhedra. Thus, it has been established that the total electron count in a condensed polyhedron is equal to the sum of the electron count for the parent polyhedron minus the electron count characteristic of the atom, pairs of atoms, or face of atoms which are common to both polyhedra. The frontier orbitals of the capping fragment are matched in

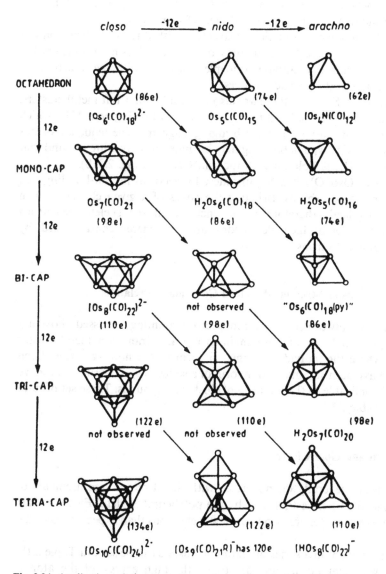

Fig. 2.34. Application of the decapping and capping principles to osmium carbonyl clusters. Reproduced with permission from Mingos DMP, May AS (1990) In: Shriver DF, Kaesz HD, Adams RD (eds), The chemistry of metal cluster complexes. VCH, New York, p 11

symmetry by orbitals of the parent clusters which are already bonding, so no changes in the number of skeletal bonding molecular orbitals result from the capping process.

Consequently, capping with a conical $M(CO)_3$ fragment with twelve electrons in non-bonding and metal-ligand bonding orbitals (isolobal for instance with a CH-group) leads to an increase in cluster valence electron count of twelve. Thus, capped deltahedral clusters are characterized by $[(14n + 2) + 12m]$ valence electrons, where n is the number of metal atoms in the parent deltahedron and m is the number of capping conical metal fragments.

A classical example of correlation of structure with valence electron count is shown in Fig. 2.34. There the structures of a series of osmium clusters are systematized by using the capping principle combined with the approach of decapping (i.e. removal of cluster vertices).

As mentioned in Sect. 2.2. there are also ring and planar raft metal clusters. Many of them may be considered to be derived from the fragment ML_4 which indeed frequently appears as an edge-bridging fragment. The tendency of this fragment to adopt this type of bonding may be understood bearing in mind that the fragment d^8-ML_4 is isolobal with the unit CH_2 (see Fig. 2.31b).

The fragment $Os(CO)_4$ has, for instance, 14 electrons in non-bonding and metal-ligand bonding orbitals. Bridging the edges of a given cluster with m of these groups leads therefore to an increase in the skeleton electrons of 14m. Hence edge-bridged deltahedra are in general characterized by $[(24n + 2) + 14m]$ electrons.

2.3.5 Rules for Cluster Structure-Electron Counting Correlations

The relationships between structure and electron counting discussed above are valid for a great number of both main group and transition metal cluster compounds where the individual atoms conform to the noble-gas rule. Such correlations have been summarized in form of a series of specific rules that have proved to be applicable to the majority of cluster compounds. These set of rules are listed in Table 2.17.

2.3.6 Platinum and Gold Clusters

Platinum clusters generally do not agree with the electronic requirements of transitions metal clusters established by the polyhedral Skeletal Electron Pair Theory. These clusters generally possess fewer electrons than expected from this theory.

The electron count in a series of platinum clusters can be seen in Table 2.18. The reason why that occurs appears to be the high energy of the atomic p_z-orbital. The molecular orbitals derived from this orbital perpendicular to the metal-ligand plane are also high-lying and unavailable for bonding.

Table 2.17. Specific rules for structure-electron counting correlations in main group and transition metal clusters

	Rule	Formula	Examples — Cluster	Geometry	No. of valence electrons S
1.	Ring clusters are characterized by 6n or 16n valence electrons.	$6n$ or $16n^a$ $n \geq 3$	$(PPh)_4$	Square	48
			S_8	Octagon	64
			$Fe_3(CO)_{12}$	Triangle	24
			$Fe_4(CO)_{11}(PC_6H_4Me)_2(P(OMe)_3)$	Square	48
2.	Three connected cluster are characterized by 5n or 15n valence electrons. They correspond to electron precise polyhedra.	$5n$ or $15n$ n even ≥ 4	P_4	Tetrahedron	20
			C_8H_8 (Cubane)	Cube	40
			$Ir_4(CO)_{12}$	Tetrahedron	60
			$Re_6(CO)_{18}(PMe)_3$	Trigonal prism	90
			$[Co_6C(CO)_{15}]^{2-}$	Trigonal prism	90
3.	Connected polyhedral molecules in which metal atoms lie approximately on a single spherical surface are characterized by 14n + 2 valence. electrons irrespective of interstitial metal atoms.	$14n + 2$	$Co_6(CO)_{16}$	Octahedron	86
			$Ru_6C(CO)_{17}$	Octahedron	86
			$[Co_8C(CO)_{18}]^{2-}$	Square antiprism	114
			$[Rh_{12}Rh^*(CO)_{24}H_3]^{2-}$ (Rh^* = interstitial atom)	Anticubo-octahedron	170
4.	Closo deltahedral clusters are characterized by 4n + 2 or 14n + 2 valence electrons.	$4n + 2$ or $14n + 2$ $n \geq 5$	$[B_6H_6]^{2-}$	Octahedron	26
			$C_2B_6H_8$	Icosahedron	34
			$Os_5(CO)_{16}$	Trigonal bipyramid	72
			$[Rh_{10}S(CO)_{22}]^{2-}$	Bicapped square antiprism	142
			$[Rh_{12}Sb(CO)_{27}]^{3-}$	Icosahedron	170
5.	Nido deltahedral clusters are characterized by 4n + 4 or 14n + 4 valence electrons.	$4n + 4$ or $14n + 4$ $n \geq 4$	B_5H_9	Pentagonal pyramid	24
			$C_2B_4H_8$	Square pyramid	28
			$Ru_5C(CO)_{15}$		74
6.	Arachno deltahedral clusters are characterized by 4n + 6 or 14n + 6 valence electrons.	$4n + 6$ or $14n + 6$ $n \geq 4$	B_4H_{10}	Butterfly	26
			$[Os_4N(CO)_{12}]^-$	Butterfly	62
			$Fe_4(CO)_{11}(PPh)_2$	Square	62

Table 2.17. (continued)

Rule	Formula	Examples		No. of valence electrons
		Cluster	Geometry	
7. The total electron count for a condensed polyhedron is the sum of the electron counts of parent polyhedra minus the electron count of the element common to both polyhedra: an atom (18 electrons), an edge (34 electrons), a triangular face (48 electrons), or a square face (62 electrons).	e.g. fusion of two *closo* metal deltahedra a) with a common atom: $14 (M_1 + M_2) + 4 - 18$	a) $Os_5(CO)_{18}$	Bow-tie (triangle + triangle)	78
		$HgRu_6(CO)_{20}(NO)_2$	Vertex-sharing butterfly (butterfly + butterfly)	106
	b) with a common edge: $14 (M_1 + M_2) + 4 - 34$	b) $[Ru_{10}C_2(CO)_{24}]^{2-}$	Two edge-sharing octahedra (octahedron + octahedron)	138
		$H_2Os_5(CO)_{16}$	Edge-bridged tetrahedron (Tetrahedron + triangle)	74
	c) with a common face: $14 (M_1 + M_2) + 4 - 48$	c) $Re_6P(CO)_{19}(PMe)(PMe_2)$	Capped square pyramid (Square pyramid + tetrahedron)	86
		$Os_7(CO)_{21}$	Capped octahedron (Octahedron tetrahedron)	92

[a] n = number of main group or transition metal cluster atoms respectively.

References

Mingos DMP, May AS (1990) In: Shriver DF, Kaesz HD, Adams RD (eds) The chemistry of metal cluster complexes. VCH, New York, p. 11
Mingos DMP, Johnston RL (1987) Struc. Bonding 68:29
MacPartlin M, DMP Mingos (1984) Polyhedron 3:1321

Table 2.18. Electron counting in platinum cluster compounds

Skeletal geometry	Platinum cluster	Electron count
Triangle	$Pt_3(CO)_3(PR_3)_3$	42
Tetrahedron	$Pt_4(PR_3)_4H_8$	56
Butterfly	$Pt_4(CO)_5(PR_3)_4$	58
Triagonal pyramid	$Pt_5(PR_3)_5H_8$	68
Edge-bridged tetrahedron	$Pt_5(CO)_6(PR_3)_4$	70
Triagonal prism	$[Pt_6(CO)_{12}]^{2-}$	90

Reference

Mingos DMP, Wardle RWM (1985) Transition Met. Chem. 10:441

The high effective nuclear charge experienced by d electrons in the elements of the group 11 or coinage metals causes them to be contracted and low-lying. Hence the d^{10} shell in gold clusters practically does not contribute to metal-metal bonding. Moreover, and by the same reason, the s-p promotion energy is very large and the valence p orbitals have a diminished role in cluster bonding. Nonetheless there are a great number of clusters containing AuL fragments. The $Au(PR_3)$ fragment, which appears to have a low-lying sp hybrid orbital with high s-character and two empty p orbitals, is indeed isolobal among others with H, CH_3, or $Co(CO)_4$. Accordingly, $Au(PR_3)$ is often found capping triangular faces in deltahedra or bridging edges in various types of clusters.

In gold clusters of the type $[Au_m(PR_3)_m]^x$, as the triphenyl phosphine derivatives shown schematically in Fig. 2.35, the electron count depends critically on both the nuclearity and geometry of the species.

In high nuclearity gold clusters with the general formula $[Au\{Au(PR_3)\}_m]^{x+}$, which are found in the two types of topologies exemplified in Fig. 2.36, not only the s gold orbitals but the p orbitals are also used for cluster bonding.

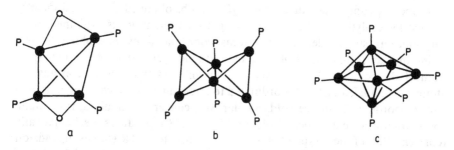

Fig. 2.35a–c. Schematic structures of gold clusters. (**a**) $Au_4I_2(PPh_3)_4$; (**b**) $[Au_6(PPh_3)_6]^{2+}$; (**c**) $[Au_7(PPh_3)_7]^+$

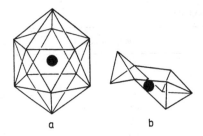

Fig. 2.36. Examples of high nuclearity gold clusters (**a**) Metal nucleus in the clusters $[Au(Au_{12}Cl_2(PR_3)_{10}]^{3+}$; (**b**) $[Au(Au_9Cl_3PR_3)_6]^+$

a   b

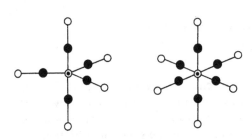

Fig. 2.37. Schemes of polyaurated species with five and six gold atoms. X = C or N; Y = PPh$_3$ [Schidbaur H (1990) Gold Bull. 23:11]

In recent developments of gold chemistry, an unexpected affinity between gold atoms, even with closed shell electronic configurations and equivalent electrical charges, has been observed. In such a context the term "aurophilicity" has been mentioned. According to this concept, gold atoms show a pronounced tendency to cluster around a given central element. Thus pyramidal species $[E(AuL)_3]^+$ with E = O, S, or Se show short Au–Au distances. A number of polyaurated species as those shown in the schemes in Fig. 2.37 have been indeed isolated.

2.3.7 Large Clusters

Gold phosphine compounds have also played a major role in the synthesis of exceptional large or "giant" clusters. From the reduction of PhAuCl with diborane, the compound $Au_{55}(PPh_3)_{12}Cl_6$ can be obtained. This stoichiometry verified by analytical and molecular weight determinations is explained as an Au_{13}-nucleus surrounded by Au-atoms coordinated by phosphine ligands giving thus an arrangement of close-packed gold atoms. The Mössbauer spectrum of the compound shows indeed four different types of gold atoms, i.e. atoms in the nucleus, atoms coordinated to phosphine and to chloride ions, and also uncoordinated surface gold. Similar M_{55} clusters of rhodium, ruthenium, and platinum have also been isolated. Figure 2.38 reproduces a schematically representation of these type of clusters. According to NMR studies, ligands on the cluster surface are mobile and change their positions very fast. Mobility of cluster surface atoms can be also deduced from such experiments. Consequently

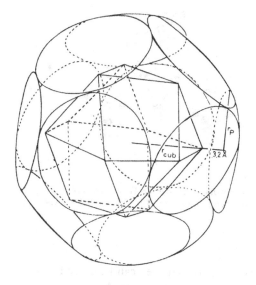

Fig. 2.38. Model for molecular cluster $M_{55}(PR_3)_{12}$. Reproduced with permission from G. Schmid (1985) Struct. Bonding 62:52

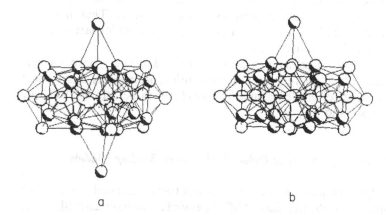

Fig. 2.39. Sideviews of metal framework of clusters (**a**) $(p\text{-Tol}_3P)_{12}Au_{18}Ag_{20}Cl_{14}$ and (**b**) $[(p-\text{Tol}_3P)_{12}Au_{18}Ag_{19}Br_{11}]^{2+}$. Reproduced with permission from Teo, BK (1988) Polyhedron 7:2317

a liquid-like behavior of the surface atoms appears to be one of the peculiar physical properties of such large clusters. It is very probable that this characteristic behavior of these giant clusters prevents the formation of the material required for crystallographic structure determinations.

Among large metal clusters there are also some heteronuclear or mixed-metal clusters that have been mentioned as "metal-alloy clusters". The bimetallic Au–Ag clusters $(p\text{-Tol}_3P)_{12}Au_{18}Ag_{20}Cl_{14}$ and $[(p\text{-Tol}_3P)_{12}Au_{18}Ag_{19}Br_{11}]^{2+}$, whose metal frameworks are illustrated in Fig. 2.39, belong to this class. A very

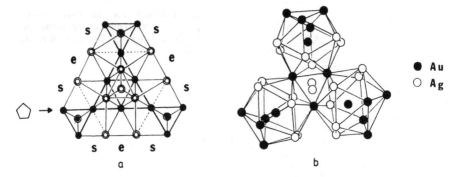

Fig. 2.40. Schematic representation of metal framework of cluster. $(p\text{-Tol}_3P)_{12}Au_{18}Ag_{20}Cl_{14}$ as three-edge-sharing icosahedra. Reproduced with permission from Teo, BK (1988) Polyhedron 7:2317

interesting structural feature in these clusters is that they can be considered as being formed by 13-atom centered icosahedral cluster units. As shown in the scheme reproduced in Fig. 2.40, the 38-atom cluster may be described as three icosahedra sharing three vertices plus two capping atoms. That has been mentioned as an example of a possible building up principle for large clusters based on the idea of "a cluster of clusters".

Electron counting rules available for small and medium-sized clusters appear to be insufficient for very large clusters with coordinatively unsaturated surfaces. These types of clusters could be considered to be actually half way between complexes and metals.

2.3.8 Quantitative Approaches to Delocalized Cluster Bonding Models

The analogy between boranes and carboranes and transition metal clusters rests on the assumption that the fragments $M(CO)_x$ as well as the fragments BH in the boron derivatives contribute with three orbitals to the formation of the metal polyhedron. This analogy between electron-deficient boron species and the electron-rich metal clusters is fundamentally empiric. However there are some theoretical calculations for specific cluster species that make some clarity about the degree of validity of such an analogy.

The nature of metal cluster networks is to some extent like that existing in bulk metals with a high degree of electron delocalization. Molecular orbital treatments are more appropriate for the description of such kinds of systems. However detailed calculations including the cluster metal atoms as well as those of the ligands are rather complex and expensive even for high symmetry systems. Accordingly, many approaches frequently consider in a first approximation only the valence atomic orbitals of skeletal metal atoms. That is equivalent to considering an isolated bare cluster. Thereafter the metal-ligand interactions can be taken into account by selecting those cluster molecular orbitals energetical and symmetrically better disposed for these interactions. Such an approach is consistent with the idea of a cluster as an independent entity.

Rather rigorous calculations have been made for the anion $[Co_6(CO)_{14}]^{4-}$. The energies as well as the symmetries of the molecular orbitals arising from calculations for the isolated cluster Co_6 are reproduced in Table 2.19a. From the combination of the 54 metal valence atomic orbitals to groups of molecular orbitals arise: 31 which can be considered as bonding or weakly bonding orbitals and 23 which have a strongly antibonding nature. From the interaction of the metal cluster molecular orbitals with the CO-ligands – with symmetries a_{1g}, e_g and t_{1u} for terminal CO groups and a_{1g}, a_{2u}, t_{1u}, and t_{2g} for the μ_2-bridging CO-groups – the molecular orbital scheme reproduced in Table 2.19b arises. According to this scheme, the 86 cluster valence electrons – 58 from the cobalt atoms and 28 from the donor CO-ligands – occupy 14 σ-metal-carbon MOs and 29 bonding or weakly non-bonding cluster MOs. Therefore, the external orbitals of the closed-shell electronic configuration of this cluster are

$$\ldots [T_{2u}(2)]^6 [T_{1g}(1)]^6 [A_{2g}]^2.$$

The HOMO as well as the LUMO in this cluster have metallic character. The HOMO-LUMO gap is about $80\,000$ cm^{-1}, so a configuration with more than 86 electrons would be rather unstable.

The fundamental assumption in the boranes and carboranes-metal clusters analogy that considers each unit of the cluster contributing only three s-p hybrid orbitals to form the metal cluster skeleton appears to be not totally confirmed by the theoretical treatment analyzed above. This shows that is not possible to define a precise set of molecular orbitals involved exclusively with the bonding in the metal network. However, the number and the symmetry of the antibonding molecular orbitals in $[Co_6(CO)_{14}]^{4-}$ (11 orbitals with symmetries T_{1u}, T_{2u}, T_{1g} and E_g) with a strong s-p character are qualitatively analogous with the equivalent orbitals calculated for the anion $[B_6H_6]^{2-}$.

A somewhat more general theoretical treatment of metal cluster has been performed through extended Hückel calculations on a series of isolated rhodium clusters with different geometries. Although the parameters (internuclear distances and AOs energies) used in such studies were always those from rhodium species, their results depend mainly on the symmetry and shape of the AOs being less sensitive to the parameters. The MO energy diagrams obtained from such calculations are therefore to be considered as characteristic for the corresponding geometries and able to be extended to other metal clusters.

According to these calculations, cluster MOs may be in general classified into two groups: 1. Valence-MOs, i.e. low-energy MOs containing metal electrons or functioning as acceptors for donor ligand electron pairs and 2. Antibonding orbitals, i.e. high-energy MOs in general not able to either hold non-bonding electrons or match ligand donor orbitals. The number of the orbitals in each group (1) or (2) depends fundamentally on the geometry of the

Table 2.19. (A) Molecular orbital energies for the octahedral cluster Co_6 and (B) a qualitative MO-scheme for the cluster anion $[Co_6(CO)_{14}]^-$

(A) Isolated Co_6 cluster		(B) The $[Co_6(CO)_{14}]^-$ anion		
Orbital type	Energy[a]	Cluster orbital type	Metal-ligand orbital type	Comments
$E_g(4)$	3.3579	$E_g(4)$		
$T_{1u}(5)$	2.1404	$T_{1u}(5)$		Antibonding
$T_{1g}(2)$	1.8640	$T_{1g}(2)$		cluster MO's
$T_{2u}(3)$	0.1614	$T_{2u}(3)$		
$T_{1u}(4)$	− 0.2241		$A_{1g}(3)^*, T_{1u}(4)^*,$	
$A_{1g}(3)$	− 0.3163		$E_g(3)^*$	Antibonding
$E_g(3)$	− 0.5065		$A_{1g}(1)^*, T_{2g}(3)^*,$	metal ligand MO's
$T_{2g}(3)$	− 0.4609		$A_{2u}^*, T_{1u}(3)^*$	
$T_{1u}(3)$	− 0.7271			
A_{2g}	− 0.9356	A_{2g}		
$T_{1g}(1)$	− 0.9430	$T_{1g}(1)$		
$T_{2u}(2)$	− 0.9530	$T_{2u}(2)$		
$E_g(2)$	− 0.9565	$E_g(2)$		
E_u	− 0.9691	E_u		Bonding and
$T_{2g}(2)$	− 0.9874	$T_{2g}(2)$		weakly antibonding
$T_{1u}(2)$	− 0.9908	$T_{1u}(2)$		cluster MOs
$T_{2u}(1)$	− 1.0167	$T_{2u}(1)$		
$E_g(1)$	− 1.0344	$E_g(1)$		
$T_{1u}(1)$	− 1.0411	$T_{1u}(1)$		
A_{2u}	− 1.0510			
$A_{1g}(2)$	− 1.0439	$A_{1g}(2)$		
$T_{2g}(1)$	− 1.0523	$T_{2g}(1)$		
$A_{1g}(1)$	− 1.0857		$A_{1g}(3), T_{1u}(4)$	
			$E_g(3)$	Bonding
			$A_{1g}(1), T_{2g}(1),$	metal ligand
			$A_{2u}, T_{1u}(3)$	MOs

[a] Energies in units of the metal d-valence orbital ionization energy.

Reference

Mingos DMP (1974) J. Chem. Soc., Dalton Trans., 133

system. In very compact arrays where the interactions among atomic orbitals are strong, the relative number of high-energy molecular orbitals increases, thus decreasing the number of electrons in the system. That occurs for example in the case of the electron-deficient compounds and pure metals, the latter being a limiting case.

In Fig. 2.41 the orbital interaction diagram for the bare cluster M_3 may be observed. The 37 atomic orbitals of the three metal atoms give rise to 27 molecular orbitals which can be classified in the groups (1) and (2) mentioned above. Orbitals with symmetry e'' and a_2' are strongly unstabilized and have an antibonding character. The energy of these orbitals is significantly higher than that of the p orbitals in the free atoms, so it can be assumed they are normally vacant orbitals. Thus, a three-atom cluster has only 24 MOs to be used for both accepting ligand electron pairs and housing non-bonding electrons up to a total of 48 electrons. Analogously to mononuclear compounds in which the metal has the tendency to have 18 valence-electrons, the cluster nucleus M_3 tend to reach 48 electrons. As can be observed on Fig. 2.42, some of the metal AOs match ligand orbitals of the same symmetry leading to normal bonding and antibonding AO-combinations. In the case of $M_3(CO)_{12}$ all cluster valence-MOs are occupied; 12 of them combined with CO orbitals and 12 remain unaltered containing 24 non-bonding electrons. It should be stressed that HOMO as well as LUMO have mainly metallic character.

In general, it is then valid to say that a cluster of three metal atoms with any given number of d-electrons will bond to the number of ligands necessary to reach a 48-electron configuration. However in certain clusters of the nickel group, specially platinum derivatives, an additional high-lying e'' orbital set remains vacant and 44-electron clusters result instead. This is perhaps not surprising since, as commented previously, platinum is known to form many mononuclear square-planar 16-electron complexes.

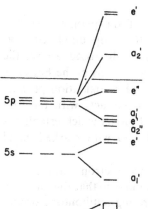

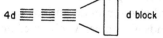

Fig. 2.41. Orbital interaction diagram for a M_3 cluster. Reproduced with permission from Lauher JW (1978) J. Am. Chem. Soc. 100:5305

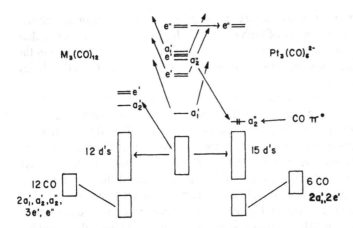

Fig. 2.42. Orbital interaction diagrams for $M_3(CO)_{12}$ and $[Pt_3(CO)_6]^{2-}$. Reproduced with permission from Lauher JW (1978) J. Am. Chem. Soc. 100:5305

Although MO-calculations do not specifically locate the metal-metal bonds, they always show the presence of three high-lying antibonding orbitals that indirectly indicate the existence of the former.

Theoretical calculations for other clusters with different nuclearities and geometries lead to qualitative similar results. Results obtained for some of most common cluster geometries are summarized in Table 2.20.

The application of this treatment appears to be successful in many cases, often explaining the differences between different geometries, for instance between tetranuclear species. Tetrahedral, square-planar, or butterfly metal atom arrays may be differentiated by considering the corresponding molecular orbital diagrams and the number of electrons in the system. However the benefits of such calculations do not go farther than those obtained from the analogy with boranes discussed above.

Theoretical descriptions may be to some extent verified experimentally. For instance, the valence-band photoelectron spectrum of the compound [Fe(CO)π-Cp]$_4$ shows that, in agreement with the calculations discussed above, the highest occupied valence orbitals are actually metallic orbitals of the Fe$_4$ core. Mössbauer experiments of the same compound and of its oxidation product [Fe(CO)π-Cp]$_4^+$ indicate that the unpaired electron interacts with the iron moiety. This feature is consistent with the theoretical model which establishes the unpaired electron would occupy a delocalized MO with marked metal character.

Electron Spin Resonance analyses of the 49-valence electron paramagnetic trinuclear compounds, $Co_3(CO)_9S$ and $Co_3(CO)_9Se$, indicate that the unpaired electron in these compounds occupy a non degenerating antibonding orbital with a symmetry a_2. The influence of this unpaired electron in an antibonding orbital on the structural properties of the compounds is notable. As observed in

Table 2.20. Bonding capabilities of transition metal clusters estimated from extended Hückel calculations of the isolated metal clusters

Cluster geometry	Atoms (N)	Atomic orbitals (9N)	High laying antibonding orbitals	Cluster valence molecular orbitals	Cluster valence electrons	Examples
Trimer	3	27	3	24	48	$Os_3(CO)_{12}$
Tetrahedron	4	36	6	30	60	$Rh_4(Co)_{12}$
"Butterfly"	4	36	5	31	62	$[Rh_4(CO)_{16}]^{2-}$
Square plane	4	36	4	32	64	$Pt_4(MeCO_2)_8$
Triagonal pyramid	5	45	9	36	72	$Os_5(CO)_{16}$
Square pyramid	5	45	8	37	74	$Fe_5(CO)_{15}C$
Bicapped tetrahedron	6	54	12	42	84	$Os_6(CO)_{18}$
Octahedron	6	54	11	43	86	$Ru_6(CO)_{17}C$
Capped square pyramid	6	54	11	43	86	$Os_6(CO)_{18}H_2$
Triagonal prism	6	54	9	45	90	$[Rh_6(CO)_{15}C]^{3-}$
Capped octahedron	7	63	14	49	98	$[Rh_7(CO)_{16}]^{3-}$
Square antiprism	8	72	15	57	114	$[Co_8(CO)_{18}C]^{2-}$
Cube	8	72	12	60	120	$Ni_8(PPh)_6(CO)_8$
Truncate hexagonal bipyramide	13	117	32	85	170	$[Rh_{13}(CO)_{24}H_3]^{2-}$

Reference

Lauher JW (1978) J. Am. Chem. Soc. 100:5305

Table 2.21. Effect of unpaired electrons on internuclear distances in tricobalt clusters

Compounds	Interatomic distances (Å)[a]			
	M–M	$\Delta(M–M)$[b]	M–X	$\Delta(M–X)$[b]
$FeCo_2(CO)_9S$	2.544	0.083	2.159	0.020
$Co_3(CO)_9S$	2.637		2.139	
$FeCo_2(CO)_9Se$	2.577	0.039	2.285	0.003
$Co_3(CO)_9Se$	2.616		2.282	

[a] Average values.
[b] Differences between interatomic distances d(M–M) and d(M–X) in the compounds $FeCo_2(CO)_9X$ and $Co_3(CO)_9X$ when X is S and Se respectively.

Reference

Strouse CE, Dahl LF (1971) J. Am. Chem. Soc. 93:6032

Table 2.21 on going from the diamagnetic species $FeCo_2(CO)_9X$ with 48 valence electrons to the paramagnetic one $Co_3(CO)_9X$ with 49, a significative increase in the average metal-metal distances is observed. This feature agrees with the predicted antibonding nature of the unpaired electron. The metallic characteristics of this last electron are also apparent in such experiments, since the enlargement effect is observed only in the metal cluster – not in the metal-ligands nor in the metal-chalcogen interatomic distances.

2.4 Synthesis of Cluster Compounds

Cluster formation is in general an agglomeration process in which labile intermediate species undergo condensation reactions for giving products with higher nuclearities. The art of synthesizing clusters is related therefore to finding proper reagents and reaction conditions for generating and condensing such kinds of intermediates.

In initial studies on transition metal cluster chemistry synthetic work was frequently based on chance discoveries. However the explosive development of this branch of chemistry in the last fifteen years has also produced an accelerated evolution of cluster synthetical methods. On going from casual synthesis to tailored cluster synthesis, the systematic preparation of a great variety of homo and heteronuclear cluster compounds is already possible.

Fundamental consideration of the type of condensation reactions involved in cluster formation shows that the methods most frequently used for metal cluster synthesis may be classified in the following three groups:

1. Condensation reactions induced by the presence of coordinately unsaturated species,
2. Condensation under reductive conditions,
3. Condensation reactions between coordinately saturated species with different oxidation states, and
4. Condensation reactions induced by the presence of special ligands.

2.4.1 Condensation Reactions Induced by Coordinately Unsaturated Species

For inducing the formation of metal-metal bonds, it is, in general, necessary for the metal atoms to exist in a rather low oxidation state. Especially suitable for the synthesis of metal clusters are therefore electron-deficient fragments in which the metal atoms have the same oxidation state than that they will have in the cluster. These fragments are normally very unstable and labile species. Electron deficiency induces a coordinative unsaturation that in absence of strong donor species will tend to form metal-metal bonds.

Reduction of Salts of Metals in Normal Oxidation States. In many cases the formation of reactive fragments can be achieved directly from salts of the metals in their normal oxidation states. In these processes the following two steps should be distinguished:

1. A reduction process to obtain the metal in adequately low oxidation state, and

2. The disruption of one or more bonds to produce unsaturated species.

Frequently both processes may be carried out simultaneously in the same trial. In other cases, however, the precursors in a low oxidation state are prepared first and then the condensation reaction is induced by thermal, photochemical, or chemical generation of unsaturate species.

Important pioneer work in the synthesis of polynuclear metal carbonyls was made by reduction of simple metal salts in the presence of carbon monoxide or similar ligands. The best results are obtained in general for the most stable clusters such as triruthenium, triosmium, tetrarhodium, and tetrairidium dodecacarbonyls. Thus, heterogeneous pyrolysis reactions of metal halides with carbon monoxide in the presence of halide acceptors such as Cu or Zn under relative drastic conditions of temperature and pressure can be used for preparing these types of three and tetranuclear carbonyl clusters:

$$4IrCl_4 + 12CO + 8Cu \xrightarrow[350 \text{ atm}]{100-200\,°C} Ir_4(CO)_{12} + 8CuCl_2$$

$$4RhCl_3 + 12CO + 12Cu \xrightarrow[200 \text{ atm}]{50-80\,°C} Rh_4(CO)_{12} + 12CuCl$$

Homogenous solution reactions often offer the possibility of carrying out these synthesis under considerably milder conditions:

$$4IrCl_4 + 20CO + 8H_2O \xrightarrow[200 \text{ atm}]{60\,°C, \text{ MeOH}} Ir_4(CO)_{12} + 16HCl + 8CO_2$$

$$6RuCl_3 + 33CO + 9H_2O \xrightarrow[5-10 \text{ atm}]{\text{MeOH}} 2Ru_3(CO)_{12} + 18HCl + 9CO_2$$

The formation of metal carbonyl clusters by reduction of the metal halides appears to occur via the formation of halo-carbonyl compounds as intermediate species. Thus, the reduction of halide derivatives is often a good alternative for the formation of clusters:

$$2RhCl_3(H_2O)_x + 6CO \xrightarrow{90\,°C} Rh_2(CO)_4Cl_2 + 2COCl_2 + 2x\,H_2O$$

$$\eta\text{: } 90\%$$

$$2Rh_2(CO)_4Cl_2 + 6CO + 4NaHCO_3$$

$$\xrightarrow[\text{hexane } (H_2O)]{25\,°C} Rh_4(CO)_{12} + 6CO_2 + 4NaCl + 2H_2O$$

$$\eta\text{: } 80-85\%$$

$$4Ir(CO)_2(Cl)L + 4CO + 2Zn \xrightarrow[90\,°C, \text{ 4 atm}]{2\text{-MeOEtOH}} Ir_4(CO)_{12} + 2ZnCl_2 + 4L$$

$$\eta\text{: } 80-85\%$$

Reduction of halides has also been applied recently in the synthesis of giant clusters:

$$Ph_3PAuCl \xrightarrow[\text{Benzene, 50–60 °C}]{B_2H_6} Au_{55}(PPh_3)_{12}Cl_6$$

$$p\text{-}Tol_3PAuCl + (p\text{-}Tol_3P)_4Ag_4Cl_4 \xrightarrow[\text{EtOH}]{NaBH_4} (p\text{-}Tol_3P)_{12}Au_{18}Ag_{20}Cl_{14}$$

Thermal Activation. Although the synthesis of carbonyl metal clusters directly from high oxidation state derivatives was historically of great importance, it is an intrinsically inefficient method because, as it will be seen in the next section, carbonyl metal clusters, in general, undergo degradation reactions in the presence of carbon monoxide. The method is therefore applicable only for the synthesis of thermodynamic and kinetic stable metal clusters. In the case of unstable compounds, for instance the carbonyl clusters of the elements of the first transition metal series, obtaining cluster products is possible only by starting from metal carbonyls:

$$2Co_2(CO)_8 \xrightarrow[\text{60–90 °C}]{\text{petroleum ether}} Co_4(CO)_{12} + 4CO$$
$$\eta: 90\%$$

Binary metal-carbonyl complexes are known for most transition metals and they are, in general, the more appropriate starting material in pyrolitic and photolytic synthetical methods.

Pyrolyses of metal carbonyls are carried out either in inert solvents or, to facilitate the elimination of CO, in the solid state under vacuum.

Although thermal induced elimination of ligands is a very convenient method to perform in the laboratory, it is very difficult to control the extent of the elimination. Such reactions are therefore rarely specific leading to the formation of a variety of products whose distribution normally depends on their relative stabilities. Reaction conditions are rather critical, hence for obtaining good yields of a given product they must be carefully selected:

$$Co_2(CO)_8 \xrightarrow{50\,°C} Co_4(CO)_{12} \xrightarrow{63-100\,°C} Co_6(CO)_{16} \xrightarrow{120\,°C} dec.$$

In the case of relatively unstable compounds frequently all ligands are lost and bulk metal is obtained:

$$Ni(CO)_4 \xrightarrow{60\,°C} Ni^0$$

$$Re_2(CO)_{10} \xrightarrow{130-170\,°C} Re^0$$

$$3Fe_2(CO)_9 \xrightarrow{70\,°C} 3Fe(CO)_5 + Fe_3(CO)_{12} \longrightarrow Fe^0$$

Table 2.22. Selected examples of metal cluster preparations via photolitic and thermolitic methods

Compound	Reagents, conditions, and yield	Ref.
$[Re_7C(CO)_{21}]^{3-}$	$[Re(CO)_4H]^-$; n-tetradecane, Δ; 70%	1
$Fe_3(CO)_{12}$	$Fe(CO)_5$; n-heptane, hν	2, 3
$[Fe_5N(CO)_{14}]^-$	$[Fe_2(CO)_8]^{2-}$ + $Fe(CO)_5$ + $NOBF_4$; diglima, Δ; 66%	4
$Ru_3(CO)_{12}$	$Ru(CO)_5$; n-heptane, 30 °C; h; 100%	2, 3, 5
$Ru_6C(CO)_{17}$	$Ru_3(CO)_{12}$ + CH_2Cl_2; Δ; 70%	6
$Os_3(CO)_{12}$	$Os(CO)_5$; n-heptane, 30 °C; sunlight	2, 3
$Os_5(CO)_{19}$	$Os_6(CO)_{18}$ + CO(90 atm); heptane, Δ; 80%	7
$Os_6(CO)_{18}$	$Os_3(CO)_{12}$; vacuum pyrolysis; 80%	8
$Os_6(CO)_{18}H_2S_2$	$Os_3(CO)_{10}H(SCH_2C_6H_5)$; h$\nu$; N_2, hexane; 8%	9
$Os_6(CO)_{16}(CPh)_2$	$Os_6(CO)_{18}$ + PhCCPh; hν	10
$Os_6(CO)_{16}S_4$	$Os_3(CO)_9S_2$, hν; N_2; 34%	11
$Os_7(CO)_{19}S$	$Os_4(CO)_{12}S$ + $Os_3(CO)_{10}(MeCN)_2$; octane, Δ; 23%	12
$[Os_8(CO)_{22}H]^-$	$Os_6(CO)_{10}$; BuOH, Δ; 30%	13
$[Os_{10}C(CO)_{24}]^{2-}$	$Os_3(CO)_{11}(C_5H_5N)$; vacuum pyrolysis; 65%	14
$[Co_6(CO)_{15}]^{2-}$	$Co_2(CO)_8$ + (i)EtOH; (ii), vacuum; (iii) + H_2O, KBr; 80%	15
$[Co_6C(CO)_{15}]^{2-}$	$[Co(CO)_4]^-$ + (i) $Co_3(CO)_9CCl$; diisopropylether, Δ; (ii) KBr; 70–80%	16
$[Co_{13}(C)_2(CO)_{24}]^{4-}$	$[Co_6C(CO)_{15}]^{2-}$; diglyme, Δ; 40–50%	17
$Rh_6(CO)_{16}$	$RhCl_3 3HO$ + CO(40 atm); MeOH; Δ; 80–90%	18
$[Rh_9P(CO)_{21}]^{2-}$	$Rh(CO)_{12}(acac)$ + CO, H_2(400 atm); 80%	19
$[Rh_{10}P(CO)_{22}]^{2-}$	$Rh(CO)_2(acac)$ + CO, H_2(400 atm) + Cs-$[PhCOO]$ + $Cs[BF_4]$ + PPh_3; diglyme; Δ; 98%	20
$[Rh_{10}As(CO)_{22}]^{2-}$	$Rh(CO)_2(acac)$ + CO, H_2 (400 atm) + Cs-$[PhCOO]$ + $AsPh_3$; diglyme, Δ; 98%	21
$[Rh_{13}(CO)_{24}H_2]^{3-}$	$Rh(CO)_2(acac)$ + CO, H_2(12 atm) + $Cs[PhCOO]$/$(CH_2OH)_2$; tetraglyme, Δ; 91%	22
$[Rh_{14}(CO)_{25}]^{4-}$	$Rh(CO_2(acac)$ + CO, H_2(200 atm) + $Cs[PhCOO]$/$(CH_2OH)_2$ + N-methylmorpholine; tetraglyme, Δ; 60%	23
$[Rh_{15}(CO)_{27}]^{3-}$	$Rh(CO)_2(acac)$ + CO, H_2 (19 atm) + $Cs[PhCOO]$/$(CH_2OH)_2$ + N-methylmorpholine; tetraglyme; Δ; 87%	24
$[Rh_{17}(S)_2(CO)_{32}]^{3-}$	$Rh(CO)_2(acac)$ + CO, H_2(300 atm) + $Cs[PhCOO]$ + H_2O + H_2S or SO_2; tetraglyme, Δ; 73%	25
$Ni_4(CNBu^t)_7$	$Ni(CNBu^t)_4$ + Ni(COD)	26
$Pt_3(CNBu^t)_6$	$Pt(COD)_2$ + $6CNBu^t$	27
$[Pt_{19}(CO)_{22}]^{4-}$	$[Pt_9(CO)_{18}]^{2-}$; MeCN, Δ; 50%	28
$CrOs_2(CO)_{11}(PMe_3)_2$	$CrOs(CO)_9(PMe_3)$; C_6F_6, hν; 31%	29
$WOs_3(CO)_{12}(PMe_2Ph)(S)_2$	$Os_3(CO)_9S_2$ + $W(CO)_5(PMe_2Ph)$ + hν; 27%	30
$W_2Os_3(CO)_{14}(PMe_2Ph)_2(S)_2$		
$[W_2Ni_3(CO)_{16}]^{2-}$	$[W_2(CO)_{10}]^{2-}$ + $Ni(CO)_4$; THF, Δ; 68%	31
$Fe_2Pt(CO)_8PR_3)_2$	$Fe_2(CO)_2$ + $Pt(PR_3)_4$	32
$[Fe_3Pt_3(CO)_{15}]^{2-}$	$[Pt_3(CO)_6]^{2-}$ + $Fe(CO)_5$; MeCN, Δ; 70%	33
$Ru_4Au_2C(CO)_{12}(PMe_2Ph)_2$	$Ru_5AuC(CO)_{14}(PMe_2Ph)_2$ + CO (80 atm); toluene, Δ; 80%	34
$[Os_3Co(CO)_{13}]^-$	$[Co(CO)_4]^-$ + $Os_3(CO)_{12}$; hν	35

Table 2.22. (continued)

Compound	Reagents conditions and yield	Ref.
$Os_3Au_2(CO)_{10}(PEt_3)_2$	$[Os_3(CO)_{11}H]^- + Au(PEt_3)Cl + TlPF_6$; $CHCl_3$, Δ; 80%	36
$[Co_4Ni_2(CO)_{14}]^{2-}$	$NiCl_2 + [Co(EtOH)_x][Co(CO)_4]_2 +$ (i) vacuum, (ii) KBr; H_2O; 60%	37
$[Rh_6IrN(CO)_{15}]^{2-}$	$[Rh_6N(CO)_{15}]^{2-} + [Ir(CO)_4]^-$; THF, Δ; 79–80%	38
$[Rh_{12}Sb(CO)_{27}]^{3-}$	$Rh(CO)_2$ (acac) + CO, H_2(400 atm) + Cs[PhCOO] + $SbPh_3$; tetraglyme; Δ; 66%	39

References

1 Ciani G, D'Alfonso G, Freni M, Romiti P, Sironi A (1982) J. Chem. Soc., Chem. Commun., 339
2 King RB (1965) Organoment. Synth. 1:93
3 Moss JR, Graham AG (1977) J. Chem. Soc., Dalton Trans., 95
4 Tachikawa M, Stein J, Muetterties EL, Teller RG, Beno, MA, Gebert E, Williams JM (1980) J. Am. Chem. Soc. 102:6648
5 Calderazzo F, L'Plattenier F (1967) Inorg. Chem. 6:1220
6 Johnson BFG, Lewis J, Sankey SW, Wong K, McPartlin M, Nelson WJH (1980) J. Organomet. Chem. 191:C3
7 Farrar DH, Johnson BFG, Lewis J, Raithby PR, Rosales MJ (1982) J. Chem. Soc., Dalton Trans., 2501
8 Eady CR, Johnson BFG, Lewis J (1975) J. Chem. Soc., Dalton Trans., 2606
9 Adams RD, Horvath IT, Mathur P, Segmüller BE (1983) Organometallics 2:996
10 Fernandez JM, Johnson BFG, Lewis J, Raithby PR, (1978) Acta Crystallogr. Sect. B 34:3086
11 Adams RD, Horvath IT (1984) J. Am. Chem. Soc. 106:1869
12 Adams RD, Foust DF, Mathur P (1983) Organometallics 2:990
13 Johnson BFG, Lewis J, Nelson WJH, Vargas MD, Braga D, Henrick K, McPartlin M, (1984) J. Chem. Soc., Dalton Trans, 2151
14 Jackson PF, Johnson BFG, Lewis J, Nelson WJH, McPartlin M (1982) J. Chem. Soc., Dalton Trans., 2099
15 Chini P (1967) J. Chem. Soc., Chem. Commun., 29
16 Martinengo S, Strumolo D, Chini P, Albano VG, Braga D (1985) J. Chem. Soc., Dalton Trans., 35
17 Albano VG, Braga D, Chini P, Ciani G, Martinengo S (1982) J. Chem. Soc., Dalton Trans., 645
18 Chaston SHH, Stone FGA (1969) J. Chem. Soc. A, 500
19 MacLaughlin SA, Taylor NJ, Carty AJ (1983) Organometallics 2:1194
20 Vidal JL, Walker WE, Schoening RC (1981) Inorg. Chem. 20:238
21 Vidal JL (1981) Inorg. Chem. 20:243
22 Vidal JL, Schoening RC (1981) J. Organomet. Chem. 218:217
23 Vidal JL, Schoening RC (1981) Inorg. Chem. 20:265
24 Vidal JL, Schoening RC (1982) Inorg. Chem. 21:438
25 Vidal JL, Fiato RA, Cosby LA, Pruett RL (1978) Inorg. Chem. 17:2574
26 Thomas MG, Pretzer WR, Beier BF, Hirsekorn FJ, Muetterties EL (1977) J. Am. Chem. Soc. 99:743
27 Green M, Howard JAK, Murray M, Spencer JL, Stone FGA (1977) J. Chem. Soc., Dalton Trans., 1509
28 Washecheck DM, Wucherer EJ, Dahl LF, Ceriotti A, Longoni G, Manassero M, Sansoni M, Chini P (1979) J. Am. Chem. Soc. 101:6110
29 Davis HB, Einstein FWB, Johnston VJ, Pomeroy RK (1988) J. Am. Chem. Soc. 110:4451
30 Adams RD, Horvath IT, Mathur P (1984) J. Am. Chem. Soc. 106:6296
31 Ruff JK, White RP, Dahl LF (1971) J. Am. Chem. Soc. 93:2159
32 Bruce MI, Shaw G, Stone FGA (1972) J. Chem. Soc., Dalton Trans., 1082
33 Longoni G, Manassero M, Sansoni M (1980) J. Am. Chem. Soc. 102:7973
34 Cowie AG, Johnson BFG, Lewis J, Raithby PR (1984) J. Chem. Soc., Chem. Commun., 1790

Table 2.22. (continued)

35 Burkhardt E, Geoffroy GL (1980) J. Organomet. Chem. 198:179
36 Burgess K, Johnson BFG, Kaner DA, Lewis J, Raithby PR, Mustaffa SNAB (1983) J. Chem. Soc., Chem. Commun., 455
37 Chini P, Cavalieri A, Martinengo S (1972) Coord. Chem. Rev. 8:3
38 Martinengo S, Ciani G, Sironi A (1984) J. Chem. Soc., Chem. Commun., 1577
39 Vidal JL, Troup JM (1981) J. Organomet. Chem. 213:351

In the synthesis of high nuclearity carbonyl clusters, small clusters are often used as starting materials:

$$Os_3(CO)_{12} \xrightarrow{210°C} Os_6(CO)_{18} + Os_5(CO)_{16} + Os_7(CO)_{21} + Os_8(CO)_{25}$$
$$80\%$$

$$Os_3(CO)_{11}(C_5H_5N) \rightarrow [Os_{10}C(CO)_{24}]^{2-} + [Os_5(CO)_{15}H]^-$$
$$65\%$$

$$+ Os_5C(CO)_{14}H(NC_5H_4)$$

Photochemical Activation. Coordinative unsaturated fragments may also be produced by photolytic reactions. In presence of UV-irradiation metal carbonyl compounds lose sequentially CO-ligands. Electron-deficient, solvent coordinated species produced in this way may combine with inactivated metal complexes via the formation of donor-acceptor metal-metal bonds. Iron, ruthenium, and osmium trinuclear carbonyl clusters may be prepared by this way:

$$3M(CO)_5 \xrightarrow{h\nu} M_3(CO)_{12} + 3CO \qquad M = Fe, Ru, Os$$

Thermolytic reactions are in general unfavorable for the synthesis of heteronuclear compounds due to the lack of specificity of the method. Photolytic procedures require instead milder conditions and they can sometimes even used for synthesizing some heteronuclear monoanions:

$$[Co(CO)_4]^- + Fe(CO)_5 \xrightarrow{h\nu} [FeCo(CO)_8]^- + CO$$

$$Os_3(CO)_{12} + [Co(CO)_4]^- \xrightarrow{h\nu} [Os_3Co(CO)_{12}]^- + 4CO$$

Selected examples of preparations of metal clusters via thermolytic and photolytic procedures are described in Table 2.22.

High-valence halide clusters of transition metals – in which the metal also has an oxidation state lower than in normal salts – are normally prepared by pyrolitic reduction of the mononuclear halide:

$$3ReCl_5 \rightarrow Re_3Cl_9 + 3Cl_2$$

$$18TaX_5 + 16Al \rightarrow 3Ta_6X_{14} + 16AlX_3 \qquad X = Cl, Br, I$$

$$MoCl_5 + Mo \rightarrow MoCl_3 \rightarrow Mo_6Cl_{12}$$

2.4.2 Condensation Between Coordinately Saturated Species with Different Oxidation States

Redox condensation of an anionic mono or polynuclear carbonyl species with a neutral, cationic, or anionic fragment provides a valuable method for the synthesis of carbonyl metal clusters.

Among the characteristics of this method are the mild reaction conditions that avoid the degradation of the products and permit the formation of high-nuclearity products. The facility with which such kinds of condensations occur is probably due to the redox character of these reactions.

Reaction of metallates with neutral metal complexes. The reaction of cluster carbonyl metallates with neutral carbonyl compounds has proved to be very useful in the preparation of homo and heteronuclear derivatives:

$$[Rh(CO)_4]^- + Rh_4(CO)_{12} \rightarrow [Rh_5(CO)_{15}]^- + CO$$

$$[Fe_3(CO)_{11}]^{2-} + Fe(CO)_5 \rightarrow [Fe_4(CO)_{13}]^{2-} + 3CO$$

$$[Ru_3(CO)_{11}]^{2-} + Ru_3(CO)_{12} \rightarrow [Ru_6(CO)_{18}]^{2-} + 5CO$$

$$[Rh_6(CO)_{15}(N)]^- + M(CO)_4 \rightarrow [Rh_6M(N)(CO)_{15}]^{2-} + 4CO$$
$$M = Cr, W$$

$$[Fe_3(CO)_9(CCO)]^{2-} + Rh(CO)_2(py)Cl \rightarrow [RhFe_3(CO)_{12}(C)]^{2-}$$

As can be seen, the chemical equations indicate that these reactions could often be considered as carbon monoxide displacement reactions. The condensations involving the square-based pyramidal dianion $[Fe_5C(CO)_{14}]^{2-}$ and some neutral species with labile ligands shown in the scheme in Fig. 2.43 may be cataloged into the same type of reactions. The same is valid in the displacement of the chlorine atom in the chloromethyl derivative $Co_3(CO)_9CCl$ by mononuclear and cluster carbonyl metallates according to the following equations.

$$Co_3(CO)_9CCl + 3[Co(CO)_4]^- \rightarrow [Co_6C(CO)_{15}]^{2-} + 6CO + Cl^-$$

$$Co_3(CO)_9CCl + 1.5[Ni_6(CO)_{12}]^{2-} \xrightarrow{-Cl^-} [Co_3Ni_9C(CO)_{20}]^{3-} + 7CO$$

Reaction of metallates with cationic complexes. The synthesis of one of the largest metal carbonyl cluster structurally characterized up to date, $[Ni_{38}Pt_6(CO)_{48}H_2]^{4-}$, has been obtained from the condensation reaction of the metallate $[Ni_6(CO)_{12}]^{2-}$ with the platinum(II) salt $PtCl_2$.

Coupling reactions of some metallates with salts of Tl, Hg, or Au often originate condensation products in which, as can be observed in the schemes in Fig. 2.44, the basic cluster geometry is preserved. As observed in Fig. 2.45, which describes the condensation of $[Rh_6C(CO)_{15}]^{2-}$ with Ag(I), sometimes the degree of condensation may be modulated by the reagents ratio.

Cationic fragments generated from complex salts of the metals copper, silver or gold appear to be a powerful method for promoting condensation reactions

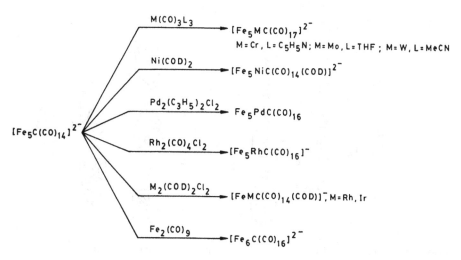

Fig. 2.43. Reactions of $[Fe_2(CO)_{14}]^{2-}$ with neutral species containing labile ligands (COD = 1,2-cyclo octadiene) [Tachikawa M, Geerts RL, Mueatterties EL (1981) J. Organoment. Chem. 213:11]

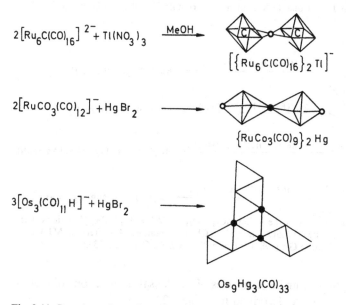

Fig. 2.44. Reactions of metallates with Tl and Hg salts [Vargas MD, Nicholl JN (1986) Adv. Inorg. Chem. Radiochem. 30:123]

with metallates. Condensation reactions of $[Ru_4(CO)_{12}H_2]^{2-}$ with cations $[M(PR_3)]^+$ described in the scheme in Fig. 2.46 exemplifies this type of reaction. These complexes with weak coordinating contraions (PF^{6-}, ClO^{4-} etc.) react readily in solution. In the case of halides, however, the presence of silver or

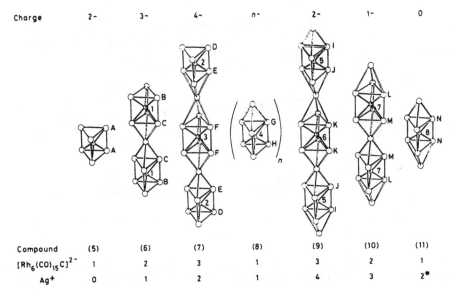

Compound	(5)	(6)	(7)	(8)	(9)	(10)	(11)
$[Rh_6(CO)_{15}C]^{2-}$	1	2	3	1	3	2	1
Ag^+	0	1	2	1	4	3	2*

Fig. 2.45. $[Rh_6C(CO)_{15}]^{2-}/Ag$ adducts at different clusters/silver molar ratios. Reproduced with permission from Heaton BT, Strona L, Martinengos, Strumulo D, Albano VG, Braga D (1983) J. Chem. Soc. Dalton Trans. 2175

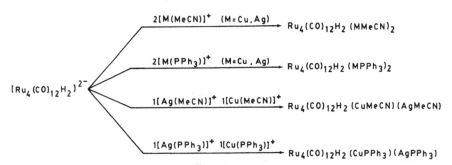

Fig. 2.46. Condensation reactions of the anion $[Ru_4(CO)_{12}H_2]^{2-}$ with $[MPR_3]^+$ cations. Ref.: Brown SSD, Salter ID, Smith BM (1985) J. Chem. Soc., Chem. Commun., 1439; Freeman MJ, Green M, Orpen AG, Salter ID, Stone FGA (1983) J. Chem. Soc., Chem. Commun. 1332

thallium salts is required. Further examples of condensation reactions of metallates with neutral species are described in Table 2.23.

2.4.3 Generation of Unsaturated Species by Chemical Methods

Ligand elimination constitutes in some case an alternative way to produce coordinative unsaturated metal species. Trimethyl oxide and also the hydroxide ion can remove carbonyl ligands converting them to CO_2.

Table 2.23. Cluster synthesis by condensation of metallates with neutral metallic species

Product	Reagents	Ref.
$[MFe_3(CO)_{13}C]^{2-}$; M = Cr, W	$Fe_3(CO)_9(CCO)]^{2-} + M(CO)_3(NCR)_3$	1
$[MFe_4(CO)_{15}C]^{2-}$; M = Cr, W	$[Fe_4(CO)_{12}C]^{2-} + M(CO)_3L_3$	2
$[M_2Ni_3(CO)_{16}]^{2-}$; M = Cr, Mo, W	$[M_2(CO)_{10}]^{2-} + 3Ni(CO)_4$	3
$[MNi_6(CO)_{17}]^{2-}$; M = Cr, Mo, W	$M(CO)_6 + (Ni_6(CO)_{12})^{2-}$	4
$[MnFe_2(CO)_{12}]^-$	$[Mn(CO)_5]^- + Fe_3(CO)_{12}$	5
$[MnNi_2(CO)_5Cp_2]^-$	$[Mn(CO)_5]^- + Ni_2(CO)_2Cp_2$	6
$[ReRu_3(CO)_{16}]^-$	$[Re(CO)_5]^- + Ru_3(CO)_{12}$	7
$[Fe_4(CO)_{13}]^{2-}$	$Fe(CO)_5 + [Fe_3(CO)_{11}]^{2-}$	8
$[M_3Co(CO)_{13}]^-$; M = Fe, Ru	$M_3(CO)_{12} + [Co(CO)_4]^-$	9
$[Fe_4Co(CO)_{14}C]^-$	$[Fe_4(CO)_{12}C]^{2-} + Co_2(CO)_8$	10
$[Fe_5Ni(CO)_{14}(COD)C]^-$	$[Fe_5(CO)_{14}C]^{2-} + Ni(COD)_2$	2
$[Fe_4M(CO)_{16}]^{2-}$; M = Pd, Pt	$3[Fe_3(CO)_{11}]^{2-} + 2[MCl_4]^{2-}$	11
$[Fe_3Pt_3(CO)_{15}]^{2-}$	$3Fe(CO)_5 + [Pt_3(CO)_6]^{2-}$	12
$[Fe_4Rh(CO)_{14}C]^-$	$[Fe_4(CO)_{12}C]^{2-} + [Rh(CO)_2Cl]_2$	2
$[RuCo_3(CO)_{12}]^-$	$RuCo_2(CO)_{11} + [Co(CO)_4]^-$	13
$[Co_8(CO)_{18}C]^{2-}$	$2[Co_6(CO)_{15}C]^{2-} + Co_4(CO)_{12}$	14
$[Rh_7(CO)_{16}]^{3-}$	$[Rh_6(CO)_{15}]^{2-} + [Rh(CO)_4]^-$	15
$[Rh_7(CO)_{16}I]^{2-}$	$2[Rh_6(CO)_{15}]^{2-} + Rh_2(CO)_4I_2$	16
$[Rh_{12}(CO)_{30}]^{2-}$	$[Rh_6(CO)_{15}]^{2-} + Rh_6(CO)_{16}$	17
$[Rh_6Ni(CO)_{16}]^{2-}$	$[Rh_6(CO)_{15}]^{2-} + Ni(CO)_4$	18

References

1 Hriljac JA, Holt EM, Shriver DF (1987) Inorg. Chem. 26:2943
2 Tachikawa M, Geerts RL, Muetterties EL (1981) J. Organomet. Chem. 213:11
3 Ruff JK, White Jr RP, Dahl LF (1971) J. Am. Chem. Soc. 93:2159
4 Hall TL, Ruff JK (1981) Inorg. Chem. 20:4444
5 Anders U, Graham WAG (1966) J. Chem. Soc., Chem. Commun., 291
6 Hsich ATT, Knight J (1971) J. Organomet. Chem. 26:125
7 Knight J, Mays MJ (1972) J. Chem. Soc., Dalton Trans., 1022
8 Hieber W, Shubert H (1965) Z. Anorg. Allg. Chem. 338:37
9 Steinhardt PC, Gladfelter WL, Harley AD, Fox JR, Geoffroy GL (1980) Inorg. Chem. 19:332
10 Hriljac JA, Swepston PN, Shriver DF (1985) Organometallics 4:158
11 Longoni G, Manassero M, Sansoni M (1980) J. Am. Chem. Soc. 102:3242
12 Longoni G, Manassero M, Sansoni M (1980) J. Am. Chem. Soc. 102:7973
13 Roland E, Bernhardt W, Vahrenkamp H (1986) Chem. Ber. 119:2566
14 Chini P, Longoni G, Albano VG (1976) Adv. Organomet. Chem. 14:285
15 Martinengo S, Chini P (1972) Gazz. Chim. Ital. 102:344
16 Martinengo S, Chini P, Giordano G, Ceriotti A, Albano VAG, Ciani G (1975) J. Organomet. Chem. 88:375
17 Albano VG, Ceriotti A, Chini P, Ciani G, Martinengo S, Anker M (1975) J. Chem. Soc., Chem. Commun., 859
18 Fumagalli A, Longoni G, Chini P, Albinati A, Brunckner S (1980) J. Organomet. Chem. 202:329

The reaction with hydroxide ions appears to yield first hydrometal carbonyl anions,

$$Os_3(CO)_{12} + OH^- \rightarrow [HOs_3(CO)_{11}]^- + CO_2$$

which under stronger conditions can condensate to produce large clusters.

$$Os_3(CO)_{12} + OH^- \xrightarrow[48\,h]{100\,°C} [H_3Os_4(CO)_{12}]^- + [HOs_6(CO)_{18}]^-$$
$$\qquad\qquad\qquad\qquad\qquad\qquad\quad 45\% \qquad\qquad 18\%$$

$$+ \text{ other products}$$

Reactions of carbonyl compounds with Me_3NO appear to depend on the ability of the medium to stabilize the intermediate products. In very poor donating solvents, condensation does occur, but relatively drastic conditions are needed; however the specificity of the reactions is poor.

$$\pi\text{-CpRh(CO)}_2 + Me_3NO \xrightarrow{80\,°C} \pi\text{-Cp}_2Rh_2(CO)_3 + \pi\text{-Cp}_3Rh_3(CO)_3$$
$$\qquad\qquad\qquad\qquad\qquad\qquad\qquad 10\text{--}25\% \qquad\quad 30\text{--}60\%$$

$$+ \pi\text{-Cp}_4Rh_4(CO)_2$$
$$5\%$$

Better results are achieved when these reactions are performed in weak donating solvents such as acetonitrile which can stabilize the decarbonylated intermediates:

$$M(CO)_x + n\,Me_3NO \xrightarrow{MeCN} M(CO)_{x-n}(MeCN)_n + n\,CO_2 + n\,NMe_3$$

These lightly stabilized complexes easily undergo substitution processes in which the weakly coordinated ligand is displaced by other uncharged 18-electron metal complexes. This kind of reaction will be referred to in the next section in discussing displacement reactions.

2.4.4 Displacement Reactions

Very useful as synthetic methods in cluster chemistry are the displacement reactions of weak coordinated ligands such as MeCN, THF, Py, alkenes and others. Complexes containing these ligands may be prepared by thermal, photo-chemical, or chemical methods.

$$Cr(CO)_6 \xrightarrow[NCMe]{h\nu} Cr(CO)_3(NCMe)_3 + 3\,CO$$

$$Os_3(CO)_{12} \xrightarrow[NCMe]{Me_3NO} Os_3(CO)_{11}(NCMe)$$

These lightly stabilized complexes may be then substituted by donor metal complexes by the formation of metal-metal bonds:

$$Os_3(CO)_{11}(NCMe) \xrightarrow{H_2Os(CO)_4} H_2Os_4(CO)_{15}$$

$$Cr(CO)_3(NCMe)_3 \xrightarrow{[Fe_5(CO)_{14}(C)]^{2-}} [Fe_5Cr(CO)_{17}(C)]^{2-}$$

The possibility of governing the stoichiometry of the reactions by controlling the quantity of added ligand, on the one hand, and the mild conditions under which the cluster-forming reaction can be performed, on the other, are the two principal advantages offered by these methods.

2.4.5 Ligand-Assisted Condensation Reactions

There are some ligands which have the property of bonding many metal atoms simultaneously – holding them together and thus inducing the formation of metal-metal bonds. Many of the more effective bridging ligands are derived from the heavier members of the main groups: GeR_2, SnR_2, PR_2, AsR_2, SR. etc.. Also some small unstable fragments such as RN, RC, RP, S, etc. can lead to the aggregation of metal atoms around themselves supporting thus metal-metal bonds. Although these fragments are very reactive species that can not exist separately, joined to metal polyhedra they form rather stable cluster compounds.

Although compounds with edge and face-bridging ligands may be formally considered as main group-transition metal mixed clusters (s. Chap. 3), in these section they will be referred to only as ligands.

The condensation processes involved in the formation of metal-metal bonds assisted by the presence of main group bridging ligands (E) are often initiated by a displacement reaction. In the case of two metal complexes $L_n ME$ and $L'_m M'$, an electron lone pair on the ligand E can displace a ligand in the second complex forming a bridged binuclear species which by further ligand eliminations finally leads to a new metal-metal bond.

The relatively high specificity of displacement mechanisms assisted by bridging ligands make such reactions useful for the synthesis of heteronuclear compounds. The reaction scheme in Fig. 2.47 shows an example in which by using the bridging and donor capacity of the group AsR_2 a new heteronuclear metal-metal bond is achieved. This corresponds to the reaction sequence used for obtaining the first cluster with four different metal atoms.

The validity of the ligand displacement mechanisms in ligand-assisted cluster synthesis mentioned above as well as the capability of the sulfur ligand to bind several atoms simultaneously may be appreciated in the synthesis scheme shown in Fig. 2.48. The reaction proceeds in two steps. In the first one, $M(CO)_5$ groups are assembled around the sulfur atom by displacement reactions on the light stabilized $M(CO)_5THF$ units. In the second step, metal-metal bonds are formed by elimination of carbon monoxide. This last step is reversible. Since these

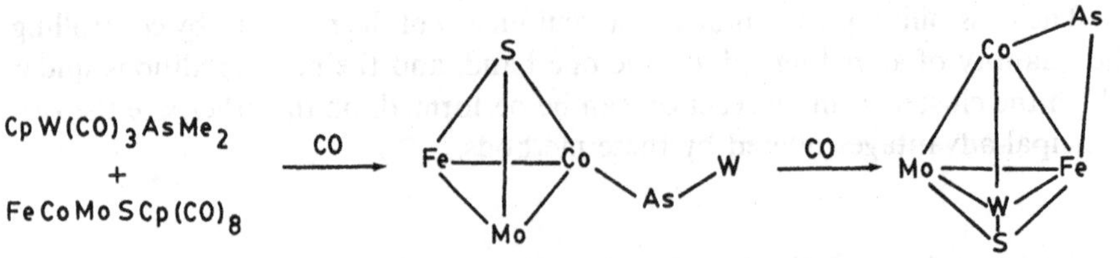

Fig. 2.47. Formation of heteronuclear metal-metal bonds by condensation reactions assisted by ligands μ_2-AsR$_2$ [Richter F, Vahrenkamp (1979) Angew. Chem. Int. Ed. Engl. 18:531]

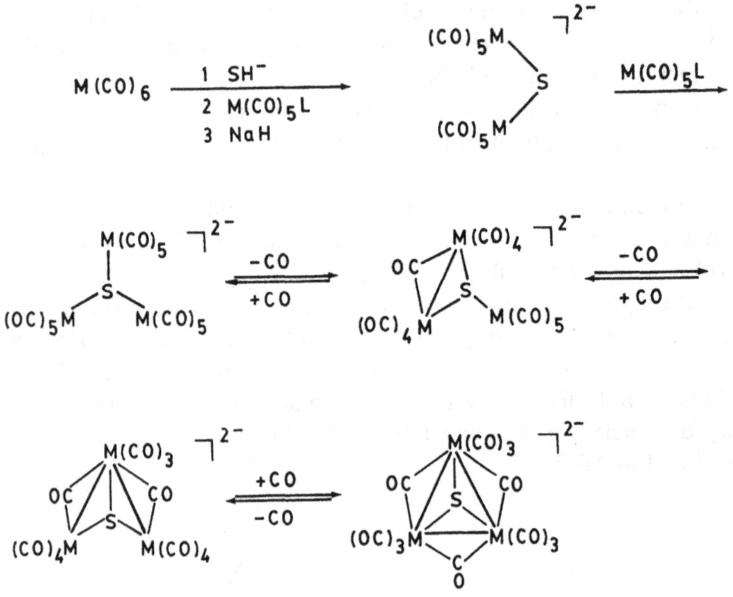

Fig. 2.48. Metal aggregation around a sulfur site [Darensbourg DJ, Zalewski DJ, Sanchez KM, Delordt T (1989) Inorg. Chem. 27:821]

clusters are synthesized step by step, it is possible to incorporate different metal atoms into the same discrete metal cluster. The three metals and the sulfur atoms form a tetrahedral mixed cluster. This trinuclear metal cluster can react with a fourth pentacarbonyl fragment to give pentanuclear derivatives. However the last metal ligand attaches itself to the sulfur but does not form new additional metal-metal bonds to the other metal centers of the cluster.

This type of bridge-assisted addition, which is in general feasible for clusters possesing ligands with electron lone pairs, can sometimes be used for obtaining high nuclearity clusters, as shown in the instances described in Fig. 2.49. In these cases, cluster enlargement occurs by self-condensation under appropriate conditions, or by addition of lightly stabilized clusters.

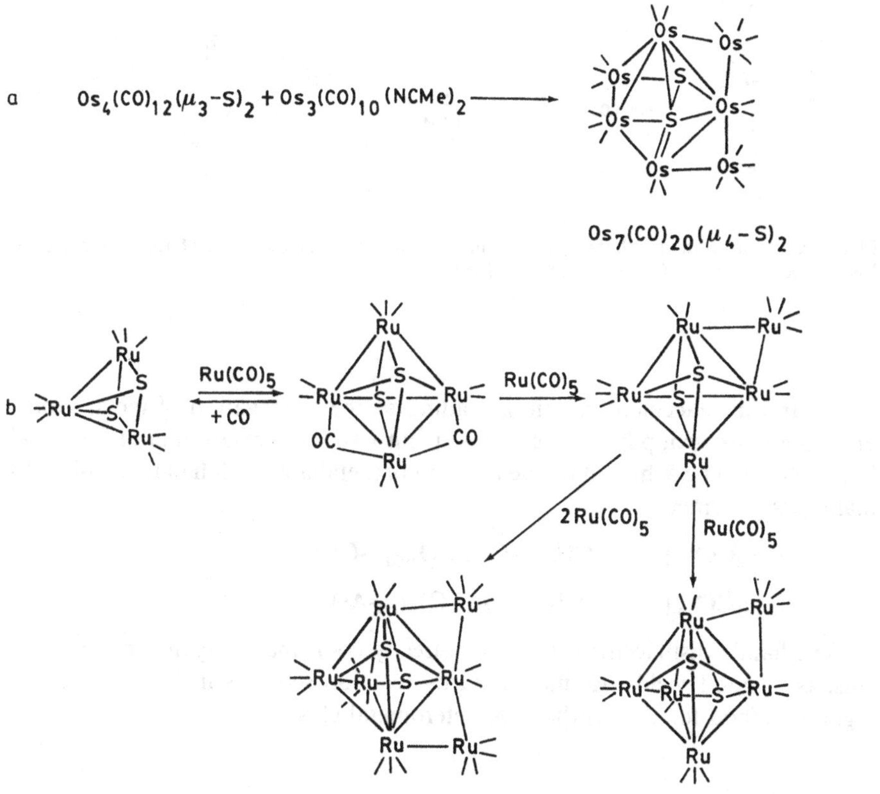

Fig. 2.49. Synthesis of large clusters by bridge assisted additions [(a) Adams RD (1985) Polyhedron 4:2003; (b) Adams RD, Babin JE, Tasi M (1987) Inorg. Chem. 26:2807]

Carbido and nitrido ligands also have been utilized as a multidentate ligand system capable of aggregating metal atoms and thus of inducing the formation of clusters. Good precursors to bridged carbido and nitrido ligands are ketenylidene (μ-CCO) and cyanate (-NCO) ligands.

The reaction scheme in Fig. 2.50 shows an example of how a ketenylidene metal complex can be used as precursor for a bridged carbido ligand that can be subsequently used for preparing a variety of new heteronuclear complexes.

Similarly, the cyanate ligand in the cluster $Ru_4(CO)_{13}(NCO)$ may be decarbonylated upon heating to yield nitrido clusters which can be enlarged by a sequence of redox condensation reactions.

$$Ru_4(CO)_{13}(\mu_2\text{-NCO}) \xrightarrow[-2CO]{68\,°C} Ru_4(CO)_{12}(\mu_4\text{-N}) \xrightarrow[+CO]{Ru(CO)_5}$$

$$\rightarrow Ru_5(CO)_{14}(\mu_5\text{-N}) \xrightarrow[+CO]{Ru(CO)_5} Ru_6(CO)_{16}(\mu_6\text{-N})$$

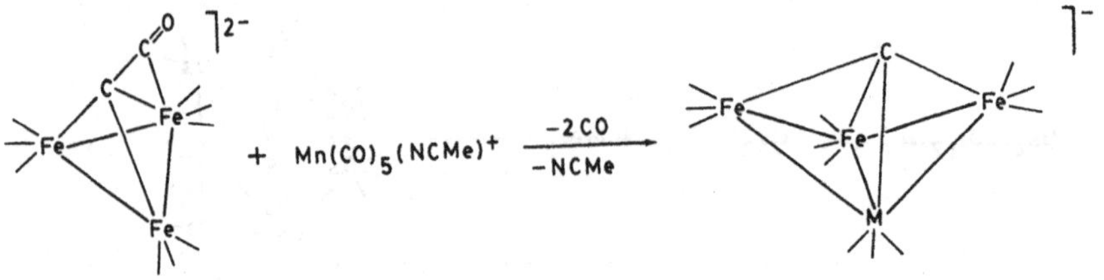
2-
Fe
Fe
Fe
C
C
O
+ Mn(CO)5(NCMe)+
-2CO
-NCMe
Fe
Fe
Fe
C
M
-

Fig. 2.50. Preparation of carbide clusters from ketenylidene metal complexes [Hriljac JA, Swepson PN, Shriver DF (1985) Organometallics 4:158]

Ligand displacement reactions leading to the aggregation of metal atoms around a main group ligand and subsequently to the formation of metal-metal bonds can also be achieved by the reaction of metallates with halides of 14 or 15 main group elements.

$$3[Co(CO)_4]^- + RCCl_3 \rightarrow Co_3(CO)_9(\mu_3\text{-}CR)$$

$$3[Co(CO)_4]^- + AsCl_3 \rightarrow Co_3(CO)_9(\mu_3\text{-}As)$$

The halide displacement from the already coordinated ligands can lead to clusters in good yields. As illustrated by reaction schemes in Fig. 2.51 this is a good method for the synthesis of heterometal clusters.

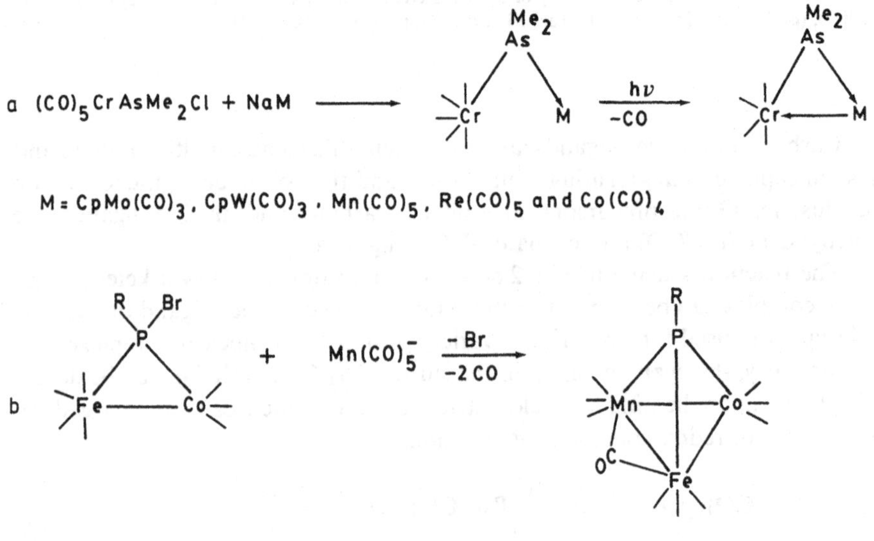
a (CO)5CrAsMe2Cl + NaM
Me2
As
Cr
M
hν
-CO
Me2
As
Cr
M
M = CpMo(CO)3, CpW(CO)3, Mn(CO)5, Re(CO)5 and Co(CO)4
b
R
Br
P
Fe
Co
+
Mn(CO)5-
- Br
-2CO
R
P
Mn
Co
OC
Fe

Fig. 2.51. Synthesis of heterometal clusters by displacement reactions on coordinated cluster ligand [(a) Ehrl W, Vahrenkamp H (1973) Chem. Ber. 106:2550; (b) Muller M, Vahrenkamp H (1983) Chem. Ber. 16:2322]

The oxidative addition of hydride and halide derivatives of the 14 and 15 main group elements to neutral polynuclear metal compounds can also be used for obtaining clusters with bridging ligands.

$$3\,Fe(CO)_5 + 2\,PhPH_2 \xrightarrow{h\nu} Fe_3(CO)_9(\mu_3\text{-}PPh)_2 + 6\,CO + 2\,H_2$$

$$3/2\,Co_2(CO)_8 + PhGeH_3 \rightarrow Co_3(CO)_9(\mu_3\text{-}GePh) + 3\,CO + 3/2\,H_2.$$

Heteronuclear metal clusters may also be produced by this method.

$$Co_2(CO)_8 + Fe(CO)_5 + PhSH \xrightarrow{h\nu} FeCo_2(CO)_9(\mu_3\text{-}S) + C_6H_6 + 4\,CO$$

$$FeCo(CO)_7(\mu_2\text{-}PHR) + Co(CO)_3(C_3H_5) \xrightarrow{-CO} FeCo_2(CO)_9(\mu_3\text{-}PR)$$
$$+ C_3H_6$$

$$Cr(CO)_5(PH_2R) + Fe_3(CO)_{12} \rightarrow CrFe_2(CO)_{11}(\mu_3\text{-}PR).$$

2.5 Cluster Reactivity

Metal atoms with low oxidation states and high polarizability of both metal-metal and metal-ligand bonds make metal clusters susceptible to a variety of chemical reactions. Depending on the reagents and reaction conditions, metal cluster species are able to undergo nucleophilic and electrophilic substitution reactions, homo and heterolytic bond dissociations, and redox processes. However, the great majority of cluster reactions appears to be complex processes in which is not always possible to recognize these fundamental processes. In general the knowledge about reaction mechanisms of cluster reactions is still lacking. Therefore, although important advances have been made in the elucidation of particular cases, any classification of cluster reactions must be still established more by their products than by mechanistic considerations.

By considering the oxidation states of metal atoms, reactions may be roughly classified into two groups: Reactions involving changes in the oxidation states of cluster metal atoms and the reactions in which metal oxidation states remain unchanged.

From the point of view of cluster reactivity, those reactions in which the cluster structure is either conserved or modified by structural and nuclearity changes but retaining its cluster nature are more interesting. However reactions leading to degradation and disruption of the clusters must be often considered.

2.5.1 Reactions With Changes in Cluster Oxidation State

Redox reactions are very frequent in the chemistry of transition metal complexes. d-orbitals are forming there many relatively low-lying occupied and unoccupied molecular orbitals which permit a variety of oxidation states. Among metal complexes, clusters are specially appropriate for redox reactions. As discussed in Sect. 2.3, the formation of a metal network gives rise to many new electron energy levels which can participate as donor or acceptor centers in electron exchange processes. In many cases metal clusters are indeed able to act as electron reservoirs. The best examples of the electron transfer versatility of cluster compounds are the iron-sulfur clusters, which because of their biological importance will be discussed separately in Chapter 5.

Because of the nature of metal clusters, assignation of oxidation numbers to individual metal atoms as it is usually made for mononuclear species does not make sense in the case of metal aggregates. Cluster polyatomic metal nucleus must be considered as a whole, so only average oxidation numbers or states can be assigned to individual atoms.

The oxidation state of a cluster may be formally defined by considering the skeleton valence electron counting as established in Sect. 2.3. An alternative method could be the same counting but without considering the electrons being involved in the valence shell of the ligands. Dividing these metal valence electrons among the metal atoms involved in the cluster, an average number of electrons per metal atom useful for comparisons between clusters independent of their nuclearities may be obtained. An increase in the number of electrons in the metal skeleton implies, in general, a reduction of cluster species whereas cluster oxidation will be associated with a diminution in electrons in the polyatomic metal nucleus. Nonetheless, there is not a unique method which could be useful and consistent in every case. As it will be often seen in this Chapter, the evaluation of changes in oxidation states is in many instances more a criterion matter than a response to a predetermined pattern.

Table 2.24 illustrates schematically some ways by which changes in the oxidation state of cluster species may be achieved. The simplest one by which a redox process can occur is the addition or substraction of electrons or protons without producing big changes in cluster structure.

Degradation processes leading to a diminution in the number of metal-metal bonds with concomitant formation of metal-ligand bonds imply in general cluster oxidation. On the other hand, the conversion of metal-ligand bonds in metal-metal bonds increasing thus the cluster nuclearity implies in general a decrease in the oxidation state of involved metal atoms and thus a reduction of the cluster species as a whole.

In the following description of cluster reactions involving changes in cluster oxidation states both electrochemical and chemical processes will be considered. In order to systematize the chemical redox processes, they have been roughly cataloged as reduction, oxidation, or additive oxidation reactions.

Electrochemical Processes. The simplest type of redox reaction in cluster chemistry is the direct addition or removal of electrons at an electrode surface. Cyclic voltammetry has proved to be a very useful electrochemical technique for studying this kind of redox process. It permits us primarily to obtain informa-

Table 2.24. Examples of redox processes in metal clusters

Type of reaction	Reaction	Comments	Examples	Ref.
Electron-Transfer	$[M\text{-}M]^{m+} \pm ne^- \rightarrow [M\text{-}M]^{(m\pm n)+}$	Oxidation and reduction by chemical or electrochemical electron-transfer.	$[Nb_6Br_{12}]^{2+} - 2e^- \xrightarrow[0.426V\ (SCE)]{} [Nb_6Br_{12}]^{4+}$	1
			$[Os_6(CO)_{18}]^{2-} + I_2 \rightarrow Os_6(CO)_{18}$	2
Proton Addition	$[M\text{-}M]^{m+} + H^+ \rightarrow [M\text{-}M\text{-}H]^{(m+1)+}$	Oxidation by proton addition yielding metal hydride cluster.	$[Ru_6(CO)_{18}]^{2-} + 2H^+ \rightarrow H_2Ru_6(CO)_{18}$	3
			$[HIr_4(CO)_{11}]^- + H^+ \rightarrow H_2Ir_4(CO)_{11}$	4
Base-Induced Degradation	$M\text{-}M + L \rightarrow M:L$	Soft Lewis-base addition induces metal-metal bond breakdown	$Os_5(CO)_{16} + P(OMe)_3 \rightarrow Os_5(CO)_{16}[P(OMe)_3]_3$	5
			$Os_5(CO)_{16} + CO \rightarrow Os_5(CO)_{19}$	5
Oxidative Addition	$M\text{-}M + L:L' \rightarrow ML + ML'$	Simultaneous addition of the fragments of a covalent species	$Os_3(CO)_{12} + X_2 \rightarrow X[Os(CO)_4]_3X$	6
			$Ru_3(CO)_{12} + HX \rightarrow (\mu_2\text{-}H)Ru_3(CO)_{10}(\mu_2\text{-}X)$	7

References

1 Espenson JH, McCarley RE (1966) J. Am. Chem. Soc. 88:1063
2 Eady CR, Johnson BFG, Lewis J (1976) J. Chem. Soc., Chem. Commun., 302
3 Eady CR, Jackson PF, Johnson BFG, Lewis J, Malatesta MC, McPartlin M, Nelson WJH (1980) J. Chem. Soc., Dalton Trans., 383
4 Angoletta M, Malatesta L, Caglio G (1975) J. Organomet. Chem. 94:99
5 Farrar DH, Johnson BFG, Lewis J, Raithby PR, Rosales MJ (1982) J. Chem. Soc., Dalton Trans., 2501
6 Cook N, Smart L, Woodward P (1977) J. Chem. Soc., Dalton Trans., 1744
7 Tachikawa M, Shapley IR (1977) J. Organomet. Chem. 124:C19

tion about the number and relative stability of the species formed on the electrode. Moreover, sometimes kinetic and mechanistic features associated with the electron transfer processes may also be investigated by this method. Coulommetric measurements and exhaustive electrolysis are also useful as complementary techniques for determining the number of electrons associated to investigate redox processes as well as a method for obtaining products of electrochemical reactions at a preparative scale.

The analysis of electrochemical reactions by cyclic voltammetry depends fundamentally on the stability of their products under experimental conditions. In general species with half-life times of the order of seconds are needed for this technique.

In Table 2.25 some selected examples of electrochemical reactions of metal clusters are described.

Table 2.25. Selected examples of electrochemical electron-transfer processes in metal clusters

Cluster	Products	Comments	Ref.
$[Ta_6Cl_{12}]^{4+}$	$[TaCl_{12}]^{3+}$ $[TaCl_{12}]^{2+}$	Acid water solutions. Reversible, rapid electron-transfers at 0.25 and 0.59 V (SCE); cluster remains intact.	1
$[CpFe(CO)]_4$	$[CpFe(CO)]_4^+$ $[CpFe(CO)]_4PF_6$	Dichloromethane 0.1 M $[NBu_4](PF_6)$ quantitative oxidation; product insoluble in reaction mixture.	2
$[CpFe(CO)]_4PF_6$	$[CpFe(CO)]_4^-$ $[CpFe(CO)]_4$ $[CpFe(CO)]_4^{2+}$	Acetonitrile, 0.1 M $[NBu_4]PF_6$. Electrochemically reversible processes: $[Fe_4]^{2+}/[Fe_4]^+$, 1.08 V; $[Fe_4]^+/[Fe_4]$, 0.32 V; $[Fe_4]/[Fe_4]^-$, -1.30 V. Monoanion $[Fe_4]^-$ not isolated because of extreme air sensitivity. Dication detected only by cyclic voltammetry (200 mvs^{-1}).	2
$[CpFeS]_4$	$[CpFeS]_4^-$ $[CpFeS]_4PF_6$ $[CpFeS]_4(PF_6)_2$ $[CpFeS]_4^{3+}$	Acetonitrile, 0.1 M Bu_4NPF_6. Four electrochemically reversible one-electron processes: $[Fe_4]^{3+}/[Fe_4]^{2+}$, 1.41 V; $[Fe_4]^{2+}/[Fe_4]^+$, 0.88 V; $[Fe_4]^+/[Fe_4]$, 0.33 V; $[Fe_4]/[Fe_4]^-$, -0.33 V. Salts of mono and dication are isolated in 60–70% yields.	3
$(Cp)_3Co_3S_2$	$[(Cp)_3Co_3S_2]^+$ $[(Cp)_3Co_3S_2]^{2+}$ $[(Cp)_3Co_3S_2]^-$	Benzonitrile, Pt-electrode; three one-electron electrochemically reversible processes: $[Co_3]^{2+}/[Co]^{3+}$, 0.57 V; $[Co_3]^+/[Co_3]$, -0.12 V; $[Co_3]/[Co_3]^-$, -1.13 V. Potentials referred to Ag/AgCl.	4
$Co_3(CO)_9S$	$[Co_3(CO)_9S]^+$ $[Co_3(CO)_9S]^{2-}$	Dichloroethane, one near-reversible, one-electron oxidation at $+0.04$ V and one irreversible two-electron reduction at -1.1 V. Potentials referred to Ferrocinium/Ferrocene (Fc^+/Fc) couple. Monocation stable only in solution.	5
$RCCo_3(Cp)_3CR'$	$[RCCo_3(Cp)_3CR']^-$ $[RCCo_3(Cp)_3CR']^+$ $[RCCo_3(Cp)_3CR']^{2+}$	R, R' = Ph, Ph; Ph, H; Me_3Si, H; Me_3Si, I. Dichloromethane or acetone. One electrochemically irreversible reduction and two oxidations, one reversible and one irreversible near the solvent limit: $[Co_3]^-/[Co_3]$, ca. -1.5 V $[Co_3]/[Co_3]^+$, ca. 0.6 V; $[Co_3]^+/[Co_3]^{2+}$, 1.3 V (vs. Ag/AgCl). For R=R'=ferrocene, redox-series is extended to $[RCCo_3(Cp)_3 CR']^{3+,2+,1+,0,1-}$.	6

References

1 Cooke NE, Kuwana T, Espenson J (1971) Inorg. Chem. 10:1081
2 Ferguson JA, Meyer TJ (1972) J. Am. Chem. Soc. 94:3409
3 Ferguson JA, Meyer TJ (1971) J. Chem. Soc. Chem. Commun., 623
4 Machach T, Vahrenkamp H (1981) Chem. Ber. 114:505
5 Honrath U, Vahrenkamp H (1984) Z. Naturforsch. B39:555
6 Colbran SB, Robinson BH, Simpson J (1984) Organometallics 4:1344

Further examples and references in Meyer TJ (1975) Progr. Inorg. Chem. 19:1; Zanello P (1988) Coord. Chem. Rev. 83:199; Lemoine P (1988) Coord. Chem. Rev. 83:169

One typical case exemplifying the ability of some cluster compounds to undergo redox processes is the cubane cluster $[\pi\text{-CpFe(CO)}]_4$ whose cyclic voltammogram is illustrated in Fig. 2.52. Observed waves corresponding to totally reversible, one-electron processes can be assigned according to following equations,

$$[\pi\text{-CpFe(CO)}]_4^{2+} \xrightarrow[1.07\ V]{e^-} [\pi\text{-CpFe(CO)}]_4^+ \xrightarrow[0.32\ V]{e^-} [\pi\text{-CpFe(CO)}]_4$$

$$\xrightarrow[-1.30\ V]{e^-} [\pi\text{-CpFe(CO)}]_4^-$$

Potentials are measured with a Saturated Calomel Electrode (SCE) as reference.

Although all these species are stable in the time scale of this technique, their separation from the reaction mixture is not always possible. The anionic species $[\pi\text{-CpFe(CO)}]_4^-$ is stable in solution for a long time but its reactivity, specially toward dioxygen producing $[\pi\text{-CpFe(CO)}]_4$, prevents its being obtained as a solid salt. The dicationic species $[\pi\text{-CpFe(CO)}]_4^{2+}$ is stable only in the time scale of the measurements. It decomposes rapidly in acetonitrile leading to the degradation products Fe_2^+ and $[\pi\text{-CpFe(CO)}_2CH_3CN]_4^+$. The monocation $[\pi\text{-CpFe(CO)}]_4^+$ is stable and can be obtained by both, chemical and electro-chemical oxidation of the neutral compound. Crystallographic characterization of the cation indicates that its structure is similar to that of the neutral compound. The oxidation produces only a small distortion of the geometry T_d of the Fe_4-core and a small diminution in the average metal-metal distance (0.036 Å).

The cyclic voltammograms of high-valence clusters $[M_6X_{12}]^{2+}$ (M = Nb or Ta; X = Cl, Br) in aqueous acid solutions show two reversible waves which correspond to two consecutive one-electron transfers. Oxidized species retain

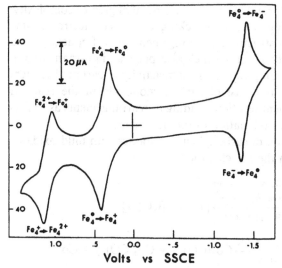

Fig. 2.52. Cyclic voltammogram of compound $[\pi\text{-CpFe(CO)}]_4$. Reproduced with permission from Ferguson JA, Meyer, TJ (1972) J. Am. Chem. Soc. 94:3409

their structures after the electron transfer so the equilibria between cations with charges $2+$, $3+$ and $4+$ are possible.

$$[M_6X_{12}]^{2+} \xrightarrow[-0.55\ V]{e^-} [M_6X_{12}]^{3+} \xrightarrow[-0.85]{e^-} [M_6X_{12}]^{4+}$$

Redox potentials (vs SCE) in the scheme above correspond to those of the bromide derivatives.

Chemical Oxidation Reactions. Atmospheric oxygen, weak oxidants such as iron(II) chloride, mercury(II) or copper(II) salts, and protonic acids are oxidation agents frequently used for oxidizing metal clusters.

Many metal clusters are air-stable compounds. However those of the first transition metal series are in general more sensitive to air. Thus, binary carbonyl of ruthenium, rhodium, and iridium are rather air-stable species while those of iron, $Fe_3(CO)_{12}$, and cobalt, $Co_4(CO)_{12}$, rapidly decompose apparently via formation of the carbonates.

Mild oxidant agents often permit oxidations with retention of the cluster structures. One example of this kind of simple chemical electron-transfer reactions are the oxidations of niobium and tantalum clusters $[M_6X_{12}]^{2+}$ to the species with charges $3+$ and $4+$. The reaction with iron(III) chloride and cerium(IV) salts leads directly to the cations $[M_6X_{12}]^{4+}$.

$$[M_6X_{12}]^{2+} \xrightarrow[\text{or Ce(IV)}]{FeCl_3} [M_6X_{12}]^{4+}$$

Meanwhile in the reactions with milder oxidation agents, e.g. Hg(II) or Cu(II) salts or iodine, intermediate oxidation states of the clusters are obtained.

$$[M_6X_{12}]^{2+} \xrightarrow[\text{Cu(II) or } I_2]{\text{Hg(II)},} [M_6X_{12}]^{3+}$$

The oxidation of carbonyl metal clusters is in general of greater complexity. Especially when strong bridging ligands are lacking, changes in electron counting are often associated with structural rearrangements and, frequently, also with nuclearity changes. Another factor which also plays an important role in the stabilization of the oxidation products is the reactivity of carbonyl clusters with carbon monoxide. The presence of carbon monoxide in the reaction mixture often induces subsequent reactions leading to rearrangements as well as to cluster degradation and recombination of the fragments.

In some cases, the oxidation of carbonyl cluster species with mild oxidants can occur without changes in their nuclearities.

$$[Fe_3Pt_3(CO)_{15}]^{2-} \xrightarrow[\text{OH}^- \text{ in MeOH}]{\substack{H_3PO_4 \text{ or } H_2SO_4 \text{ or} \\ Cu(I) \text{ or } Ag(I) \text{ salts}}} [Fe_3Pt_3(CO)_{15}]^-$$

$$[Co_6(CO)_{15}]^- \xrightarrow[25°,\ H_2O]{[HgCl_4]^{2-} \text{ or } FeCl_3} Co_6(CO)_{16} + Co_4(CO)_{12} + Co^{2+}$$

However, in spite of the formation of $Co_6(CO)_{16}$ in the last reaction, the diversity of products observed indicates a reaction mechanism much more complex than a simple electron transfer.

As mentioned above carbonyl clusters are in general labile to restructuration. This property is normally enhanced by electron transfer reactions which indeed promote processes leading to new structures that agree better with the electron counting. Molecular-orbital studies like those for the cluster $[\pi\text{-}C_5H_5FeS]_4$ illustrated in Fig. 2.53 permit to understand such a behavior. In metal clusters the HOMO and the LUMO are usually metal-metal bonding and antibonding molecular orbitals respectively. Electron transfer to clusters often results therefore in a weakening of metal-metal interactions favoring cluster fragmentation or, at least, lowering the activation barriers to structural rearrangements via metal-metal bond rupture. In general, this tendency to undergo polyhedron rearrangements increases with increasing cluster nuclearity.

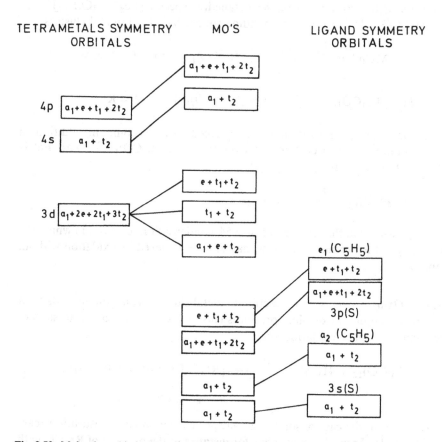

Fig. 2.53. Molecular orbital scheme for $[\pi\text{-}CpFES]_4$ [Trinh–Toan, Teo BK, Ferguson JA, Meyer TJ, Dahl LF (1977) J. Am. Chem. Soc. 99:408]

In the chemical oxidation of the anion $[Os_6(CO)_{18}]^{2-}$ by iodine leading to the neutral species the cluster nuclearity is retained.

$$[Os_6(CO)_{18}]^{2-} \xrightarrow{I_2} [Os_6(CO)_{18}]$$

However, a reorganization of the metal polyhedron from an octahedral to a bicapped tetrahedral geometry is observed.

Iron(III) salts are often used as oxidant agents in carbonyl cluster chemistry. However, depending on the nature of the metal cluster as well as on the reaction conditions, different processes may occur.

The oxidation of $[Co_6C(CO)_{15}]^{2-}$ (with 90 valence electrons) by $FeCl_3$ with the loss of one mole carbon monoxide is also accompanied by cluster rearrangement from the original trigonal prismatic to an octahedral geometry.

$$[CoC(CO)_{15}]^{2-} + FeCl_3 \rightarrow [CoC(CO)_{14}]^- + FeCl_2 + Cl^- + CO$$

The same oxidant $FeCl_3$ may in many instances cause cluster degradations as occurs in the conversion of the octahedral species $[Fe_5MC(CO)_{17}]^{2-}$ or $[Fe_5MC(CO)_{16}]^{2-}$ to the corresponding square-based pyramidal clusters.

$$[Fe_5MC(CO)_{17}]^{2-} \xrightarrow{FeCl_3} Fe_4MC(CO)_{16} \quad M = Cr, Mo, W$$

$$[Fe_4MC(CO)_{16}]^{2-} \xrightarrow{FeCl_3} [Fe_4MC(CO)_{14}]^- \quad M = Rh$$

In some cases, the oxidation may proceed with concomitant addition of carbon monoxide as occurs in the oxidation of $[Os_5(CO)_{15}]^{2-}$ with $FeCl_3$ under CO-atmosphere.

$$[Os_5(CO)_{15}]^{2-} \xrightarrow[CO]{FeCl_3} Os_5(CO)_{16}$$

As illustrated in the scheme in Fig. 2.54, condensation reactions in which the nuclearity of the products increases may also be achieved by oxidation with an iron(III) salt.

Acids as Oxidant Agents. The reaction of metal clusters with oxidant acids such as nitric or sulfuric acids often produces cluster breakdown leading to salts of the metals in higher oxidation states.

$$Rh_4(CO)_{12} + 4H_2SO_4 \xrightarrow[CH_3CH]{25\,°C} 4[Rh(CH_3CN)_2(CO)_2]HSO_4$$

$$+ 2H_2 + 4CO$$

However in the case of some high-valence cluster species substitution reactions have also been observed. Thus for instance in the reaction of Re_3Cl_9 with

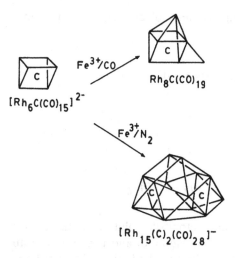

Fig. 2.54. Condensations reactions by oxidation of $[Rh_6C(CO)_{15}]^{2-}$ by Fe(III)-salts [Albano VG, Chini P, Martinengo S, Sansoni M, Strumolo DJ (1974) J. Chem. Soc., Chem. Commun., 299]

nitric acid where, beside the formation of the rhenium(IV) salt $Cs(ReCl_6)$, cluster ligand exchange of chloride to nitrate to an extent of ca. 17% is achieved.

$$Re_3Cl_9 \xrightarrow[HNO_3, \, 0\,°C]{CsNO_3} \begin{array}{l} Cs(ReCl_6) \\ \\ (Cs_3Re_3Cl_9)(NO_3)_3 \end{array}$$

Oxidation of metal clusters may also be performed by reaction with Bronsted-acids through straightforward addition of protons to metal backbone. Thus, carbonyl clusters of ruthenium, osmium and iridium are stable in acids and may be protoned without decomposition. The ^{1}H-NMR spectra of these carbonyls in concentrated sulfuric or trifluoroacetic acid indicate the formation of cationic metal hydrides:

$$M_3(CO)_{12} + H^+ \xrightarrow[90\%]{H_2SO_4} [HM_3(CO)_{12}]^+ \qquad M = Ru, Os$$

$$Ir_4(CO)_8L_4 + 2H^+ \xrightarrow[\text{or } CF_3COOH]{H_2SO_4} [H_2Ir_4(CO)_8L_4]^{2+} \qquad L = Co, PR_3$$

Affinity of cluster species to protons increases with increasing electron density in the cluster. Thus many anions, generally obtained by reduction of carbonyl clusters in basic media (vide infra), are straightforwardly transformed to the corresponding hydrides by addition of protons from acids or solvents.

$$[Fe_4(CO)_{12}]^{2-} \rightarrow [HFe_4(CO)_{13}]^- + H_2Fe_4(CO)_{13}$$

In many cases, by properly choosing acid and solvent, sequential protonation of the anions may be achieved.

$$[Os_{10}C(CO)_{24}]^{2-} \begin{array}{c} \xrightarrow[\text{H}_2\text{SO}_4]{\text{CH}_2\text{Cl}_2 \text{ or THF}} [HOs_{10}C(CO)_{24}]^- \\[1em] \xrightarrow[\text{CH}_3\text{CN}]{\text{H}_2\text{SO}_4} H_2Os_{10}C(CO)_{24} \end{array}$$

$$[HOs_{10}C(CO)_{24}]^- \xrightarrow{X^-} H_2Os_{10}C(CO)_{24}$$

Similarly, the addition of acid to an acetonitrile solution of the cluster $[Ni_{38}Pt_6(CO)_{48}]^{6-}$ leads to a mixture of the mono, di and trihydrides in an equilibrium that may be easily governed by controlled addition of acid or base.

$$[Ni_{38}Pt_6(CO)_{48}]^{6-} \xrightarrow[\text{OH}^-]{\text{H}_2\text{O}} [Ni_{38}Pt_6(CO)_{48}H]^{5-}$$

$$\xrightarrow[\text{CO}_3{}^{2-}]{\text{H}^+} [Ni_{38}Pt_6(CO)_{48}H_2]^{4-} \xrightarrow[\text{MeCN}]{\text{H}^+} [Ni_{38}Pt_6(CO)_{48}H_3]^{3-}$$

The mono and dihydrides are stable enough for isolating crystalline salts with voluminous cations. The acidity of the trihydride prevents us,-however, from obtaining pure compounds.

Protonation reactions are often complicated by further chemical transformations of the unstable hydride derivatives. Thus in many instances the reaction of anionic clusters with acids leads to oxidation products with elimination of molecular hydrogen. That occurs for example in the reaction of the anion $[Ir_6(CO)_{15}]^{2-}$ with acetic acid from which, depending of the reaction conditions, two isomers of the metal cluster $Ir_6(CO)_{16}$ can be isolated.

$$[Ir_6(CO)_{15}]^{2-} \begin{array}{c} \xrightarrow[\text{CO}]{\text{MeCOOH, CH}_2\text{Cl}_2} Ir_6(CO)_{12}(\mu_2\text{-CO})_4 \\[1em] \xrightarrow[\text{MeCOOH}]{\text{CO}} Ir_6(CO)_{12}(\mu_3\text{-CO})_4 \end{array}$$

The elimination of hydrogen depends to a great extent upon the presence of nucleophiles in the reaction medium. Thus, the dihydride $H_2Ir_4(CO)_{11}$ can be prepared by protonation of the anionic monohydride in a nitrogen atmosphere, while in the presence of carbon monoxide, the hydrogen elimination described above occurs.

The complexity of subsequent and secondary reactions in the protonation of anionic clusters may be observed in the following equations that illustrate the protonation of the tetraanion $[Rh_6(CO)_{14}]^{4-}$ under different reaction conditions.

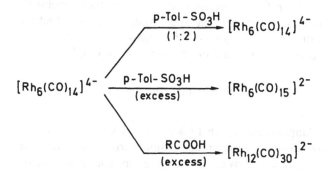

Cluster oxidation by reaction with Bronsted acids may also occur by ligand elimination. Thus, species Re_3^{7+} stabilized in the presence of tertiary amine groups in the cluster coordination sphere are oxidized by acid to Re_3^{9+}

$$1/n[Re_3Cl_7L_2]_n \xrightarrow[\text{HCl}]{\text{MeOH}} [LH]_2[Re_3Cl_{11}]$$

Both, ligand and molecular hydrogen elimination processes probably occur by a similar mechanism with hydride species as intermediates. Isolation of species like $[Os_5(CO)_{15}H_2I]^-$ and NMR detection of intermediates like $[Os_7(CO)_{20}H_2Cl]^-$ in reactions of the corresponding hydrides with iodide and chloride ions agree with this kind of mechanism.

$$H_2Os_5(CO)_{15} + I^- \longrightarrow [H_2IOs_5(CO)_{15}]^-$$
$$H_2Os_7(CO)_{20} + Cl^- \longrightarrow [H_2ClOs_7(CO)_{20}]^-$$

Protonation and deprotonation reactions of carbonyl cluster species are in general rather slow. A screening effect of the carbonyl groups around the metal nucleus which makes the direct interaction of the reactants difficult is considered to be the cause of large activation barriers.

The same size effects appear to prevent the addition of other electrophiles to cluster metal atoms. The attack of other nucleophiles normally leads to their attachment to the ligand sphere at the high electron density point. Thus for instance the reaction of $FeCo_2(\mu_3\text{-}S)(CO)_9$ with the electrophile $Cr(CO)_5$ which attacks at the μ_3-sulfur ligand may be explained.

$$FeCo_2(\mu_3\text{-}S)(CO)_9 + Cr(CO)_5THF \longrightarrow Cr(CO)_5(\mu_3\text{-}S)FeCo_2(CO)_9$$

Analogously, the coordination of main group element electrophiles normally occurs at the oxygen centers of bridged CO-ligands.

Cluster Reduction Reactions. Alkali metals, sodium borhydride, and alkali hydroxides are frequently used as reduction agents in cluster chemistry.

Straightforward electron transfers to clusters can be often achieved by reaction with metal atoms. Transferred electrons replace either metal-metal or metal-ligand bonds in the cluster. One example illustrating a simple reduction process is the reaction of the anion $[Rh_{12}(CO)_{30}]^{2-}$ with sodium in which the transfer of two electrons causes the rupture of one metal-metal bond.

$$[Rh_{12}(CO)_{30}]^{2-} + 2Na \xrightarrow[25\,°C]{THF} 2[Rh_6(CO)_{15}]^{2-} + 2Na^+$$

However this class of simple reductions is rather unusual. Cluster reduction with alkali metals is almost always a rather more complex process in which the initial electron transfer is usually followed by degradation, condensation, or rearrangement processes.

In the case of carbonyl metal clusters, the presence of free carbon monoxide notoriously affects the reduction processes inducing condensations as well as degradation of the clusters species. Thus the reduction of the tetranuclear cobalt cluster $Co_4(CO)_4$ with lithium, sodium, or potassium in which the final product is the hexanuclear cluster $[Co_6(CO)_{15}]^{2-}$ is an illustrative example of the complexity of such reactions.

$$Co_4(CO)_{12} \xrightarrow[THF]{Na} [Co_4(CO)_{11}]^- \xrightarrow[THF]{CO} [Co_6(CO)_{15}]^{2-}$$

Initial electron transfer induces the rupture of the metal-CO bond. The subsequent reaction of the monoanion with the ejected CO produces the augmented nuclearity. An excess of alkali metal leads to the formation of $[Co_6(CO)_{14}]^{4-}$. The formation of this tetraanion by this method is, however, not straightforward because the anion itself undergoes degradation in the presence of carbon monoxide:

$$11[Co_6(CO)_{15}]^{2-} + 22Na \longrightarrow 11[Co_6(CO)_{14}]^{4-} + 22Na^+ + 11CO$$

$$2[Co_6(CO)_{14}]^{4-} + 11CO \longrightarrow [Co_6(CO)_{15}]^{2-} + 6[Co(CO)_4]^-$$

Because of the π-acceptor properties of carbon monoxide, the tendency to cluster degradation increases with increasing negative cluster charge. Therefore, an excess of an alkali metal in a CO-atmosphere often produces the total degradation of the cluster.

$$M_4(CO)_{12} + 4Na + 4CO \xrightarrow[CO]{Na(THF)} 4[M(CO)_4]^- + 4Na^+ \quad M = Rh, Ir$$

In liquid ammonia reduction processes are normally drastic.

$$M_3(CO)_{12} + 6Na \xrightarrow{\text{liq. NH}_3} 3Na_2[M(CO)_4] \qquad M = Ru, Fe$$

$$Co_4(CO)_{12} + 3Na \xrightarrow{\text{liq. NH}_3} 3Na[Co(CO)_4] + Co$$

Reactions with Hard Bases. Cluster degradation produced by the presence of CO in cluster reduction reactions may be avoided by using hard nucleophiles as OH^-, OR^-, or R^- as reduction agents. Under such conditions oxidation of carbon monoxide occurs leading to the formation of anionic species.

$$Rh_6(CO)_{16} + 8OH^- \longrightarrow [Rh_6(CO)_{14}]^{4-} + 2CO_3^{2-} + 4H_2O$$

$$Co_4(CO)_{12} + 4OR^- + 2H_2O \longrightarrow [Co_4(CO)_{11}]^{2-} + CO_3^{2-} + 4ROH$$

Isolation of derivatives containing groups CO_2X from reactions of carbonyl clusters with nucleophiles OR^- or X^- as those illustrated in the scheme in Fig. 2.55 point out to a mechanism via formyl derivatives.

$$M_m(CO)_n + OH^- \longrightarrow [M_m(CO)_{n-1}(CO_2H)]^-$$

$$M_m(CO)_{n-1}(CO_2H)]^- \longrightarrow [M_m(CO)_{n-1}]^{2-} + CO_2 + H^+$$

In contrast to the reaction of $[Os_5(CO)_{16}]$ with the base OH^- that also produces the dianion

$$Os_5(CO)_{16} + 2OH^- \longrightarrow [Os_5(CO)_{15}]^{2-} + CO_2 + H_2O$$

the high nuclearity clusters Os_6, Os_7, and Os_8 lead, under similar conditions, to fragmentation resulting in the formation of the dianions $[Os_5(CO)_{15}]^{2-}$, $[Os_6(CO)_{18}]^{2-}$, and $[Os_7(CO)_{20}]^{2-}$ respectively. The different behavior on reduction observed for these compounds appears to be related to the electron structures of the clusters. Thus the cluster $Os_5(CO)_{16}$ is an electron-precise cluster compound with six skeletal electron pairs and a regular trigonal bipyramidal geometry meanwhile those of the other member of the series are

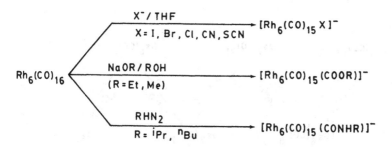

Fig. 2.55. Reduction of reactions of $Rh_6(CO)_{16}$ by hard Lewis bases [Chini P, Martinego S, Giordano G (1972) Gazz. Chim. Ital. 102:330]

electron-deficient compounds with capped polyhedral geometries. As illustrated on Fig. 2.56, the electron-deficient Os_6, Os_7, and Os_8 cluster derivatives are transformed by ejection of $Os(CO)_3$ capping groups in electron-precise species or, in the case of $[Os_7(CO)_{20}]^{2-}$, in one with lower deficiency than the original neutral cluster $Os_8(CO)_{23}$.

By using milder nucleophiles than OH^- such as nitriles and halides, it is also possible to reduce high nuclearity osmium carbonyl clusters to the corresponding dianions. Thus, for instance, the reaction of $Os_6(CO)_{18}$ with different nitriles or with iodide leads to the dianion $[Os_6(CO)_{18}]^{2-}$

$$Os_6(CO)_{18} \xrightarrow{\text{Y}} [Os_6(CO)_{18}]^{2-} \qquad Y = I^-, RCN, \text{ or } P_y$$

The reduction in this case involves a reorganization of the metal framework with a change of its geometry from a capped trigonal geometry to a regular octahedral one.

Examples of the reduction of cluster species with neutral Lewis bases are also found in the chemistry of the high-valence rhenium clusters. Thus the action of tertiary amines on halide clusters Re_3X_9 leads to rhenium(II) cluster compounds.

$$Re_3X_9 + 6NR_3 \xrightarrow{\text{H}_2\text{O}} 1/n[Re_3X_6(NR_3)_3]^{n+} + 3R_3NHX$$

Reactions with Soft Bases. Metal-metal bonds are often weaker than those the same metal can form with ligands. Thus the reaction of clusters with nucleophiles frequently leads to cleavage of metal-metal bonds and to cluster breakdown. The normally low oxidation state of metal centers on clusters makes their reaction with soft bases such as CO, PR_3, olefins etc. very favorable. Clusters of lighter transition metals undergo mainly fragmentation reactions. Heavier element derivatives however are likely to participate in addition and substitution reactions.

Proper cluster reduction reactions can occur only when the clusters are electron-deficient species.

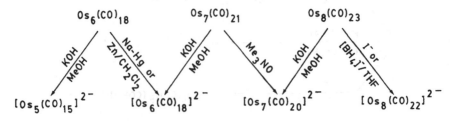

Fig. 2.56. Reduction reactions of osmium carbonyl clusters [John GR, Johnson BFG, Lewis J (1979) J. Organomet. Chem. 186:C69; Vargas MD, Nicholls JN (1986) Adv. Inorg. Chem. Radio Chem 30:123]

Although unsaturation is a rather uncommon phenomenon there are some examples of such compounds, for instance, the clusters $(\mu_2\text{-}H)_2Os_3(CO)_{10}$ and $Fe_4(CO)_{11}(\mu_4\text{-}PR)_2$ which contain a total of 46 and 62 electrons respectively i.e. clusters with two electrons less than those required by the effective atomic number rule. As mentioned earlier in Sect. 2.3, in these cases metal-metal double bonds are supposed to be present. This class of clusters can add several different donor ligands to form electronic saturated species with 48 and 64 electrons respectively.

$$(\mu_2\text{-}H)_2Os_3(CO)_{10} + L \longrightarrow H(\mu_2\text{-}H)Os_3(CO)_{10}L$$

$$L = CO, PR_3, P(OR)_3, CNR.$$

$$Fe_4(\mu_2\text{-}CO)(CO)_{10}(\mu_4\text{-}PR)_2 + L \longrightarrow Fe_4(CO)_{11}L(\mu_4\text{-}PR)_2$$

$$L = CO, P(OMe)_3$$

As mentioned above, addition of electron donors to electronic saturated species normally results in cluster degradation. However the presence of special multi-bridging substituents sometimes permits the addition of neutral Lewis bases without cluster degradation. Nonetheless the addition always implies the rupture of metal-metal bonds. The fragments remain, however, close by the bridging ligands. That is illustrated schematically for the reaction of the mixed-metal cluster $MnFe_2(CO)_8Cp(\mu_3\text{-}PR)$ with various electron donors in Fig. 2.57.

In the absence of ligands bridging the cluster structure, the addition of Lewis bases to clusters often produces metal-metal bond cleavages inducing rearrangements of the cluster metal framework. Many interesting examples of this class of reactions are found in the chemistry of osmium carbonyl clusters. Thus, the addition of three CO-groups to the trigonal-bipyramidal compound $Os_5(CO)_{16}$ produces the cleavage of three metal-metal bonds yielding the cluster $Os_5(CO)_{19}$ with a "bow-tie" structure. The reaction can be reversed when carbon monoxide is removed by heating.

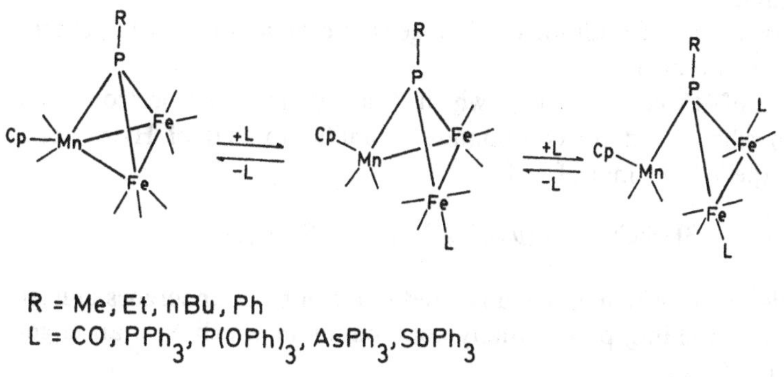

R = Me, Et, nBu, Ph
L = CO, PPh$_3$, P(OPh)$_3$, AsPh$_3$, SbPh$_3$

Fig. 2.57. Reactions of the mixed clusters $MnFe_2(CO)_8Cp(\mu_3\text{-}PR)$ with electron pair donors [Schemide J, Huttner G (1983) Chem. Ber. 116:917]

Oxidative Additions. Many covalent compounds (XY) such as hydrogen, halogens, mercury and silver halogenides can react with clusters resulting in an homolytic cleavage of the X–Y bond and in the oxidative addition of the two fragments to a metal-metal bond thus increasing the formal oxidation state of the metal atoms.

$$X-Y \; + \; -M-M- \; \longrightarrow \; -M\overset{\displaystyle Y}{\underset{\displaystyle X}{|}}M-$$

This kind of reaction implies that there is a transfer of electrons to the clusters. Therefore, if there is no cluster unsaturation before the reaction, these oxidative additions will involve ligand substitution or metal-metal cleavage.

The oxidative addition of hydrogen to cluster is the best known reaction of this type. They are widely used for preparing hydride metal clusters.

The electron-deficient cluster $H_2Os_3Pt(CO)_{10}P(cyclo-C_6H_{11})_3$ adds one mole H_2 yielding the 60 electron compound $H_4Os_3Pt(CO)_{10}(cyclo-C_6H_{11})$. This reaction which occurs under a pressure of 200 bar H_2, can be reversed in a N_2-atmosphere.

$H_2Os_3Pt(CO)_{10}P(cyclo-C_6H_{11})_3$

$$\xrightarrow[N_2]{H_2, \; 200 \text{ atm}} H_4Os_3Pt(CO)_{10}(cyclo-C_6H_{11})$$

The H_2-addition to the 60-electron precise cluster $Os_3(CO)_{12}$ yielding $H_3Os_3(CO_{10})$ appears to be preceded by formation of vacant sites on the metal atoms through loss of CO ligands.

$$Os_3(CO)_{12} + H_2 \xrightarrow{125\,°C} H_2Os_3(CO)_{10} + 2CO$$

Further examples of additive oxidation of hydrogen and other reagents are described in Table 2.26.

Analogously, oxidative additions of halogens and halides follow in general a similar pattern to that of H_2.

The addition of HX compounds in which X has no unshared electron pairs as it occurs in silicon and tin compounds is similar to that of H_2 i.e. one CO-ligand is expelled per mol of HX.

$$Os_3(CO)_{12} + 3HSiCl_3 \longrightarrow (\mu_2\text{-}H)_3Os_3(CO)_9(SiCl_3)_3$$

Compounds HX in which X has unshared electron pairs undergoes cluster addition reactions yielding preferentially compounds in which X is an edge-bridging electron donor

$$Ru_3(CO)_{12} + HX \longrightarrow (\mu_2\text{-}H)Ru_3(CO)_{10}(\mu_2\text{-}H)$$

Table 2.26. Selected examples of oxidative additions to metal clusters

Reactions	Comments	Ref.
$Os_3(CO)_{12} + H_2 \rightarrow H_2Os_3(CO)_{10} + 2CO$	H_2 addition proceeds by unsaturation achieved by CO ejection.	1
$Co_2Ru_2(CO)_{13} + H_2 \rightarrow H_2Co_2Ru_2(CO)_{12} + CO$		2
$H_2Os_3Pt(CO)_{10}PR_3 + H_2 \rightarrow H_4Os_3Pt(CO)_{10}PR_3$	R = cyclo-C_6H_{11}. Because of unsaturation in starting complex, addition can occur without ligand ejection.	3
$Os_5C(CO)_{15} + H_2 \rightarrow H_2Os_5C(CO)_{15}$		4
$Os_3(CO)_{12} + X_2 \rightarrow X[Os(CO)_4]_3X$	X = Cl, Br, I. Product is a linear cluster.	5
$Ru_5C(CO)_{15} + X_2 \rightarrow Ru_5C(CO)_{15}X_2$		6
$Os_6(CO)_{20} + O_2 \rightarrow Os_6(CO)_{18}(\mu_3\text{-CO})(\mu_3\text{-O})$		7
$H_2Ru_3(CO)_9(\mu_3\text{-S}) + SnCl_4$ $\rightarrow H_2Ru_3(CO)_8(SnCl_3)(\mu_3\text{-S})(\mu\text{-Cl})$	Product is an open triangular cluster.	8
$Ru_5(CO)_{15}(\mu_5\text{-C}) + HX$ $\rightarrow HRu_5(CO)_{15}(X)(\mu_5\text{-C})$	X = Cl, Br, Product is an opened cluster in which a Ru–Ru bond to apical vertex is lacking.	9, 10

References

1 Knox SA, Koepke JW, Andrews AA, Kaesz HD (1975) J. Am. Chem. Soc. 97:3942
2 Roland E, Vahrenkamp H (1983) Organometallics 2:183
3 Farrugia LJ, Green M, Hankey DR, Orpen AG, Stone FGA (1983) J. Chem. Soc., Chem. Commum., 310
4 Johnson BFG, Lewis J, Nelson WJH, Nicholls JN, Puga JN, Raithby PR, Rosales MJ, Schöder MJ, Vargas MD (1983) J. Chem. Soc., Dalton Trans., 2447
6 Johnson BFG, Lewis J, Nelson WJH, Nicholls JN, Puga J, Whitmire KH (1983) J. Chem. Soc., Dalton Trans., 787
8 Adams RD, Katahira DA, (1982) Organometallics 1:53
9 Oxton IA, Powell DB, Farrar DH, Johnson FBG, Lewis J, Nicholls JN (1981) Inorg. Chem. 20:4302
10 Johnson BFG, Lewis J, Nicholls JN, Puga J, Raithby PR, Rosales MJ, McPartlin M, Clegg W (1983) J. Chem. Soc., Dalton Trans., 227

Further examples and reference can be found in Adams RD, Horvath IT (1985) Progr. Inorg. Chem. 33:127; and Vargas MD, Nicolls JN (1986) Adv. Inorg. Chem. Radiochem. 30:123

As expected for a mechanism in which the oxidative addition is preceded by ligand elimination, light stabilized clusters add straightforwardly different substances HX.

$$Os_3(CO)_{10}(NCMe)_2 \xrightarrow{\text{HX}} (\mu_2\text{-H})Os_3(CO)_{10}(\mu_2\text{-X})$$

In contrast to the H_2-addition in which the preliminary CO-ligand elimination permits to retain the cluster structure, the addition of halogens occurs preferentially through metal-metal bond cleavage without CO-elimination. This mechanism which implies the formation of X^- ions in a preliminary adduct

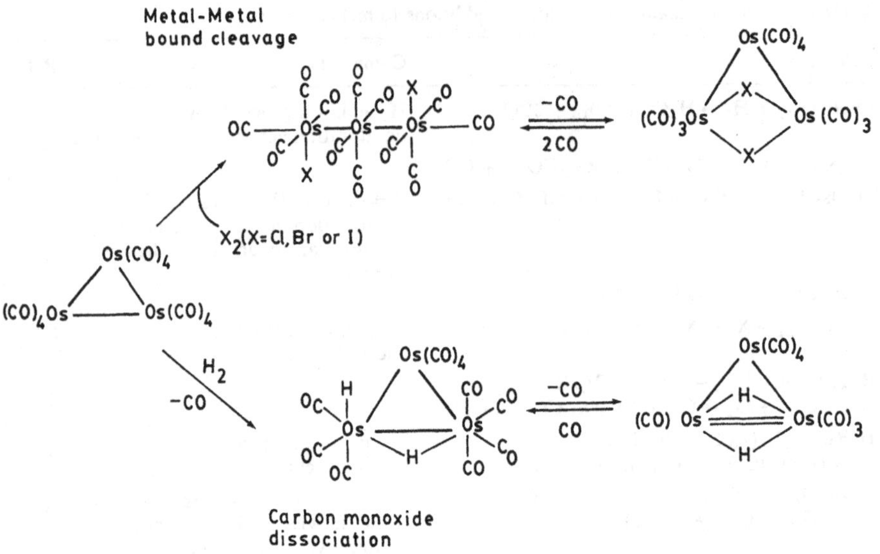

Fig. 2.58. Comparison of mechanisms of hydrogenation and halogenation [Deeming AJ (1980) In: Johnson BFG (ed) Transition metal clusters. John Wiley Chichester, p 391]

formation or the formation of a coordinatively unsaturated intermediate are energetically unfavorable when X = H.

A comparison of both types of mechanism is illustrated in the scheme in Fig. 2.58. Further decarbonylation of the primarily reaction product leads, in both hydrogenation and halogenation, to products with equivalent geometries but different electronic structures.

From the point of view of the application of cluster species to catalysis studies, oxidative addition involving the fission of C–H bonds are specially important. However most of these reactions are intermolecular processes following ligand exchange reactions. This theme will be therefore analyzed in a following section after the discussion of ligand substitution reactions.

2.5.2 Cluster Reactions Without Changes in the Cluster Oxidation State

The action of Lewis bases on metal clusters leads to very many reactions. In the last section we analyzed a series of reactions in which the addition of bases induces different kind of oxidation-reduction processes. Nevertheless, Lewis bases can also induce other kind of reactions which do not imply a change in the formal oxidation state of the cluster metal nucleus. They are normally classified as ligand exchange or ligand substitution reactions. Specially interesting are those reactions in which the cluster structure remains unaltered in the substitution process. However, it should be also considered that ligand exchange reactions are often accompanied by degradation processes, cluster rearrangements and coupled reactions which make it very difficult to establish the limits between one or other type of reaction. Most redox reactions are indeed preceded by Lewis acid-base interactions which lead to substitution products being intermediate in the electron-transfer processes.

Ligand Exchange in Metal Halide Clusters. Rhenium, molybdenum, tantalum and niobium halides are high-valence metal clusters that formally behave as Lewis acids towards halide ions forming anionic complexes that retain the structure of the original cluster.

$$Re_3X_9 + 3X^- \longrightarrow [Re_3X_{12}]^{3-}$$

$$[M_6X_8]^{4+} + 6X^- \longrightarrow [M_6X_{14}]^{2-} \qquad M = Mo, W.$$

$$[M_6X_{12}]^{2+} + 6X^- \longrightarrow [M_6X_{18}]^{4-} \qquad M = Nb, Ta.$$

Anionic cluster complexes undergo ligand exchange reactions in which some of the ligands may be easily substituted by other anionic or neutral Lewis bases.

$$Nb_6Cl_{14} \cdot 8H_2O + 4L \longrightarrow (Nb_6Cl_{12})Cl_2L_4$$

$$L = Py, R_3PO, DMSO, DMF, etc..$$

In the anions $[Mo_6X_{14}]^{2-}$ only 6 of the halide ligands may be substituted with relative facility, namely those bonded to the metal as terminal ligands. The other 8 face-bridging ligands are, however, very inert to exchange reactions.

$$[Mo_6Cl_{14}]^{2-} \xrightarrow{L} (Mo_6Cl_8)Cl_4L_2$$

In the case of rhenium(III) species it is also apparent that the terminal halides are more liabile to exchange than those in bridging positions. Furthermore, terminal ligands in equatorial position are exchanged more easily than those out of the molecular plane. Thus for Re_3Cl_9 the liability of the chloride ions follows the series

Cl terminal in plane > Cl terminal out of plane $\gg$ Cl bridged

2.5.3 Ligand Substitution Reactions in Carbonyl Metal Clusters

Although the mechanisms of cluster reactions in general have been not extensively studied, there are a couple of studies dealing with the mechanism of ligand substitution processes.

Three classes of reaction pathways may be distinguished in cluster substitution reactions:

1. Associative mechanisms (A). The reaction proceeds through the association of the reactants via a demonstrable intermediate in which the cluster metal

nucleus displays a coordination number greater than in the original compound.

2. Dissociative mechanism (D). The reaction proceeds via the dissociation of the cluster leading to a demonstrable intermediate in which the coordination number of the metal cluster is lower than in the original compound.

3. Interchange mechanism (I). This corresponds to reaction pathways for which the nature of the intermediate is not directly demonstrable.

Demonstration of the nature of the intermediates is in general a very difficult experimental task, therefore most ligand substitution reactions are assumed to be interchange processes. Nonetheless, kinetic studies often permit us to establish the intimate reaction mechanism which assigns to the interchange pathways either an associative or a dissociative character.

The determination of activation parameters, normally strongly associated with the nature of the transition state complex, may also be very useful in distinguishing plausible mechanisms in ligand substitution reactions. Dissociative mechanisms will be associated with a relatively high enthalpy change as well as to positive $\Delta S^{\ddagger}$ values. Associative processes will indeed show a not very high endothermic effect but a considerable decrease in entropy.

Coordinatively saturated cluster compounds, specially when they are electron-precise species, undergo ligand substitutions via dissociative pathways, meanwhile unsaturated cluster species will prefer to proceed via associative mechanisms.

Ligand substitution reactions of the iron-group trimetal dodecacarbonyl clusters and derivatives from them have been extensively studied. $M_3(CO)_{12}$ cluster, as it frequently occurs with stable small tri and tetranuclear clusters, is constituted by saturated 18-electron metal atoms. Accordingly dissociative pathways for substitutions in these compounds are expected.

$$M_3(CO)_{12} \longrightarrow [M_3(CO)_{11}] \xrightarrow{\text{L}} M_3(CO)_{11}L$$

Specially straightforward are the mechanistic studies of the acetonitrile derivative of the triosmium carbonyl $Os_3(CO)_{11}(NCMe)$ with strong osmium-osmium bonds, inert CO-ligands, and one labile acetonitrile ligand. As mentioned in Sect. 2.4. this kind of "light-stabilized" intermediate is very useful in the synthesis of substituted derivatives.

As shown in Table 2.27 the rate constants for displacement reactions of acetonitrile in $Os_3(CO)_{11}(NCMe)$ are independent of the nature as well as of the concentration of the substituting ligand. The rate constants determined for the substitution of acetonitrile by triphenylphosphin at different ratios $[MeCN]/[PPh_3]$ illustrated in the plots in Fig. 2.59 agree with the rate equation.

$$1/k_{obs} = 1/k_1 + (k_{-1}/k_1k_2)[MeCN]/[PPh_3]$$

which corresponds to a single dissociative mechanism where both acetonitrile

Table 2.27. Limiting first-order rate constants for the substitution reactions of $Os_3(CO)_{11}(NCMe)$ with phosphorus and arsenic donor ligands

$$Os_3(CO)_{11}(NCMe) + L \xrightarrow{30\,°C} Os_3(CO)_{11}L + MeCN.$$

L	k_{obs} (s^{-1}) (average values)
PPh_3	1.45 ± 0.04
$AsPh_3$	1.52 ± 0.04
$P(OPh_3)_3$	1.59 ± 0.04

Reference

Dahlinger K, Poë AJ, Sayal PK, Sekhar VC (1986) J. Chem. Soc., Dalton Trans., 2145

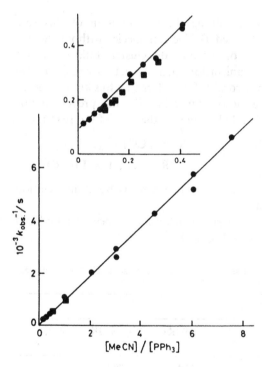

Fig. 2.59. Dependence of the first-order rate constants of ligand substitution reactions of $Os_3(CO)_{11}(NCMe)$ on the $[MeCN]/[PPh_3]$ ratio. Reproduced with permission from Dahlinger K, Poe AJ, Sayal PK, Sekhar VC (1986) J. Chem. Soc. Dalton Trans. 2145

and triphenylphosphin compete for the reaction with the unsaturated intermediate $Os_3(CO)_{11}$.

$$Os_3(CO)_{11}(NCMe) \underset{k_{-1}}{\overset{k_1}{\rightleftharpoons}} [Os_3(CO)_{11}] + MeCN$$

$$[Os_3(CO)_{11}] + PPh_3 \xrightarrow{k_2} Os_3(CO)_{11}(PPh_3)$$

In Table 2.28, the activation parameters observed for the reaction of $Os_3(CO)_{10}(NCMe)$ with PPh_3 as well as those obtained for substitution reactions of the trimetal dodecacarbonyls of iron, osmium, and ruthenium are shown. The relatively high activation enthalpies and positive $\Delta S^{\ddagger}$ values observed in this Table agree with the dissociative mechanism described above. The features in Table 2.28 also reflect the much greater liability of $Os_3(CO)_{11}(NCMe)$ compared with $Os_3(CO)_{12}$. This probably arises from a weaker bond strength of osmium with acetonitrile than with carbon monoxide.

Substitution reactions of some low nuclearity carbonyl clusters such as $Ru_3(CO)_{12}$ and $Ir_4(CO)_{12}$ lead however to more complex rate constant expressions than those expected for the dissociation process. Namely to expressions containing both ligand-dependent and ligand-independent terms.

$$k_{obs} = k_1 + k_2[L]$$

An associative mechanism for explaining the second term of this rate equation is very improbable for saturated 18-electron species without ligands like NO, cyclopentadienyl, or arenes onto which the cluster could delocalize part of its charge. A plausible mechanism for such a reaction could involve metal-metal dissociations. Specially favorable for undergoing this kind of mechanism are compounds with strong face-bridging ligands that prevent major rearrangements after metal-metal bond cleavage. In the case of the reaction

$$(\mu_3\text{-PR})Fe_3(CO)_{10-n}L_n + L \longrightarrow (\mu_3\text{-PR})Fe_3(CO)_{10-n-1}L_{n+1}$$

$$R = \text{aryl}, \ L = P(OR')_3$$

the exchange of $P(OR')_3$ ($L = L'$) L should occur via M-M bond dissociation as shown schematically in Fig. 2.60.

The M-M bond-opened intermediate of this reaction has been isolated and identified by X-ray diffraction analysis (Fig. 2.61).

Table 2.28. Activation paramenters for ligand substitution reactions in some Group 8 trinuclear carbonyl clusters

$M_3L_n + L' \rightleftarrows M_3L_{n-1}L' + L$

Cluster	Dissociated ligand	$\Delta H^{\ddagger}$ (KJ mol^{-1})	$\Delta S^{\ddagger}$ (JK^{-1} mol^{-1})	
$Fe_3(CO)_{12}$	CO	123.4	79.5	1
$Ru_3(CO)_{12}$	CO	133.1	84.5	2
$Os_3(CO)_{12}$	CO	137.5	31.8	3
$Os_3(CO)_{11}(NCMe)$	MeCN	112.4	92.2	4

References

1 Shojaie A, Atwood JD (1985) Organometallics 4:187
2 Poe AJ, Twigg MV (1974) J. Chem. Soc., Dalton Trans., 1860
3 Poe AJ, Sheckar VE (1985) Inorg. Chem. 24:4376
4 Dahlinger K, Poe AJ, Sayerla PK, Seckar, VC (1989) J. Chem. Soc., Dalton Trans., 2145

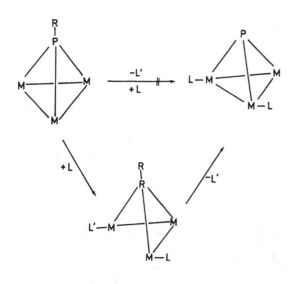

Fig. 2.60. Diagram for a probable dissociative mechanism of ligand interchange reactions in $M_3(CO)_9L$ (M = Ru, Fe) [Darenbourg DJ (1990) In: Shriver DF, Kaez HD, Adams RD (eds) The chemistry of metal cluster complexes. VCH New York, p 171]

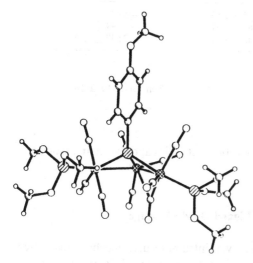

Fig. 2.61. Molecular structure of (p-TolP)Fe$_3$(CO)$_{10}$[P(OMe)$_3$]$_2$ Reproduced with permission from Knoll K, Huttner I, Zsolnai L, Jibril I, Wasincionek M (1985) J. Organoment. Chem. 294:91

As discussed in Sect. 2.4, the reaction of carbonyl metal clusters with N-oxide trimethylamine may often be a convenient method for synthesizing cluster compounds via ligand substitution. Kinetic studies of the substitution of CO-ligands by phosphines and arsines in group-18 trimetal dodecarbonyls

$$M_3(CO)_{12} + Me_3NO + L \longrightarrow M_3(CO)_{11} + Me_3N + CO_2$$

indicate that the formation of $M_3(CO)_{11}L$ is first order in $[M_3(CO)_{12}]$ and $[Me_3NO]$ but zero order in the concentration of L. One mechanism in accord with these features is that described in the scheme in Fig. 2.62. Activation parameters obtained for these reactions are reproduced in Table 2.29. Moderate

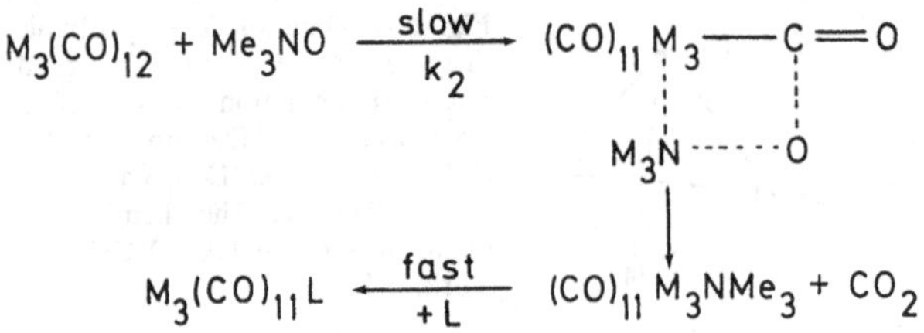

$$M_3(CO)_{12} + Me_3NO \xrightarrow[k_2]{slow} (CO)_{11}M_3 \!\!-\!\! C\!=\!O$$

$$M_3N \cdots\cdots O$$

$$M_3(CO)_{11}L \xleftarrow[+L]{fast} (CO)_{11}M_3NMe_3 + CO_2$$

Fig. 2.62. Interchange mechanism for CO-substitutions in $M_3(CO)_{12}$ (M = Fe, Ru, Os) in the presence of Me_3NO [Shen J-K, Shi Y-L, Gao Y-C, Shi Q-Z, Basolo F (1988) J. Am. Chem. Soc. 110:2414

Table 2.29. Ligand substitution reactions in Group 8 trimetal carbonyls. Activation parameters for the deplacement of carbon monoxide by triphenylphosphine in the presence of trimethylamine oxide

$$M_3(CO)_{12} + Me_3NO + L \rightarrow M_3(CO)_{11}L + Me_3N + CO_2$$

Cluster	L	$\Delta H^{\ddagger}$ (KJ mol^{-1})	$\Delta S^{\ddagger}$ (JK^{-1} mol^{-1})
$Fe_3(CO)_{12}$	PPh_3	52.26 ± 2.55	-58.37 ± 8.79
$Ru_3(CO)_{12}$	PPh_3	50.67 ± 0.67	-79.99 ± 2.26
$Os_3(CO)_{12}$	PPh_3	71.46 ± 1.21	-24.77 ± 4.06

Reference

Shen JK, Shi, Y-L, Gao Y-C, Shi Q-Z, Basolo F (1988) J. Am. Chem. Soc. 110:2414.

activation enthalpies and negative activation entropies agree with an associative process.

2.5.4 Oxidative Additions Involving Ligand Bond Cleavage

Oxidative additions may also occur via intramolecular mechanisms, often leading to substitution reactions. Thus, in some carbonyl clusters the substitution of CO-groups by phosphine, arsine, or anilines originates species in which the proximity of the new ligands to metal cluster nucleus often activate bonds in the latter inducing their cleavage. The stabilization of the products frequently occurs by bonding the ligand fragments to the metal cluster framework. Most examples of this class of reactions are found in the chemistry of osmium and ruthenium.

The substitution of triosmium dodecacarbonyl by arsine and phosphine derivatives under mild conditions yields a series of simple substitution products.

$$Os_3(CO)_{12} + nL \longrightarrow Os_3(CO)_{12-n}L_n + nCO \qquad L = PR \text{ or } AsR3$$

Thermal treatment of these products often leads to a series of processes such as rearrangements, metalations, ligand fragmentations, etc. in which the Os_3 skeleton is retained. However these reactions are not always predictable and are often difficult to control. Thus for instance, from the pyrolysis of compounds of the series $Os_3(CO)_{12-n}(PPH_3)_n$ it is possible to isolate and to identify the compounds described in Fig. 2.63.

However in many instances it is not possible to isolate the products of primarily substitution reactions. Thus the reaction of $Os_3(CO)_{12}$ with aniline leads, by loss of two CO-ligands, to the formation of $HOs_3(CO)_{10}(\mu\text{-NHPh})$. Heating the products at 150 °C causes the ejection of another CO-ligand and to a new fragmentation of the aniline. As illustrated in Fig. 2.64, a competition between two types of ligand oxidative additions is observed, namely between the addition of the other N–H bond yielding $H_2Os_3(CO)_9(NPh)$ and the addition of a C–H bond in position *ortho* to the phenyl group with the formation of $H_2Os_3(CO)_9(HNC_6H_7)$. This kind of oxidative addition is known as orthometallation.

Another example of orthometallation is the series of reactions shown in Fig. 2.65. The thermal treatment of the compound $Os_3(CO)_{11}(PPh_2R)(R = Ph$ or Me) induces loss of carbon monoxide providing cluster unsaturation which promotes in turn the activation of a C–H bond in the phosphine giving finally $Os_3(CO)_9(\mu_3\text{-}C_6H_4)(\mu_3\text{-PR})$. The metallation of two carbon atoms observed in

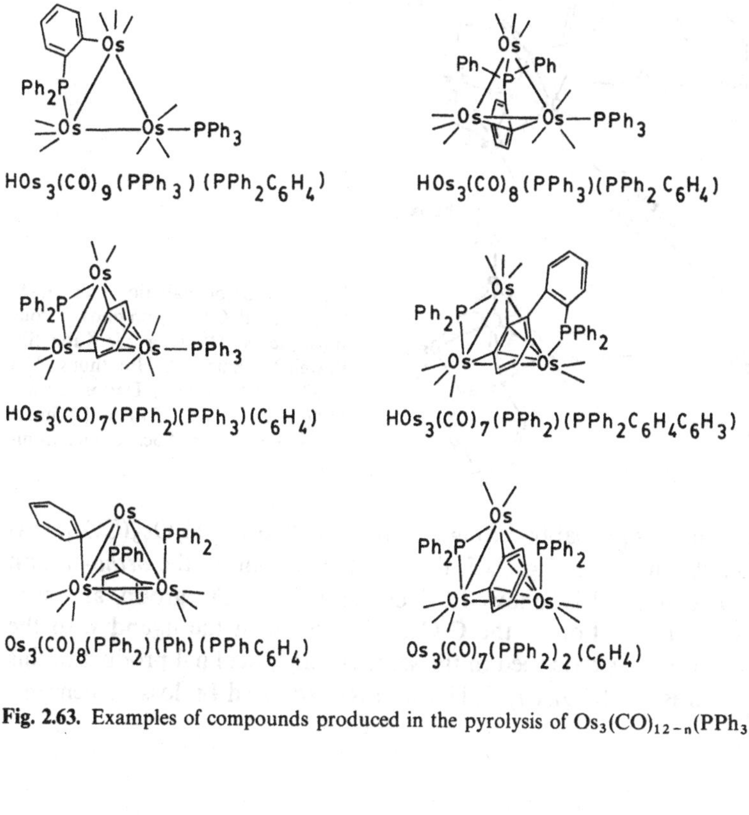

$HOs_3(CO)_9(PPh_3)(PPh_2C_6H_4)$

$HOs_3(CO)_8(PPh_3)(PPh_2C_6H_4)$

$HOs_3(CO)_7(PPh_2)(PPh_3)(C_6H_4)$

$HOs_3(CO)_7(PPh_2)(PPh_2C_6H_4C_6H_3)$

$Os_3(CO)_8(PPh_2)(Ph)(PPhC_6H_4)$

$Os_3(CO)_7(PPh_2)_2(C_6H_4)$

Fig. 2.63. Examples of compounds produced in the pyrolysis of $Os_3(CO)_{12-n}(PPh_3)_n$

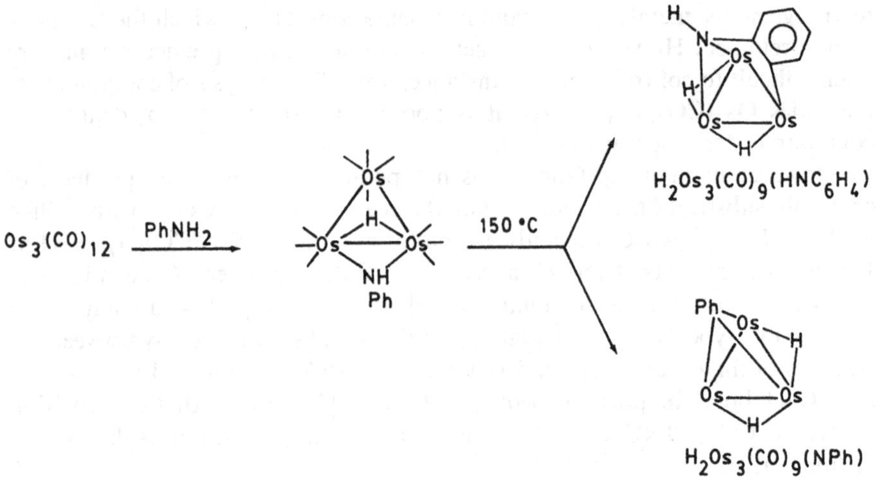

Fig. 2.64. The reaction of triosmium dodecacarbonyl with aniline

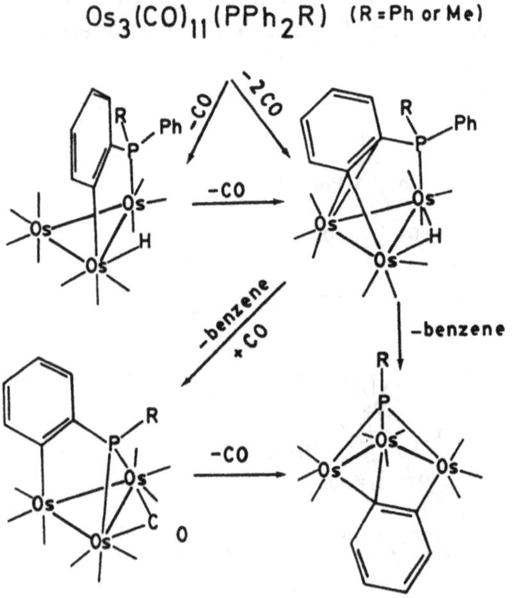

$$Os_3(CO)_{11}(PPh_2R) \quad (R = Ph \text{ or } Me)$$

Fig. 2.65. Ortometallation of C–H bonds and P–C bonds on phosphorous substituents [Deeming AJ, Kabir SE, Powell NI, Bates PA. Hursthouse MB (1987) J. Chem. Soc., Dalton Trans., 1529, Brown SC, Evans J, Smart L (1980) J. Chem. Soc., Chem. Commun., 1021]

this compound occurs probably through the intermediate $(\mu_2\text{-H})Os_3(CO)_{10}(\mu_2\text{-P}(C_6H_4)PhR)$. As shown in Fig. 2.65, the first metallation of the *ortho*-carbon forming an Os-C σ-bond is followed by loss of another CO group, and subsequently by the interaction of the C-C double bond in the ligand with the adjacent metal atom. As described in the same scheme, the final product of this thermal reaction is $Os_3(CO)_9(\mu_3\text{-}C_6H_4)(\mu_3\text{-PR})$ produced by loss of benzene.

The orthometallation process and the possibility of stabilizing fragments RP by bonding to the metal framework make this oxidative addition with cleavage of the strong phosphorus-carbon bond possible.

These oxidative additions to $Os_3(CO)_{12}$ which can occur with a variety of substrata may be described in a simple generalized way by the following equation.

$$Os_3(CO)_{12} + H_2Y \longrightarrow H_2Os_3(CO)_9Y + 3CO$$

where H_2Y may be aliphatic or aromatic hydrocarbons, alkyl or aryl phophines, aniline, H_2S etc., and the products have molecular structures as those shown in Fig. 2.66.

Especially interesting are oxidative additions of unsaturated hydrocarbons because they are the basis of many processes as hydrogenations and isomerizations which can be catalyzed by metal clusters with similar and often better results than those obtained with mononuclear catalysts (see Sect. 2.6).

The behavior of metal clusters in the presence of olefins is in general similar to those toward other donors. Reactions involving clusters of lighter group elements normally produce degradation of the cluster structure yielding mononuclear compounds in which olefin double bond acts as an one-electron pair donor.

$$Fe_3(CO)_{12} + C_2F_4 \longrightarrow (C_2F_4)_2 Fe(CO)_4$$

Nevertheless, with heavier elements as ruthenium or osmium it is possible the isolation of species in which the M_3 framework is retained.

The reaction conditions necessary for the substitution of CO ligands by olefines are often not mild enough for isolating the substitution products without inducing oxidative additions processes. Thus, the reaction of $Os_3(CO)_{12}$

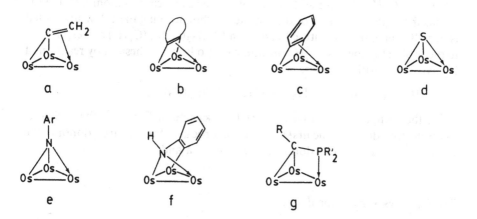

Fig. 2.66. Examples of products from oxidative addition reactions of different substrates to $Os_3(CO)_{12}$: (**a**) Ethylene, (**b**) cyclic olefine; (**c**) benzene; (**d**) H_2S; (**e**) arylamine; (**f**) aniline; (**g**) alkyl or aryl phosphine

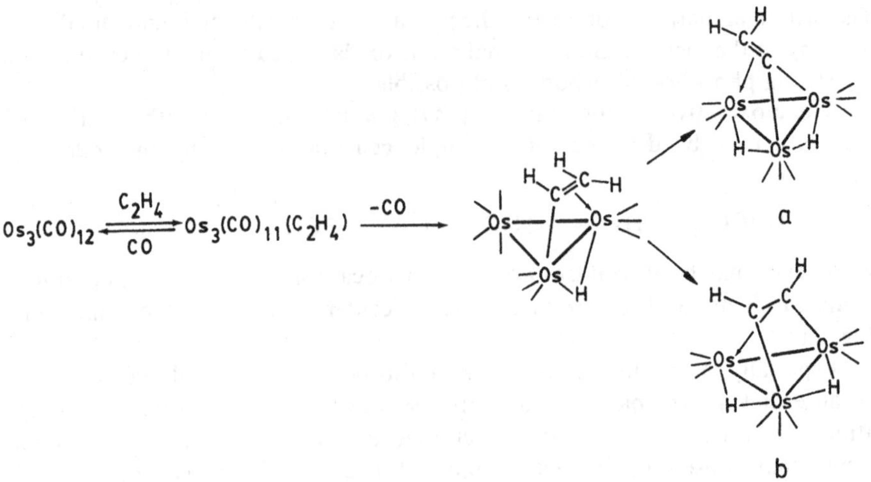

Fig. 2.67. Oxidative addition of ethylene to $Os_3(CO)_{12}$ [Ferriar RP; Vaglio GA, Gambino O, Valle M, Cetini G (1972) J. Chem. Soc., Dalton Trans., 1998]

with ethylene yields two isomers of the compound $H_2Os_3(CO)_9(C_2H_2)$ produced by oxidative addition of two C–H bonds.

Nevertheless, the intermediate substitution compounds may be prepared by alternative ways under milder conditions.

The monosubstitution of $Os_3(CO)_{12}$ by ethylene may be indeed afforded using trimethylamine oxide for displacing the carbonyl group.

$$Os_3(CO)_{12} + Me_3NO + C_2H_2 \longrightarrow Os_3(CO)_{11}(C_2H_2) + Me_3N + CO_2$$

$Os_3(CO)_{11}(C_2H_2)$ should therefore correspond to the intermediate in the oxidative addition described in the scheme in Fig. 2.67. To the substitution follow two C–H oxidative additions, both in the same carbon atom yielding the 1,1-disubstituted isomers A or one on each carbon forming the 1,2-disubstituted isomer B. The monosubstituted compound $HOs_3(CO)_{10}(C_2H_3)$ corresponding to the product of the first C–H addition has also been synthesized by reaction of $H_2Os_3(CO)_{10}$ with ethylene.

$$H_2Os_3(CO)_{10} + (C_2H_2) \longrightarrow HOs_3(CO)_{10}(C_2H_3)$$

In the compounds involved in reaction scheme in Fig. 2.67, the olefine is a two electron donor in the first substitution product, three electron donor in the second, and four electron donor in the isomers a and b.

2.6 Catalysis by Metal Clusters

The potential of metal clusters as homogeneous catalysts has been one important component in the driving force that has permitted the spectacular development of this chemistry observed in the last 15 years.

The nature of cluster compounds, which, as commented before, lies between that of mononuclear compounds and that of pure metals, makes them specially attractive from the point of view of catalysis. The characteristic metal atom distribution in clusters does indeed offer the possibility of accomplishing catalytic properties similar to those displayed by fine divided metals in many well known heterogeneous processes. However, clusters, due to their molecular nature, have the further advantage of displaying catalytic activity in homogeneous media. Actually, the metal polyatomic nuclei in cluster species simulate to some extent the metal surface. Thus they can, similarly to the pure metals, induce changes in the chemical species bonded to them, activating processes that often involve the rupture of highly energetic bonds. Since there is a great variety of homo and heteronuclear cluster species with different structures and chemical behaviours the possibilities of getting some catalysts with special properties from such metal-like species is much greater than from the metals themselves. Metal cluster chemistry may be therefore specially interesting for designing catalysts with high activity but also with a high degree of selectivity. That is specially relevant today when, because of ecological and pollution problems, it is necessary to optimize the industrial production of chemicals diminishing energetic requirements and avoiding the formation of secondary waste products.

Catalytic activity of cluster species has been detected and investigated in a number of processes. There are some detailed studies on mechanisms involved in such processes and many of them correspond indeed to reactions and mechanisms described in the previous Section. However, it is in general difficult to establish with precision which are the species actually responsible for the catalytic phenomenon.

In order to illustrate the potential of cluster chemistry in catalysis, the catalytic activity of cluster species in a number of selected processes will be described. The processes to be considered – hydrogenation and isomerization of unsaturated hydrocarbons, hydroformylation of alkenes, hydrogen reductions of carbon monoxide, and water-gas shift reaction – are interesting not only from the point of view of fundamental knowledge but also because of their great industrial and economical relevance.

2.6.1 The Water-Gas Shift Reaction

Hydrogen is required in copious quantities for the production of light fuels and a great variety of other organic chemicals. Actually the production of hydrogen for hydrogenation of hydrogen-deficient substrates, Fischer-Tropsch processes, or ammonia synthesis is principally based on steam reforming or partial oxidation of petroleum derivatives. There is however an alternative procedure for the preparation of hydrogen, namely the reaction of coal with water at high temperatures producing "water-gas", a mixture of H_2O, H_2, CO_2, and CO. Composition of this mixture may be governed by the water shift reaction.

$$H_2O + CO \rightleftharpoons CO_2 + H_2$$

Thus, after removing CO_2 and H_2O, hydrogen or hydrogen mixtures adequate for different applications can be obtained.

The water-gas shift reaction may be carried out at 360 °C by using the metal oxide Fe_3O_4–Cr_2O_3 as a heterogeneous catalyst. However, because of the thermodynamic parameters of the reaction – $\Delta G^0 = -6.82\ \text{kcal mole}^{-1}$, $\Delta H^0 = -9.84\ \text{kcal mole}^{-1}$ and $\Delta S^0 = -10.1$ eu at 298 K – its efficiency and thermal input requirements are more favorable at lower temperatures.

Some organometallic compounds and among them some metal cluster carbonyls have proved to be interesting homogeneous catalysts for the water-gas shift reaction at relatively low temperatures. Selected examples of catalysis of this reaction by cluster are described in Table 2.30.

Detailed studies involving cluster carbonyls have been carried out with ruthenium clusters. The catalytic effect of the cluster species $Ru_3(CO)_{12}$, $H_2Ru_4(CO)_{13}$, and $H_4Ru_4(CO)_{12}$ are very similar. In all these cases, whatever the ruthenium carbonyl added originally, the principal ruthenium species present under catalysis conditions are the cluster anions $[HRu_3(CO)_{11}]^-$ and $[H_3Ru_4(CO)_{12}]^-$. However, the relative amount of tri and tetranuclear species appears to depend on the relative concentration of H_2 and CO in the reaction mixture. $[H_3Ru_4(CO)_{12}]^-$ – which is present in the mixture only when H_2 is accumulated in the catalytic system – is converted in the trinuclear species by reaction with carbon monoxide. This interconversion process is described in the scheme in Fig. 2.68. ^{1}H-NMR spectroscopic studies on the mechanism of the interconversion between $[H_3Ru_4(CO)_{12}]^-$ and $[HRu_3(CO)_{11}]^-$ point out to the species $[HRu_4(CO)_{13}]^-$ as an intermediate formed by displacement of H_2 by CO:

$$[H_3Ru_4(CO)_{12}]^- + CO \xrightarrow[\text{glyme}]{60-80\,°C} [HRu_4(CO)_{13}]^- + H_2$$

Table 2.30. Selected examples of metal cluster catalysis of the water-gas shift reaction[a]

$H_2O + CO \rightarrow H_2 + CO_2$

Metal cluster	Catalytic activity[b]	
	H_2	CO_2
$Ru_3(CO)_{12}$	2.8	2.7
$H_2Ru_4(CO)_{13}$	4.4	4.0
$H_4Ru_4(CO)_{12}$	3.7	3.3
$H_2FeRu_3(CO)_{13}$	10.3	10.9
$Ir_4(CO)_{12}$	5.6	6.6
$Rh_6(CO)_{16}$	0.8	1.3

[a] Reaction Conditions: [metal cluster] = 0.012 M; $[OH^-]$ = 0.6 M; $[H_2O]$ = 6 M in 2-ethoxyethanol; P_{CO} = 0.9 atm; temp. = 100 °C.
[b] Activity in [mol gas (mol complex)$^{-1}$ d^{-1}].

Reference

Ungerman C, Landis V, Moya SA, Cohen H, Walker H, Pearson RG, Rinker RG, Ford PC (1979) J. Am. Chem. Soc. 101 : 5922.

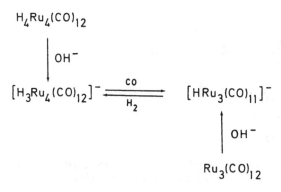

Fig. 2.68. Interconversion between tri and tetranuclear ruthenium species in the catalysis of the water-gas shift reaction [Bricker JC, Nagel CC, Bhattacharyya AA, Shore SG (1985) J. Am. Chem. Soc. 107:377]

However, the intermediate concentration in the reaction mixture is always low, probably due to its high reactivity with carbon monoxide.

$$[HRu_4(CO)_{13}]^- + 2CO \xrightarrow{60°C} [HRu_3(CO)_{11}]^- + 1/3 Ru_3(CO)_{12}$$

The catalytic cycle illustrated in Fig. 2.69 has been proposed as a possible mechanism for the catalysis of the water shift reaction by the metal cluster $Ru_3(CO)_{12}$.

The second step in this mechanism assumes the opening of a three-center/two-electron H-bridge in the anion $[HRu_3(CO)_{11}]^-$ leading thus to the electron-rich intermediate $[HRu_3(CO)_{12}]^-$ with an hydrogen bonded terminally in a two-center/two-electron linkage as described in the reaction schemes illustrated in Fig. 2.70. The assumption of the intermediate $[HRu_3(CO)_{12}]^-$ is supported by the reaction pathway inferred from kinetic studies on $^{13}CO-^{12}CO$ exchange in $[HRu_3(^{12}CO)_{11}]^-$.

$$[HRu_3(^{12}CO)_{11}]^- + {}^{13}CO \longrightarrow [HRu_3(^{13}CO)(^{12}CO)_{10}]^- + {}^{12}CO$$

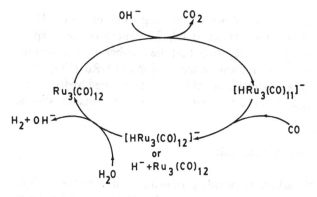

Fig. 2.69. Proposed mechanism for the catalysis of the water-gas shift reaction by the $[HRu_3(CO)_{12}]^--Ru_3(CO)_{12}$ system [Bricher JC, Nagel CC, Bhattacharyya AA, Shore SG (1985) J. Am. Chem. Soc. 107:377].

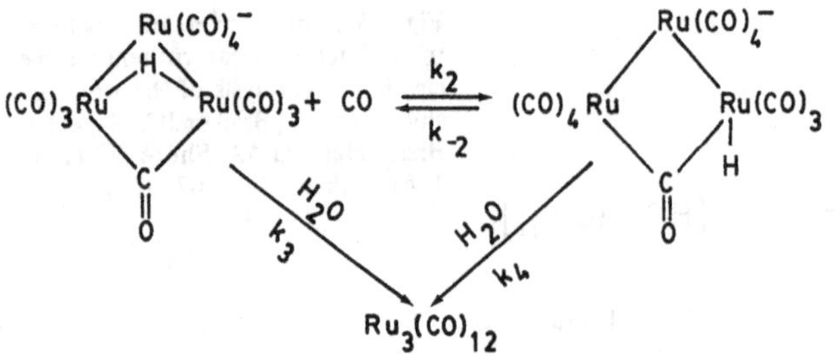

Fig. 2.70. Details of the addition of carbon monoxide to $[HRu_3(CO)_{11}]^-$ in water-gas shift reaction catalysis [Payne MW, Lenssing DC, Shore SG, (1987), J. Am. Chem. Soc. 109:618]

Table 2.31. Effect of the accumulation of hydrogen on the catalysis of the water-gas shift reaction by metal clusters

| Cluster | Conditions[a] | Turnover[b] after day | | | Mol% ruthenium species | |
		1	3	5	$[H_3Ru_4(CO)_{12}]^-$	$[HRu_3CO)_{11}]^-$
$H_4Ru_4(CO)_{12}$	H_2-removed[c]	3.4	2.6	2.8	0	100
$H_4Ru_4(CO)_{12}$	H_2-accumulated[d]	3.1	2.0	2.0	10	90
$Ru_3(CO)_{12}$	H_2-removed[c]	3.8	2.8		0	100
$Ru_3(CO)_{12}$	H_2-accumulated[d]	3.0	1.9	2.1	10	90
$Ru_3(CO)_{12}$	H_2-accumulated[e]	–		1.2	15	85

[a] 100 °C; 3 mol ethoxyethanol; 0.36 mL H_2O; 2.0 mol KOH; [cluster] = 0.04 mol; [CO] = 0.9 atm.
[b] Turnovers = nmol H_2 (nmol Ru)$^{-1}$ d^{-1}.
[c] H_2 removed continuously through Pd thimble.
[d] H_2 removed every 24 h.
[e] H_2 was added after day 4 and H_2 was allowed to accumulate through day 5.

Reference

Bricker JC, Nagel CC, Bhattacharyya AA, Shore SG (1985) J. Am. Chem. Soc. 107:377

According to the scheme in Fig. 2.68 with increasing accumulation of H_2 in the reaction mixture increases the $[H_3Ru_4(CO)_{12}]^-$ concentration. Experiments with variable accumulation of H_2 show that the number of turnovers for ruthenium atom increases with increasing concentration of $[HRu_3(CO)_{11}]^-$.

In Table 2.31 some results illustrating the effect of accumulated hydrogen on the catalysis of the water-gas shift reaction with ruthenium clusters are shown.

2.6.2. Hydrogenation of Carbon Monoxide

The catalytic reduction of carbon monoxide producing hydrocarbons, olefins, and oxygenated compounds generally known as Fischer-Tropsch chemistry is of great industrial and economical interest. The possibility of substituting petroleum by other carbonaceous sources such as coal in the production of organic compounds useful as fuels, as well as in chemical synthesis is especially appealing. Although traditionally this kind of synthesis has been carried out by using heterogeneous catalysis, there is increasing interest in the development of homogeneous catalysts applicable to this class of process. As mentioned before, homogeneous processes have in general some advantages over the heterogeneous ones. The former occurs in general under relatively milder conditions than the latter, thus permitting better control of the properties of the catalyst and making it possible to achieve a higher selectivity toward desired products.

Metal clusters have also proved to have great possibilities as homogeneous catalysts for the reaction of the reduction of carbon monoxide. Selected examples of catalytic hydrogenation of carbon monoxide by metal clusters are shown in Table 2.32.

Rhodium species as $Rh_6(CO)_{16}$ and $[Rh_5(CO)_{15}]^-$ catalyze the reduction of carbon monoxide to light alcohols, specially to ethylene glycol.

Table 2.32. Examples of catalytic hydrogenation of carbon monoxide by metal clusters

Cluster	Products	Reaction conditions	Ref.
$Os_3(CO)_{12}$	CH_4 (29%); C_2H_6 (4%) C_3H_8 (20%); CH_3Br (51%) C_2H_5Br (13%)	180 °C, 2 atm $H_2:CO = 3:1$ solvent: liq-BBr_3 sealed glass tube rate: ca. 20 turnover/h	1
$Rh_6(CO)_{16}$ $[Rh_5(CO)_{15}]^-$ $[Rh_6C(CO)_{15}]^{2-}$ $[Rh_9P(CO)_{21}]^{2-}$ $[Rh_{17}S_2(CO)_{32}]^{3-}$	Polyols, primarily ethylene glycol. MeOH, EtOH, as side products.	200–290 °C; 500–1000 atm; $H_2:CO = 1:1$. Activity is lower for compounds with heteroatoms. High-nuclearity species appear to be stable under catalysis conditions.	2, 3
$Ir_4(CO)_{12}$	$C_2H_5:CH_4 = 10:1$ (t < 3 h); $1:2$ (t > 12 h)	Sealed glass tube 180 °C 1.5 atm; 12–24 h; 100% $H_2:CO = 3:1$ solvent; molten NaCl + 3AlCl$_3$	4
$Ir_4(CO)_{12}$	After 4 h: CH_4 (14%), C_2H_6 (49%), CH_3Cl (32%)	Flow conditions, 175 °C, 1 atm. $H_2:CO = 3:1$; solvent; molten NaCl + 3AlCl$_3$	5
$Ir_4(CO)_{12}$	After 3 h, 1 atm, flow 1 and 20 ml min^{-1} CH_3 (33; 13%); C_2H_6 (57; 7%); C_3H_8 (10; 30%), C_4H_{10} (3; 50%)	Flow conditions, 170–180 °C, 1 atm $[CO] = 0.25$ atm; solvent: molten NaCl + 3 AlCl$_3$	6
$Ir_4(CO)_{12}$	CH_4 less than quantitative in presence of $P(OMe)_3$.	Sealed tube; 160–180 °C; toluene 5d.	7

References

1 Choi HW, Mutterties EL (1981) Inorg. Chem. 20:2664
2 Vidal JL, Walker WE (1980) Inorg. Chem. 19:896
3 Heaton BT, Strona L, Jonas J, Eguchi T, Hoffman GA (1982) J. Chem. Soc., Dalton Trans., 1159
4 Demitras GC, Mutterties EL (1977) J. Am. Chem. Soc. 99:2796
5 Collman JP, Brauman JI, Tustin G, Wang III GS (1983) J. Am. Chem. Soc. 105:3913
6 Wang HK, Choi HW, Muetterties EL (1981) Inorg. Chem. 20:2661
7 Schu RA, Demitras GC, Choi HW, Muetterties EL (1981) Inorg. Chem. 20:4023

Polyol formation is also catalyzed by some rhodium clusters, namely by a series of cluster species as $[Rh_6C(CO)_{15}]^{2-}$, $[Rh_9P(CO)_{21}]^{2-}$, and $[Rh_{17}S_2(CO)_{32}]^{3-}$ containing strong bridging ligands. Because of the nature of these ligands, they are specially appropriate for holding the cluster atoms together in spite of possibly unfavorable electron counting. It is therefore probable that the mechanism of such catalysis reactions involves intact rhodium clusters.

Although most of known homogeneous reductions of carbon monoxide with hydrogen catalyzed by carbonyl complexes produce a mixture of oxygenated products, the iridium carbonyl species $Ir_4(CO)_{12}$ has proved to catalyze the following homogeneous methanation reaction in toluene solutions.

$$3H_2 + CO \longrightarrow CH_4 + H_2O$$

Mononuclear complexes are, however, inactive under similar reaction conditions.

In spite of such catalytic effects of the metal cluster, the rate of this reaction remains very low. Nevertheless, this reaction is an example of a very interesting type of homogeneous catalysis. Here the activation of carbon monoxide appears to be achieved by interaction of both the carbon and oxygen atoms with the metal cluster atoms in a similar way to the CO-chemisorption on metals in heterogeneous Fischer-Tropsch processes.

The catalytic effect of $Ir_4(CO)_{12}$ on the methanation reaction is notoriously enhanced by the presence of strong Lewis acids, probably due to a cooperative activation of the carbon-oxygen bond. Thus the catalysis with $Ir_4(CO)_{12}$ in a $NaCl/AlCl_3$ melt converts at very reasonable rate carbon monoxide and hydrogen into light hydrocarbons, primarily ethane, without producing oxygenated compounds.

$$CO + 3H_2 \xrightarrow[\substack{180\,°C,\ 1–2\,atm \\ 100\%\ 12–24\,h}]{Ir_4(CO)_{12},\ AlCl_3–NaCl} C_nH_{2n+2}$$

$n = 1$ 50–90%
$n = 2$ 10–50%
$n > 2$ traces

Kinetic studies of this reaction under flow conditions point out to a mechanism with formation of chloromethane as reaction intermediate. Nonetheless, as observed in Table 2.32, the results are strongly affected by the conditions of the experiments. Temperature appears not to affect the product distribution very much but this is dramatically influenced by contact time. The variety of product distributions as well as discrepancies in possible intermediates inferred from diverse investigations could be also explained by different catalysts arising from the same precursor.

The catalysis of hydrogenation of carbon monoxide with $Os_3(CO)_{12}$ as catalyst precursor also produces a mixture of hydrocarbons with high methane content. Moreover, in this case, the formation of bromine derivatives is observed. During the catalysis, however, $Os_3(CO)_{12}$ is converted into a halogenated dinuclear species which appears to be the enduring catalyst.

2.6.3 Olefin Hydrogenation and Isomerization

Hydrogenation of unsaturated hydrocarbons is a process of great economical importance. Many transition metal cluster species have displayed interesting catalytic effects on such kind of reactions. Selected examples of hydrogenation of olefins are described in Table 2.33.

As it can be appreciated in Table 2.33, there are different kinds of hydrogenation and isomerization reactions. Fundamental differences between these processes arise from the way the catalyst precursor is converted in either coordinately or electron unsaturated species able to interact with the olefin beginning thus a catalytic cycle.

Thermal Activation. As often discussed in previous sections, metal clusters, in general, undergo relatively easy thermal decomposition with carbon monoxide ejection. From these reactions highly reactive coordinately unsaturated derivatives which can be active catalysts arise.

A relatively simple example of this class of cluster catalysis is the hydrogenation of ethylene catalyzed by the tetranuclear ruthenium cluster $H_4Ru_4(CO)_{12}$ described by the following equations.

$$H_4Ru_4(CO)_{12} \xrightarrow{\ K_1\ } H_4Ru_4(CO)_{11} + CO$$

$$H_4Ru_4(CO)_{11} + H_2 \xrightarrow{\ K_2\ } H_6Ru_4(CO)_{11}$$

$$H_4Ru_4(CO)_{11} + C_2H_4 \xrightarrow{\ K_3\ } H_3Ru_4(CO)_{11}(C_2H_5)$$

$$H_3Ru_4(CO)_{11}(C_2H_5) + H_2 \xrightarrow{\ k\ } H_6Ru_4(CO)_{11} + C_2H_6$$

Kinetic studies of this process do indeed show the catalysis being first order in the concentration of the ruthenium cluster and inversely dependent on the partial pressure of carbon monoxide indicating thus that the most important intermediate in the reaction is the coordinately unsaturated species $H_4Ru_4(CO)_{11}$. Coordinately unsaturated sites in the catalyst would be achieved, in this case, by the dissociation of a CO-ligand. The catalytic effect of the unsaturated intermediate depends on the reversible insertion of ethylene into the Ru–H bond. However there is competition between this insertion and the addition of carbon monoxide or H_2 for obtaining the starting compound or the hexahydride $H_6Ru_4(CO)_{11}$ respectively.

The hydrogenolysis of the ethyl group in $H_3Ru_4(CO)_{11}(C_2H_5)$ closing the catalytic cycle is the rate-determining step in this process.

Photochemical Activation. An alternative way of inducing the formation of coordinately non-saturated species able to participate in catalytic processes is the selective displacement of CO-ligands by photolysis. The mechanism of addition of H_2 to ethylene catalyzed by $H_4Ru_4(CO)_{12}$ is similar to that for

Table 2.33. Selected examples of catalytic hydrogenation by metal clusters

Metal Cluster	Reaction Type/Hydrocarbon	Catalysis activation	Reaction conditions and catalysis activity	Ref.
$Fe_3(CO)_{12}$	Isomerization 1-pentene	Photochemical	$25\,°C$; 30 s laser irradiation, 515 nm; $[Fe]_3 = 5 \times 10^{-3}$ M, $[R] = 9.14$ M (neat); rate: 900 turnovers min^{-1}	1
$H_4Ru_4(CO)_{12}$	Hydrogenation C_2H_4	Thermal	$72\,°C$; $P(H_2) = P(C_2H_4) = 100$.	2
$H_4Ru_4(CO)_{12}$	Hydrogenation C_2H_4	Photochemical	$25\,°C$; irradiation 366 nm, ca. 10^{-7} einstein min^{-1}; $P(C_2H_4) = P(H_2) = 0.13$ bar $[Ru_4] = 10^{-3}$ M.	3
$H_4Ru_4(CO)_{12}$	Isomerization 1-pentene	Photochemical	$25\,°C$; 2 h irridiation 355 nm, ca. 1.2×10^{-6} einstein min^{-1}; $[Ru_4] = 5 \times 10^{-4}$ M $[R] = 2$M; rate: 250 turnoverr h^{-1}; trans-to cis- 2-penetene isomerization ratio = 1.8.	4
$[HRu_3(CO)_{11}]^-$	Hydrogenation	Anion-promoted	$25\,°C$, dimethylpropylene, formanide; $[Ru_3] = 0.25$ mmol, $[R] = 0.06$ mol, $[H_2] = 20$ bar; rate: 0.3 turnover min^{-1}	5
$Ru_3(\mu_3\text{-}NPh)_2(CO)_2$	Hydrogenation 3,3-dimethylbutene	Thermal	$98\,°C$; heptane; $[Ru_3] = 7 \times 10^{-3}$ mmol $[R] = 0.35$ mmol, $[H_2] = 3$ bar; rate: 0.5 turnover min^{-1}	6
$H_2Ru_3(CO)_9(\mu_3\text{-}PR)$	Hydrogenation styrene	Ligand-assisted	$60\,°C$; $[R = Ph] = 9$ mmol; $[H_2] = 60$ bar; rate: ca. 2 turnover h^{-1}	7
$[Ru_3(\mu_2\text{-}NCO)(CO)_{10}]^-$	Hydrogenation	Anion-promoted	$25\,°C$; $[Ru_3] = 1.4$ mmol, $[R] = 0.38$ mol, $[H_2] = 0.74$ bar	8
$H_3Ru_4\{Au(PPh_3)\}(CO)_{12}$	Isomerization 1-penetene	Ligand effect	$35\,°C$; CH_2Cl_2; $[Ru_4Au] = 2$ mmol, $[R] = 54$ mmol, $[H_2] = 1$ atm; trans to cis-penetene ratio: 2.8 (after 24 h); rate: ca. 20 turnover d^{-1}	9
$NiRu_3(\mu\text{-}H)_3(CO)_9(Cp)$	Hydrogenation 1,3-pentadiene	Thermal	$120\,°C$; n-octane, cis-1,3-pentadiene; $[NiRu_3] = 1.7 \times 10^{-4}$ M, $[R] = 0.21$ M; products (%): pentane (1.4); 1-pentene (7.3), trans-2-pentene (18.4), cis-2-pentene (13.1); rate: 11.3 turnover min^{-1}	10
$H_4Os_4(CO)_{12}$	Hydrogenation styrene	Thermal	$1140\,°C$; $[Os_4] = 0.112$ M, 11 estyrene $[R] = 1.2$ M, $[H_2] = 1.05$ bar; rate: 1.45 turnover min^{-1}	11
$H_2Os_3(CO)_{10}$	1-hexene	Unsaturation	$50\,°C$; octane; $[H_2] = 3.4$ bar; $[Os_3]$: [hexane] = 1:100; 35 h; rate: < 1 turnover h^{-1}.	12
$[Os_3(NCO)(CO)_{11}]^{-1}$	Hydrogenation maleic anhidride anion	Anion-promoted	$60\,°C$; $[H_2] = 3.4$ bar; 3 equivalent succinic anhydride are produced after 24 h.	13

$\{(\mu\text{-H})Rh[P(O\text{-}i\text{-}C_3H_7)_3]_2\}_2$	Hydrogenation 2-butyne	Ligand-assisted	20 °C, toluene; $[Rh_2] = 0.02$ mmol, $[H_2] = 0.8$ bar; rate: ca. 1 turnover min^{-1}; reaction totally selective to trans alkene; catalyst life-time: 5 min	14
$Ni_4[CNC(CH_3)_3]_7$	Hydrogenation diphenylacetylene	Ligand effect	20 °C; product: *cis*-stilbene	15

References

1 Mitchener JC, Wrighton MS (1981) J. Am. Chem. Soc. 103:975
2 Doi Y, Koshizuka K, Keii T (1982) Inorg. Chem. 21:2732
3 Doi Y, Tamura S, Koshizuka K (1983) J. Mol Catal. 19:213
4 Graff JL, Wrighton MS (1980) J. Am. Chem. Soc. 102:2123
5 Süss-Fink G, Reiner J (1982) J. Mol. Catal. 16:231
6 Smieja JA, Gozum JE, Gladfelter WL (1986) Organometallycs 5:2154
7 Mani D, Varenkamp H (1985) J. Mol. Catal. 29:305
8 Zuffa JL, Blohm ML, Gladfelter WL (1986) J. Am. Chem. Soc. 108:552
9 Evans J, Jingxing G (1985) J. Chem. Soc., Chem. Commun., 39
10 Castiglioni M, Giordano R, Sappa E (1987) J. Organomet. Chem. 319:167
11 Sanchez-Delgado RA, Andriollo A, Puga J, Martin G (1987) Inorg. Chem. 26:1867
12 Keister JB, Shapley JR (1976) J. Am. Chem. Soc. 98:1056
13 Zuffa JJ, Galdfelter WL (1986) J. Am. Chem. Soc. 108:4669
14 Burch RR, Shusterman AJ, Muettertier EL, Teller RE, Wiliams JM (1983) J. Am. Chem. Soc. 105:3546
15 Mutterties EL, Band E, Kokorion A, Pretzer WR, Thomas MG (1980) Inorg. Chem. 9:1552

thermal activation discussed above but in which the substitution of the CO group by the olefin is activated photochemically.

$$H_4Ru_4(CO)_{12} + L \xrightarrow{h\nu} H_4Ru_4(CO)_{11}L + CO$$

In this case the rate determining step of the reaction is again the oxidative addition of H_2. Thus, for instance, the photolytic addition of 1-pentene to the cluster $H_4Ru_4(CO)_{12}$ leads to olefin isomerization with reaction rates of ca. 2000 turnovers per hour at 25 °C. The hydrogenation rate however is considerably lower. Thus with an increase in the H_2 pressure of $6.89 \cdot 10^4$ Pa in the same system a rate of only 60 turnovers per hour is achieved. The yield in terms of isomerization cycles per photon is notoriously high, about 300 turnovers per photon. An analogous behavior is observed in the isomerization catalyzed by

$Fe_3(CO)_{12}$ where a laser irradiation of 30 seconds originates an isomerization rate of about 10^3 turnovers per minute. That is one of the most rapid among known processes in homogeneous catalysis. However in this last case, contrasting to the isomerization with $H_4Ru_4(CO)_{12}$, the reaction occurs via metal-metal bond cleavage.

$$Fe_3(CO)_{12} \xrightarrow{h\nu} 3Fe(CO)_4$$

In general, the photolytic activation of clusters is easier than that of the equivalent mononuclear species because the former absorbs energy at lower frequencies, often in the visible spectrum.

Electron Unsaturated Catalyst Precursors. As analyzed in previous Sections, the trinuclear cluster $H_2Os_3(CO)_{10}$ is an electronically as well as coordinately unsaturated species. The hydrogenation of olefins in the presence of $H_2Os_3(CO)_{10}$ thus constitutes a typical example of catalysis induced by intrinsic unsaturation of the catalyst precursor. In these cases, the activation energy for the formation of the complex catalyst-substrate is rather low. The mechanism proposed for this kind of catalysis is illustrated in the scheme in Fig. 2.71.

Anionic Catalyst Precursors. Excellent results in the hydrogenation of olefins are obtained by labilization of the cluster by the formation of anionic species. As discussed in Sect. 2.4.2, this method is often used for the activation of clusters in substitution reactions.

The anionic cluster $[Ru_3(\mu_2\text{-}NCO)(CO)_{10}]^-$ catalyzes the hydrogenation of olefins with reaction rates of ca. 5 turnovers per minute at room temperature. According to kinetic studies, the determinant reaction step is the oxidative addition of hydrogen to the anion $[M_3(CNO)(CO)_9]^-$ ($M = Ru$, Os). The

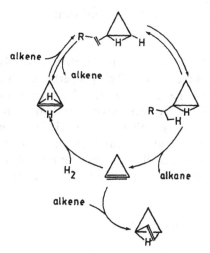

Fig. 2.71. Catalysis mechanism by unsaturated cluster species. The reaction of $H_2Os_3(CO)_{10}$ with alkenes [Keister JB, Shapley JR (1976) J. Am. Chem. Soc. 98 : 1056]

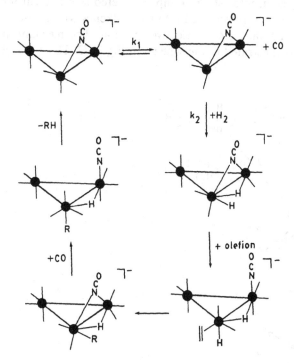

Fig. 2.72. Olefin hydrogenation catalysed by $[Ru_3(\mu_2\text{-}NCO)(CO)_{10}]^-$. [Zuffa JL, Gladfelter WL (1986) J. Am. Chem. Soc. 108 : 4669]

mechanism proposed for this reaction is illustrated in the scheme in Fig. 2.72 which has been confirmed not only by kinetic investigations but also by the synthesis and crystallographic characterization of the intermediate species $[HOs_3(NCO)R(CO)_9]^-$ with R = succinoyl anhydride. Similar studies with the ruthenium derivatives has been not possible because of considerably higher rates of the catalytic processes.

Special Ligands. As discussed in Sect 2.5.3, the substitution of carbonyl ligands in the cluster often enhances the liability of the clusters. This property has also been used in catalytic processes. Thus for instance, the nickel cluster with isocyanide ligands $Ni_4(CNR)_7$ is an efficient catalyst precursor in the hydrogenation of acetylenes. Actually, $Ni_4(CNR)_7$ reacts with acetylenes yielding the adduct $Ni_4(CNR)_4$ (acetylene)$_3$ which in turn is able to catalyze, under rather mild conditions, the hydrogenation of diaryl and dialkylacetylenes at rates of ca. one turnover per minute. Mononuclear complexes do not catalyze this reaction under the same conditions.

$$RC-CR \xrightarrow[\underset{Ni_4[CN(CH_3)_3]_4[\mu_3(\eta^2)-PhC-CPh]_3}{}]{H_2} cis \; RHC=CHR$$

The hydrogenation of acetylenes catalyzed by the dinuclear compound $\{HRh[P(O-i-C_3H_7)_3]_2\}_2$ is an interesting example of stereo selective cluster catalysis. At 20 °C acetylenes are transformed into *trans*-alkenes with a rate of 1 turnover per minute in a totally stereo-selective reaction. The mechanism proposed for this process is illustrated in the scheme in Fig. 2.73. This kind of

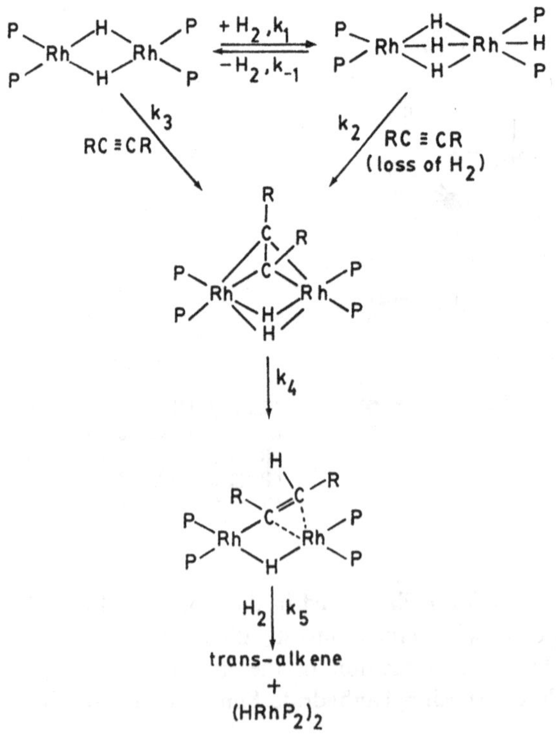

Fig. 2.73. Hydrogenation of alkynes to *trans*-alkenes catalysed by $\{(\mu-H)Rh[P(O-i-C_3H_7)_3]_2\}_2$ [Burch RR, Shusterman AJ, Muetterties EL, Teller RG, Williams JM (1983) J. Am. Chem. Soc. 105:3546]

stereo selectivity is privative of the dinuclear compound. The catalysis by mononuclear species used directly or from the degradation of dinuclear compounds gives rise primarily to the *cis*-isomer.

2.6.4 Hydroformylation

Reactions of simultaneous addition of carbon monoxide and hydrogen to olefins leading to carboxylic compounds are known as hydroformylation reactions.

$$C = C + CO + H_2 \rightarrow H\text{–}C\text{–}C\text{–}CHO$$

This class of reaction is specially interesting in the conversion of hydrocarbons in compounds with functional groups useful as material for organic synthesis. Cluster compounds have also proved useful in the catalysis of this kind of process. Selected examples of this kind of catalytic hydroformilation are shown in Table 2.34.

In many cases, olefin hydrogenation reactions lead under relatively more drastic conditions, and in the presence of carbon monoxide, to the hydroformylation of the olefin. Thus, for instance, the cluster hydride $[HRu_3(CO)_{11}]^-$, which as discussed above catalyzes under mild conditions (25 °C, 20 bar H_2) the hydrogenation of ethylene, leads, at higher temperatures and in the presence of carbon monoxide, to the formation of propanal according to the scheme illustrated in Fig. 2.74. The mechanism involved in this scheme agrees with the isotopic distribution observed in propanal by using molecular deuterium D_2. Preferential deuterations of the formyl hydrogen (98%) as well as of one

Table 2.34. Selected examples of catalytic hydroformylation of olefines by metal clusters.

Metal Cluster	Olefin	Reaction conditions	Ref.
$[HRu_3(CO)_{11}]^-$	ethylene	100 °C; $[CO]$ = 2.6 bar; $[H_2]$ = $[C_2H_4]$ = 13 bar; rate: 1 turnover per min.	1
$Co_3(CO)_9(\mu_3\text{-}CPh)$	1 pentene	90 °C; $[CO]$ = $[H_2]$ = 57 bar; 104 h; 57.5%; selectivity = 5.4.[a]	2
	2 pentene	110 °C; $[CO]$ = $[H_2]$ = 30.6 bar; 775 h; 99.8%; selectivity = 1.0.	
$Co_3(CO)_8(\mu_2\text{-}CO)_2(\mu_4\text{-}PPh)_2$	1 pentene	130 °C; $[CO]$ = $[H_2]$ = 57.8 bar; 23 h; 100%; selectivity = 2.7.	2
	2 pentene	130 °C; $[CO]$ = $[H_2]$ = 32.3 bar; 72 h; 87.7%; selectivity = 1.2.	

[a] Selectivity: Product distribution ratio of Normal : Branched hydrocarbons.

References

1 Süss-Fink G (1980) J. Organomet. Chem. 193:C20
2 Ryan RC, Pittman Jr, CU, O'Connor JP (1977) J. Am. Chem. Soc., 99:1986

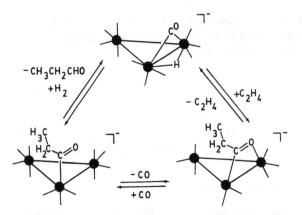

Fig. 2.74. Catalytic hydroformylation of ethylene by $[HRu_3(CO)_{11}]^-$ [Süss-Fink G, Hermann G (1985) J. Chem. Soc., Chem. Commun., 735]

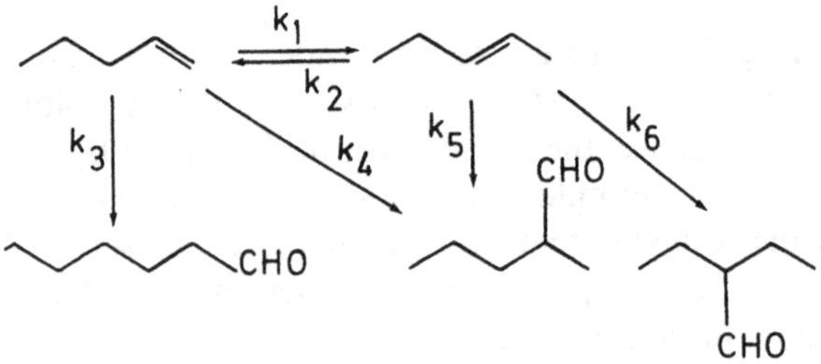

Fig. 2.75. Selectivity in hydroformylation of 1 and 2-pentene [Ryan RC, Pittman Jr. CU, O'Connor JP (1977) J. Am. Chem. Soc. 99 : 1986]

hydrogen in the methyl group (70–80%) are observed, meanwhile the methylene group shows low deuterium incorporation (5–12%).

As analyzed in a previous Section (see Sect. 2.4.5), there are some clusters containing strong face-bridging ligands which have the property of holding to cluster structures in spite of cleavage of the metal-metal bonds arising for instance from an excess of cluster skeleton electrons. This property may also be useful for cluster catalyst precursors.

The cobalt clusters $Co_3(CO)_9(\mu_3\text{-}CC_6H_5)$ and $Co_4(CO)_8(\mu_2\text{-}CO)_2 \cdot (\mu_4\text{-}PC_6H_5)_2$ also catalyze, under mild conditions, the hydroformylation of 1- and 2-pentene to aldehydes in relatively high yields. Moreover, this process can be achieved with a fairly high normal-to-branched selectivity (Fig. 2.75). The clusters are recovered unchanged with high yields from these reactions and the hydroformylations are performed at milder conditions than those required for the catalysis with $Co_2(CO)_8$. These features point out to catalysis by the intact clusters, assisted probably by the presence of the strong bridging ligands. The effect of the ligands is specially strong in the case of the phosphinidene compounds $Co_4(\mu_4\text{-}PPh)_2(CO)_{10}$ with two bridging ligands.

Chapter 3

Main Group-Transition Metal Mixed Clusters

In the chemistry of transition metal clusters outlined in Chapter 2, main group elements have been, in general, considered as electron donor ligands which are stabilizing electron-deficient transition metal conglomerates. However, there are some cases in which such an approach appears to be insufficient to explain all features related to their structures and bonding. This chapter deals mainly with compounds in which the main-group elements may be considered as cluster vertexes of the polyhedron. Nevertheless, considering the importance of clusters containing hydrogen, some aspects of the chemistry of hydride metal clusters will also be discussed briefly in this chapter.

3.1 Hydride Clusters

Since metal clusters containing hydrogen directly bonded to the metallic nucleus are often involved in the synthesis, reactivity, and catalytic properties of cluster species, they have been frequently mentioned in the chemistry discussed in Chapter 2. Nonetheless, some features related to the structure of and bonding in these compounds will be discussed separately in this Section.

3.1.1 The Structure of Hydride Clusters

The determination of the structure of hydride-transition metal clusters is in general a difficult task that frequently involves indirect approaches for locating the hydrogen atom positions. Thus, hydrogen atom locations are often inferred from the geometry of the molecule obtained from X-ray data. Unusual distances and angles or special orientations of the ligands can reveal the presence of hydrogen atoms in determinate positions.

Thus for instance in the draft of the equatorial plane of the anion $[H_2Re_3(CO)_{12}]^-$ reproduced in Fig. 3.1 it can be observed that distance Re(2)–Re(3) is shorter than those to including Re(1). This feature, in addition to the distortion of the C-Re-Re-angles from their normal values of 105° indicating the equatorial Re(CO)$_2$ groups on Re(2) and Re(3) are bending away from Re(1), point out to a location of the hydrogen atoms near the edges Re(1)–Re(2) and Re(1)–Re(3) of the triangle.

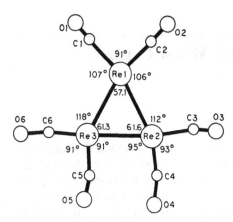

Fig. 3.1. Equatorial plane in the structure of the anion $[H_2Re_3(CO)_{12}]^-$. Reproduced with permission from Churchill MR, Bud PH, Kaesz HD, Bau R, Fontal B (1968) J. Am. Chem. Soc. 90:7135

X-ray diffraction methods may also be applied to detect the location of hydrogen atoms by evaluating electron density and obtaining difference-Fourier maps from low-angle reflections. Figure 3.2 illustrates a classic application of this method, namely the determination of the molecular structure of $H_3Mn_3(CO)_{12}$. This compound consists of a triangular array of metal atoms with one hydrogen atom bridging each of the sides.

However, the most accurate methods for determining hydrogen positions in crystalline compounds are based on neutron diffraction determinations which can provide true nuclear positions. Since neutrons are diffracted by most elements with more or less the same efficiency, the method is very appropriate in determining accurately light atom positions in the presence of heavy atoms. Since X-ray and neutron diffraction data contain fundamental information about heavy and light atom positions respectively, a method considering combined X-ray/neutron least-squares refinement appears to be very useful. The practical advantage of carrying out the least-squares refinement with both sets of data simultaneously is that it permits us to optimize the ratio benefit to

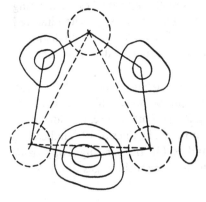

Fig. 3.2. Difference Fourier map calculated through the equatorial plane of $H_3Mn_3(CO)_{12}$. Reproduced with permission from Kirtley SW, Olsen JP, Bau R (1973) J. Am. Chem. Soc. 95:4532

Table 3.1. Examples of transition metal hybride clusters in which hydrogen atoms are located directly

Cluster	Type of H-Bond[a]	Distance M-H (Å)	Technique of location[b]	d(M-M) Å and remarks	Ref.
H₄W₄(CO)₁₂(OH)₄4Ph₂EtPO	t			—	1
HWOs₃(CO)₁₂(Cp)	e-b	1.6-1.9	X	d(Os-W) = 3.078	2
H₃WOs₃(CO)₁₁(Cp*)	e-b	1.92 (W) 1.68 (Ru)	X	d(W-Ru) = 3.078 d(Ru-Ru) = 2.917	3
H₃Mn₃(CO)₁₂	e-b	1.72	X	d(Mn-Mn) = 3.111 Å	4
HRe₃(CO)₁₄	e-b	—	X	d(Re-Re) = 3.295	5
[H₄Re₃(CO)₁₀]⁻	e-b	—	X	[NEt₄]⁺ salt. One Re(μ₂-H)₂Re bond, d(Re-Re) = 2.821, two Re(μ₂-H) Re bond d(M-M) = 3.184	6
[H₅Re₄(CO)₁₄]⁻	e-b	1.86	X	d(Re-Re) = 2.23	7
[H₄Re₄(CO)₁₃]²⁻	e-b	—	X	[NEt₄]⁺ salt. d(Re-Re) = 3.094	8
[H₆Re₄(CO)₁₂]²⁻	e-b	—	X	[NMe₃Bz]⁺ salt. d(Re-Re) = 3.157	9
HRe₆Cl₆(CH₂SiMe₃)₉	t		X		10
HFe₄(CO)₁₂BH₂	e-b	1.67	X	(d(Fe-Fe) = 2.65	11
[HFeRu₃(CO)₁₃]⁻	e-b	1.821	N	[(Ph₃P)₂N]⁺ salt d(M-M) = 2916 Ru(μ₂-H)	12
HRu₃(CO)₁₀(CNMe₂)	e-b	1.85	X	d(Ru-Ru) = 2.801	13, 14
H₃Ru₃(CO)₉(CMe)	e-b	1.81	X	data complemented by ¹H-NMR analysis. d(Ru-Ru) = 2.842	15
HRu₃(CO)₉(PhNCO)	e-b	1.74	X	d(Ru-Ru) = 2.777	16
HRu₃(CO)₉(MeNC₄H₄)	e-b	1.50	X	d(Ru-Ru) = 2.968	17
H₂Ru₄(CO)₁₃	e-b		X	d(Ru-Ru) = 2.930	18
H₄Ru₄(CO)₁₂	e-b	1.76	X	d(Ru-Ru) = 2.950	19
H₄Ru₄(CO)₈[P(OMe)₃]₄	e-b	1.773	N	d(Ru-Ru) = 2.978	12
HRu₅C(CO)₁₄(Py)	e-b		X	d(Ru-Ru) = 2.98	20
H₂Os₃(CO)₁₁	e-b		X	d(Os-Os) = 2.989	21, 22
H₂Os₃(CO)₁₀Br	e-b		N	Os(μ₂-H)Os; d(Os-Os) = 2.683	23-25
HOs₃(CO)₁₀	e-b	1.845	X	d(Os-Os) = 2.851	26
H₂Os₃(CO)₁₀(CH₂)	e-b	2.03	N	d(Os-Os) = 3.053	27
H₂Os₃(CO)₉S	e-b	1.883	N	d(Os-Os) = 2.915	28
H₃Os₄(CO)₁₂I	e-b	1.819	X	d(Os-Os) = 3.010	29
H₃Os₄(CO)₁₁(C₆H₉)	e-b		X	d(Os-Os) = 2.918	30
H₂Os₅(CO)₁₆	e-b		X	d(Os-Os) = 2.962	31
H₂Os₆(CO)₁₈	e-b		X	d(Os-Os) = 2.928	32-34
H₂Os₆(CO)₁₉	t		X	—	35

Table 3.1. (continued)

Cluster	Type of H-Bond[a]	Distance M-H (Å)	Technique of location[b]	d(M-M) Å and remarks	Ref.
$H_3Rh_3[P(OCH_3)_3]_6$	e-b	1.765	N	d(Rh–Rh) = 2.813	36, 37
$[H_4Rh_4Cp_4]^{2+}$	e-b		X	d(Rh–Rh) = 2.829	38
$\{H_7Ir_3[Ph_2P(CH_2)_3PPh_2]_3\}^{2+}$	t, e-b, f-b		X	$[BF_4]$ salt. d(Ir–Ir) = 2.772	39
$[H_2Ir_4(CO)_{10}]^-$	t		X	$[Ph_3P_2N]^+$ salt.	40
$[HPt_4(CO)_2(Ph_2PCH_2PPh_2)_4]^+$	e-b		X	d(Pt–Pt) = 2.705	41
$H_6Cu_6(PPh_3)_6$	e-b		X	d(Cu–Cu) = 2.655	42
HNb_6I_{11}	i	2.01	N	d(Nb–Nb) = 2.84	43
$H_4Re_4(CO)_{12}$	f-b	1.77	X	d(Re–Re) = 2.913	44
$[HFe_3Ni(CO)_{12}]^-$	f-b	1.57	X	$[NMe_3Bz]^+$ salt; μ-H on Fe_2Ni face	12
$H_2Ru_6(CO)_{18}$	f-b		X	d(Ru–Ru) = 2.954	45, 46
$[HRu_6(CO)_{18}]^-$	i	2.04	X, N	$[AsPh_4]^+$ salt. d(Au–Au) = 2.87	47, 48
$[HOs_6(CO)_{18}]^-$	f-b		X	$[(Ph_3P)_2N]^+$ salt. d(Os–Os) = 2.973	49
$H_4Co_4Cp_4$	f-b	1.67	N	d(Co–Co) = 2.467	50
$[HCo_6(CO)_{15}]^-$	i	1.824	N	d(Co–Co) = 2.579	51
HRh_3Cp_4	f-b		X	d(Rh–Rh) = 2.725	52
$[H_2Rh_{13}(CO)_{24}]^{3-}$	i		X	$[NBu_4^1]^+$ salt. d(Rh–Rh) = 2.823	12
$H_3Ni_4CP_4$	f-b	1.691	X, N	d(Ni–Ni) = 2.469	53–55
$[H_2Ni_{12}(CO)_{21}]^{2-}$	i	1.72 2.22	N	d(Ni–Ni) = 2.682, 2.425	56

a t = terminal; e–b = edge-bridging; f–b = face-bridging; i = interstitial
b X = X-ray; N = Neutron diffraction

References

1 Albano VG et al. (1972) J. Organomet. Chem., 34:353
2 Peng SM, Lee GL, Chi Y, Peng CL, Huang LS (1989) J. Organomet. Chem., 371:197
3 Chi T, Cheng CY, Wang SL (1989) J. Organomet. Chem., 378:45
4 Kirtley SW, Olsen JP, Bau R (1973) J. Am. Chem. Soc. 98:4532
5 Frenz BA, Ibers JA (1971) in: Muetterties EL (ed) Transition metal hydrides. Marcel Dekker, New York, p 33
6 Ciani G et al. (1977) J. Organomet. Chem. 136:C49
7 Beringhelli T, Ciani G, D'Alfonso G, Malde VD, Sironi A, Freni M (1986) J. Chem. Soc., Dalton Trans, 1051
8 Bertolucci A et al. (1976) J. Organomet. Chem. 117:C37

9 Ciani G, Sironi A, Albano VG (1977) J. Organomet. Chem. 136:339
10 Mertis K et al. (1980) J. Chem. Soc., Chem. Commun., 654
11 Fehlner TP, Housecroft CE, Scheidt WR, Wong KS (1983) Organometallics 2:825
12 Teller GR, Bau R (1982) Struct. Bonding 44:1
13 Churchill MR, DeBoer BG, Rotella FJ (1976) Inorg. Chem. 15:1843
14 Churchill MR et al. (1975) J. Am. Chem. Soc. 97:7158
15 Sheldrick GM, Yesinowski JP: (1975) J. Chem. Soc., Dalton Trans., 873
16 Bhaduri S, Khwaja H, Jones PG (1988) J. Chem. Soc., Chem. Commun., 194
17 Aime S, Osella D, Deeming AJ, Arce AJ, Hursthouse MB, Dawes HM (1986) J. Chem. Soc., Dalton Trans.,1459
18 Yawney DBW, Doedens RJ (1972) Inorg. Chem. 11:838
19 Wilson RD et al. (1978) Inorg. Chem. 17:1271
20 Conole G. McPartlin M, Powell HR, Dutton T, Johnson BFG, Lewis J (1989) J. Organomet. Chem. 379:C1
21 Shapley JR et al. (1975) J. Am. Chem. Soc. 97:4145
22 Churchill MR, DeBoer BG (1977) Inorg. Chem. 16:878
23 Orpen AG et al. (1978) J. Chem. Soc., Chem. Commun., 723
24 Orpen AG et al. (1978) Acta Cryst. B34:2466
25 Broach RW, Williams JM (1979) Inorg. Chem. 18:314
26 Churchill MR, Lashewycz RA (1979) Inorg. Chem. 18:3261
27 Schultz AJ et al. (1979) Inorg. Chem. 18:319
28 Johnson BFG et al. (1979) J. Chem. Soc., Dalton Trans., 616
29 Johnson BFG et al. (1978) J. Chem. Soc., Dalton Trans., 673
30 Bhaduri S et al. (1979) J. Chem. Soc., Dalton Trans., 562
31 Guy JJ, Sheldrick GM (1978) Acta Cryst. B34:1725
32 Orpen AG, (1978) J. Organomet. Chem. 159:C1
33 Lippard SJ, Melmed KM (1967) Inorg. Chem. 6:2223
34 Lippard SJ, Melmed KM (1967) J. Am. Chem. Soc. 89:3929
35 Johnson BFG, Khattar R, Lewis J, McPartlin M, Morris J, Powell GL (1986) J. Chem. Soc., Chem. Commun., 507
36 Brown RK et al. (1979) Proc. Nat. Acad. Sci. USA 76:2099
37 Brown RK et al. Inorg. Chem. (1980) 19:370
38 Espinet P et al. (1979) Inorg. Chem. 18:2706
39 Wang HH, Pignolet LH (1980) Inorg. Chem. 19:1470
40 Ciani G et al. (1978) J. Organomet. Chem. 150:C17
41 Douglas G, Manojlovic-Muir L, Muir KW, Jennings MC, Lloyd BR, Rashidi M, Puddephatt RJ (1988) J. Chem. Soc., Chem. Commun., 149
42 Churchill MR et al. (1972) Inorg. Chem. 11:1818
43 Simon A (1967) Z. Anorg. Allg. Chem. 355:311
44 Wilson RD, Bau R (1976) J. Am. Chem. Soc. 98:4687
45 Churchill MR, Wormald J (1971) J. Am. Chem. Soc. 93:5670
46 Churchill MR et al. (1970) J. Chem. Soc., Chem. Commun., 458

Table 3.1. (continued)

47 Eady CR et al. (1976) J. Chem. Soc., Chem. Commun., 945
48 Jackson PF et al. (1980) J. Chem. Soc., Chem. Commun., 295
49 McPartlin M et al. (1976) J. Chem. Soc., Chem. Commun., 883
50 Huttner G, Lorenz H (1975) Chem. Ber. 108:973
51 Hart DW et al. (1979) Angew. Chem. Internat. Edit. 18:80
52 Mills OS, Paulus EF (1968) J. Organomet. Chem. 11:587
53 Huttner G, Lorenz H (1974) Chem. Ber. 107:996
54 Müller J et al. (1973) Angew. Chem. Internat. Edit. 12:1005
55 Koetzle TF et al. (1979) J. Am. Chem. Soc., 101:5631
56 Broach RW et al. (1978) Adv. Chem. Ser. 167:93

data-set magnitude. Examples of compounds in which hydrogen atoms have been located directly are shown in Table 3.1.

Hydrogen is found in transition metal cluster directly attached to metal atoms, in some few cases as terminal ligand but more frequently bridging two, three, or more metal atoms. Examples of the uncommon terminal hydrogen ligands are found in the complex $H_2Os_3(CO)_{11}$ illustrated in Fig. 3.3.

The edge-bridging position is the most common site for hydrogen in metal clusters. Besides the examples illustrated in Figs. 3.2 and 3.3 in which mono hydrogen bridged edges are observed, there are also compounds in which double hydrogen bridged are formed as occurring for instance in $H_2Os_3(CO)_{10}$ (s. Fig. 3.4).

As mentioned above, the presence of hydrogen bridges implies, in general, larger internuclear distances. An interesting example of isomerization

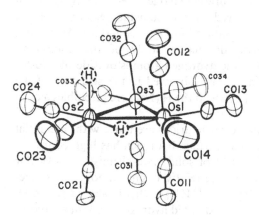

Fig. 3.3. The structure of $H_2Os_3(CO)_{11}$. Both terminal and bridging H-ligands are attached to the same metal atom. Reproduced with permission from Churchill MR, De Boer BG (1977) Inorg. Chem. 16:878

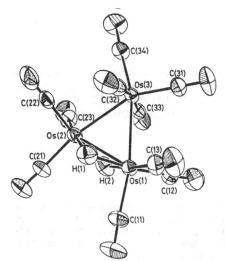

Fig. 3.4. Molecular structure of $H_2Os_3(CO)_{10}$. Reproduced with permission from Orpen AG, Rivera V, Bryan EG, Sheldrick GM, Rouse KD (1978) J. Chem. Soc., Chem. Commun., 723

among edge-bridged structures is constituted by the hydride $H_4Ru_4(CO)_{12}$ and its derivatives. $H_4Ru_4(CO)_{12}$, $H_4Ru_4(CO)_{11}[P(OMe)_3]$, and $H_4Ru_4(CO)_{10}(PPh_3)_2$ show a characteristic array of four long and two short M-M distances corresponding to M-H-M and M-M bonds respectively as illustrated in scheme (a) in Fig. 3.5. However, the derivative $H_4Ru_4(CO)_{10}(Ph_2PCH_2CH_2PPh_2)$ shows a pattern like that illustrated in Fig. 3.5, scheme (b). The existence of variety of geometrical forms for similar compounds indicates that the energy differences among the various structures are small. This is dramatically exemplified by the case of the anion $[H_3Ru_4(CO)_{12}]^-$ which in its salt with the cation $[(Ph_3P)_2N]^+$ is found under similar conditions in the two isomers schematized in Fig. 3.6. NMR studies of this kind of compounds show indeed that the hydrogen changes easily between the different edge-bridging positions.

Although not so frequent as the edge-bridging mode, face-bridging hydrogen bonding to metal clusters is also observed. Well documented examples of this mode of bonding are the structures reproduced in Figs. 3.7 and 3.8. As demonstrated by a single-crystal neutron diffraction study of the compound $HFeCo_3(CO)_9\{P(OMe)_3\}_3$ (Fig. 3.7), the hydrogen atom is outside the cluster capping the Co_3 face. μ_3-H bonding is also observed in the structure of the tetrahedral cluster $H_3Ni_4(\eta^5\text{-}C_5H_5)_4$ in which the H_3Ni_4 core resembles a cube with a missing corner (Fig. 3.8).

A very interesting class of metal cluster hydrides are those with interstitial hydrogen atoms. Although compounds in which hydrogen has high coordination number are well known in the field of solid state chemistry, molecular species with interstitial hydrogen atoms are relatively uncommon. The single-crystal neutron diffraction study of the anion $[HCo_6(CO)_{15}]^-$ shown in Fig. 3.9 provides proof of the existence of a six-coordinated hydrogen atom in a molecu-

D_{2d}

a

C_s

b

Fig. 3.5. Hydrogen bridge distributions in $H_4Ru_4(CO)_{12}$ derivatives (a) $H_4Ru_4(CO)_{12}$, $H_4Ru_4(CO)_{11}P(OMe)_3$ and $H_4Ru_4(CO)_{10}(PPh_3)_2$; (b) $H_4Ru_4(CO)_{10}(Ph_2PCH_2CH_2PPh_2)$

C_{3v}

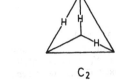

C_2

Fig. 3.6. Schematic distribution of hydrogen bridges into two isomers of the anion $[H_3Ru_4(CO)_{12}]^-$ isolated under the same conditions [Jackson, PF (1978) J. Chem. Soc. Chem. Commun 920]

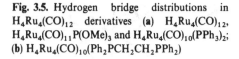

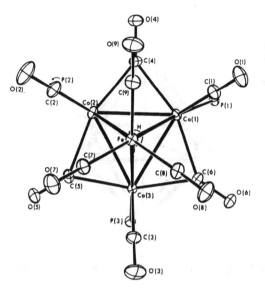

Fig. 3.7. Molecular structure of $HFeCo_3(CO)_9(POMe)_3)_3$ (MeO-groups are omitted). Reproduced with permission from Teller RE, Wilson RD, McMaller RK, Koetzle TF, Bau R (1978) J. Am. Chem. Soc. 100:3071

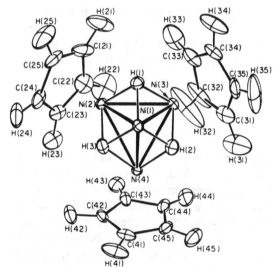

Fig. 3.8. Molecular structure of $H_3Ni_4Cp_4$ determined by neutron diffraction study. Reproduced with permission from Koetzle TF, Müller J, Tipton DI, Hart DW, Bau R (1979) J. Am. Chem. Soc. 101:5631

lar species. Interestingly, interstitial atoms in these clusters undergo easy interchange with proton-accepting solvents. This feature contrasts with the behavior of other interstitial compounds e.g. carbide clusters. The high mobility of hydrogen in these clusters is apparently similar to that within metal lattices.

3.1.2 Bonding in Hydride Cluster Compounds

From the point of view of bonding, terminal H-ligands form two-center/two-electron bonds meanwhile edge-bridging and face-bridging H-ligands are

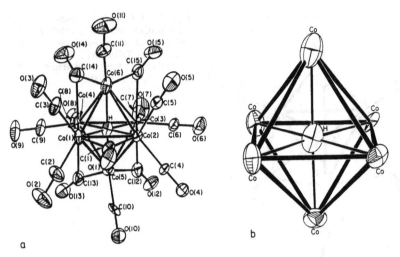

Fig. 3.9. Molecular structure of the anion $[HCO_6(CO)_{15}]^-$. (**a**) ORTEP plot; (**b**) view of the core

forming three-center/two-electron and four-center/two-electron bonds respectively.

Formally three-center M-H-M bonds can be considered as formed by the donation of an electron pair from the metal-metal bond to the proton.

$$
\begin{array}{ccc}
M & & M \\
| & & | \\
| & + \; H^+ \longrightarrow & | \!\!-\!\!-\!\!-\!\! H^+ \\
| & & | \\
M & & M
\end{array}
$$

The total number of electrons in the original cluster remains unchanged after the formation of the bridge. Such a description of hydrogen edge-bridging in metal clusters agrees with the lengthening of metal-metal bonds induced by this process, thus agreeing with the structural features commented above for the cluster $[H_2Re_3(CO)_{12}]^-$. The cluster $[H_2Os_3(CO)_{10}]$ appears to be an exception. As shown in Fig. 3.4, in this molecule, which has two edge-bridging protons on the same edge of the triangle, the Os-Os distance supporting the proton bridges is shorter. That agrees with the description of this molecule as a 46-electron unsaturated anionic species, $[Os_3(CO)_{10}]^{2-}$, containing a double bond to which two protons are added.

Analogously, the bonding of a μ_3-hydrogen to a triangular face of the cluster corresponds to the donation of an electron pair from a face-localized orbital to the 1s orbital of the proton. Clusters with face-bridging hydrogen ligands will have the same number of valence electrons than the parent anion.

Although the electron counting in cluster hydrides is straightforward, the prediction of the placement of hydrogen atoms is in general difficult. Thus, for instance, the clusters $[H_4Co_4(Cp)_4]$ and $[H_4Ru_4(CO)_{12}]$ have the 60 electrons characteristic of a tetrahedral geometry, but in the former the hydride ligands are face-bridging while in the latter they are edge-bridging.

Unsaturated species such as the cluster $[H_4Re_4(CO)_{12}]$ with 56 valence electrons can be rationalized by considering a parent hypothetical anion $[Re_4(CO)_{12}]^{4-}$ with four face-delocalized, three-center/two-electron metal-metal bonds, each of which is able to coordinate one proton in the hydride.

The anionic clusters $[HRu_6(CO)_{18}]^-$ and $[HCo_6(CO)_{15}]^-$ equally to both the dianion and the neutral hydride, have 86 electrons with seven electron pairs for skeletal bonding each, so an octahedral geometry is expected. As mentioned above, in these monoanions, hydrogen occupies an octahedral site in the center of the metal cluster. However in the neutral hydrides as well as in the isoelectronic osmium species, hydrogen atoms appear to occupy other locations not always predictable. In $[H_2Ru_6(CO)_{18}]$ each hydrogen atom is bridging a face on opposite sides of the octahedron of metal atoms, but the isoelectronic cluster $[H_2Os_6(CO)_{18}]$ shows a capped square pyramidal structure in which the hydrogen atoms are believed to be bridging two opposite edges of the base of the square pyramid.

3.1.3 Proton Magnetic Resonance

^{1}H-NMR is a valuable tool for detecting the presence and properties of hydrogen in soluble cluster species. In general, the ^{1}H-NMR signal for hydrogen in edge and face-bridging positions is observed at high field around $\delta = 25\text{-}33$ ppm. For terminally bonded hydrogen, the resonance signals are observed at lower field in the range $\delta = 10\text{--}25$ ppm. Thus the spectrum of the adduct $H_2Os_3(CO)_{10}L$ which has one terminal and one bridging hydrogen atom at $-60\,^\circ C$ shows two signals in the regions 27–30 ppm and 19–20 ppm corresponding to the bridging and terminally bonded hydrogens respectively. At higher temperature these signals are observed to broaden and coalesce due to an exchange between the two types of hydrogen.

Interestingly, in the cases of coordinately unsaturated compounds $[H_2Os_3(CO)_{10}]$ and $[H_4Re_4(CO)_{12}]$ discussed above, the former shows a chemical shift of $\delta = 21.36$ ppm for the hydrogen atoms occupying bridging positions and the latter $\delta = 15.08$ ppm for the face-bridging hydrogen atom. Although the hydrogen atom in $[HCo_6(CO)_{15}]^-$ shows a chemical shift at $\delta = -6.41$ ppm it does not correspond properly to the atom in an intersticial position (see Fig. 3.9) but probably to an hydrogen attached by exchange to an electronegative atom. Hydrogen atoms located interstitially usually show a rather high-field resonance in the range $\delta = 28\text{--}39$ ppm.

3.2 Main Group Element-Transition Metal Mixed Compounds

In this section, a series of cluster species in which one or more vertexes are constituted by a main group element will be analyzed. Although many of these compounds have been discussed to some extent in the previous Chapter, it is convenient to include them in this description. In this Section, the effect of gradually replacing transition metals by main group elements in metal clusters will be analysed, i.e. through the description of compounds with stoichiometries E_3M, M_4E, M_5E, M_6E, ME_3, M_2E_3, M_2E_5, and M_2E_6 where M is a transition metal and E a main group element.

3.2.1 Cluster Species with $M_3(\mu_3\text{-}E)$ Units

The simplest and also the commonest among main group-transition metal mixed compounds are those in which the main group element caps a triangular array of metal atoms in any triangular or more complex deltahedral metal cluster thus forming units of the type $(\mu_3\text{-}E)M_3$.

In Fig. 3.10 three types of $(\mu_3\text{-}E)M_3$ units which differ in the coordination state of the main group element E are shown schematically. Atom E always has an electron pair that may be bonded to any external one-electron group such as a halide, alkyl, or aryl; or that may act as an electron pair donor toward a transition metal, fragment M; or that can also remain as a lone pair thus giving "naked" species. For elements of the groups 13 and 14 as well as for nitrogen compounds of type A and B are formed almost exclusively. However naked species are frequently formed by heavier elements of group 15 and always for those of the group 16. Apparently the formation of naked species is possible only when the activity of s-electrons on the main group element is relatively low. For instance in group 15: Nitrogen is always bonded to an external group; phosphorus can form some naked derivatives e.g. $PCo_3(CO)_9$, but they are very reactive unless they have high steric hindrance e.g. $Fe_3(CO)_6\{P(O\text{-}iC_3H_7)_3\}$-$(\mu_3\text{-}PMn(CO)_2Cp)(\mu_3\text{-}P)$; arsenic forms instead the three kind of compounds $(\mu_3\text{-}AsR)M_3$, $(\mu_3\text{-}AsM')M_3$ and $(\mu_3\text{-}As)M_3$.

3.2.2 Butterfly M_4E Species

Although in the formation of compounds with $(\mu\text{-}E)M_3$ geometry the size of the main group element appears to be not very important, such a parameter can

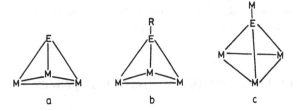

Fig. 3.10. Types of $(\mu_3\text{-}E)M_3$ units. E = main group element, M = transition metals, and R = halide or organic radical

play an important role for the stabilization of other compounds. In general, the smaller the main group element, the stronger its tendency to induce the formation of M-M bonds. Butterfly compounds M_4E are indeed known only as borides, carbides, nitrides or oxides i.e. only for elements of the first row.

3.2.3 Carbido-Clusters

M_5 and M_6E are typically encapsulated structures and similarly to M_4E species are accessible only when E is a first row element. Larger main group elements are not appropriate for these cavities. Selected examples of clusters with stoichiometries M_4E, M_5E, and M_6E are shown in Table 3.2. The most important family of clusters with interstitial or encapsulated atoms are by far the carbides.

Rhenium carbides can be afforded by pyrolysis of simple rhenium carbonyl compounds.

$$[Et_4N][H_2Re(CO)_4] \xrightarrow{235\,^\circ C} [Et_4N]_2[(\mu_3\text{-}H)Re_6(CO)_{18}C]$$

$$\xrightarrow{250\,^\circ C} [Et_4N]_3[Re_7(CO)_{21}C]$$

Both compounds contain the carbide atom in an octahedral cavity. The hydride cluster is a simple octahedron with three terminal carbonyls on each rhenium atom. The heptanuclear cluster can instead be considered as the same octahedron in which the two hydrides are replaced by a $[Re(CO)_3]^-$ fragment leading to a monocapped octahedral cluster. Both structures have seven skeletal electron pairs.

The chemistry of iron-group carbides has been extensively studied. Among iron derivatives, the species $Fe_5(CO)_{15}C$ and $[Fe_6(CO)_{16}C]^{2-}$ are specially relevant. As shown in the sequence of reactions illustrated in Fig. 3.11, these compounds can be prepared from $Fe(CO)_5$ and $[Fe(CO)_4]^{2-}$. In the same scheme several reactions involving carbido-iron clusters are described. As observed, a number of other encapsulated as well as butterfly species may be prepared from both the hexa and pentairon carbide mentioned above.

The chemistry of both ruthenium and osmium carbides is also well developed. Some representative examples of them are also included in Table 3.2.

Cobalt and rhodium also form a series of high nuclearity carbido-cluster compounds. Contrasting with carbido compounds of the iron group which have a clear tendency to put the carbide atom in octahedral cavities, carbides of the group 9 often place it in trigonal prismatic cavities. As shown in scheme in Fig. 3.12, the parent compound $[Co_6(CO)_{15}C]^{2-}$ of a series of encapsulated carbido-cobalt species may be prepared by the reaction of $Co_3(CO)_9CCl$ with $[Co(CO)_4]^-$. The same scheme also describes some carbido-cobalt cluster interconversions. Rhodium carbide clusters are in general similar to the cobalt ones.

Table 3.2. Selected examples of main group transition metal mixed clusters with stoichiometries M_4E, M_5E, and M_6E.

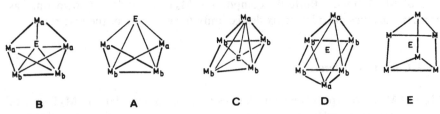

Compound	Interatomic distances (Av.) [Å]			Ref.
	$M_b–M_b$	$M_a–E$	$M_b–E$	
Butterfly Clusters (A)[a]				
$[Fe_4(CO)_{12}(CCO_2Me)]^-$	2.55	2.02	1.96	1
$Fe_4(CO)_{12}\{C=C(OMe)_2\}$	2.53	2.03	1.97	2
$Fe_4(CO)_{13}C$	2.54	1.80	1.99	3
$W_4(OPr^i)_{12}(NMe)C$	2.79	1.91	2.25	4
$HFe_4(CO)_{12}(BH_2)$	2.64	1.97	2.05	5
$[HFe_4(CO)_{12}C]^-$	2.61	1.79	1.99	6
$HFe_4(CO)_{12}CH$	2.61	1.84	1.94	7
$[Fe_4(CO)_{12}N]^-$	2.51	1.71	1.90	8
$HFe_4(CO)_{12}N$	2.80	1.93	2.11	9
$H_3Ru(CO)_{11}N$	2.81	1.96	2.06	10
$Ru_4(CO)_{12}(NCO)N$	3.24	1.90	2.13·	11
$[Os_4(CO)_{12}N]^-$	2.73	1.95	2.12	10
Bridged Butterfly (B)				
$[Fe_5(CO)_{12}(Br)_2C]^{2-}$	2.53	1.84	1.96	12
$HFe_4(CO)_{12}(AuPPh_3)C$	2.62	1.85	1.95	13
$Ru_5(CO)_{15}(MeCN)C$	2.72	1.96	2.07	14
$Ru_5(CO)_{14}(Br)(AuPPh_3)C$	2.95	1.98	2.04	15
$HRu_5(CO)_{15}(SEt)_3C$	2.85	1.96	2.13	16
$Os_5(CO)_{16}C$	2.75	1.98	2.12	17
$Os_5(CO)_{15}\{Ph_2P(CH_2)_2PPh_2)C$	2.76	1.98	2.11	18
$[Os_5(CO)_{15}(I)C]^-$	2.75	1.98	2.11	19
$HOs_5(CO)_{14}(NC_5H_4)C$	2.93	1.99	2.09	20
Square Pyramid (C)				
$Fe(CO)_{15}C$	2.66	1.96	1.89	21, 22
$[Fe_5(CO)_{14}C]^{2-}$	2.63	2.00	1.87	23
$[Fe_5(CO)_{14}N]^-$	2.61	1.92	1.84	22
$HFe_5(CO)_{14}N$	2.58	1.91	1.84	8
$Ru_5(CO)_{15}C$	2.85	2.09	2.03	14
$Ru_5(CO)_{14}(PPh_3)C$	2.87	2.12	2.04	14
$Ru_5(CO)_{13}(PPh_3)_2C$	2.88	2.16	2.05	14
$Os_5(CO)_{15}C$	2.88	2.06	2.05	19
$[Os_5(CO)_{14}C]^{2-}$	2.88	2.22	2.05	17
Octahedron (D)[b]				
$[H_2Re_6(CO)_{18}C]^{2-}$	3.02	2.14		24
$[Re_8(CO)_{24}C]^{2-}$	3.01	2.13		25
$[Fe_6(CO)_{16}C]^{2-}$	2.67	1.89		26
$Fe_6(CO)_{11}(NO)_4C$	2.68	1.90		27
$Ru_6(CO)_{17}C$	2.90	2.05		28
$[Ru_6(CO)_{16}C]^{2-}$	2.86	2.05		29

Table 3.2. (continued)

Compound	Interatomic distances (Av.) [Å]			Ref.
	M_b-M_b	$Ma-E$	M_b-E	
$Ru_6(CO)_{14}(NO)_2C$	2.88	2.04		30
$Ru_6(CO)_{15}(C_6H_{10})C$	2.90	2.06		31
$[Os_{10}(CO)_{24}C]^{2-}$	2.88	2.04		32
$[Os_{10}(CO)_{23}(NO)C]^-$	2.91	2.06		33
$[Os_{10}(CO)_{24}(I)C]^-$	2.90	2.05		34
$[Co_6(CO)_{13}C]^{2-}$	2.64	1.87		35
$[Co_6(CO)_{14}C]^-$	2.66	1.90		36
$[Rh_{15}(CO)_{28}C_2]^-$	2.90	2.04		37
Trigonal Prismatic $(E)^{b,c}$				
$[Co_6(CO)_{15}C]^{2-}$	2.56	1.95		38
$[Co_6(CO)_{15}N]^-$	2.54	1.94		39
$[Rh_6(CO)_{15}C]^{2-}$	2.80	2.13		40
$[Rh_6(CO)_{15}N]^-$	2.79	1.86		41
$Rh_8(CO)_{19}C$	2.81	2.13		37
$[Co_{13}(CO)_{24}C_2]^{3-d}$	2.57	1.97		42
$[Rh_{12}(CO)_{24}C_2]^{2-}$	2.78	2.12		43

[a] Ma–E–Ma angles in general are in the range 170–180°. Smaller angles are observed only for the clusters $W_4(OPr^i)_{12}(NMe)C$ (163.5°), $Fe_4(CO)_{12}BH_4$ (162°), $Fe_4(CO)_{12}\{C=C (OMe)\}$ (148°) and $[Fe_4(CO)_{12}(CO_2Me)C]^-$ (148°).
[b] For high nuclearity clusters, average distances refers only to the M_6E unit.
[c] Internuclear N–M distances correspond to an average of the distances within the triangular faces and between those faces.
[d] Average of the distances in the two cavities.

References

1 Bradley JS, Ansell GB, Hill EW (1979) J. Am. Chem. Soc. 101:7417
2 Bradley JS, (1982) Phil. Trans. R. Soc. London A 308:103
3 Bradley JS, Ansell GB, Leonowicz ME, Hill EW (1981) J. Am. Chem. Soc. 103:4968
4 Chisholm MH, Folting K, Huffman JC, Leonelli J, Marchant NS, Smith CA, Taylor LCE (1985) J. Am. Chem. Soc. 107:3722
5 Fehlner TP, Housecroft CE, Scheidt WR, Wong KS (1983) Organometallics 2:825
6 Holt EM, Whitmire KH, Shriver DF (1981) J. Organomet. Chem. 213:125
7 Beno MA, Williams JM, Tachikawa M, Muetterties EL (1980) J. Am. Chem. Soc. 102:4542
8 Tachikawa M, Stein J, Muetterties EL, Teller RG, Beno MA, Gebert E, Williams JM (1980) J. Am. Chem. Soc. 102:6648
9 Braga D, Johnson BFG, Lewis J, Mace JM, McPartlin M, Puga J, Nelson WJH, Raithby PR, Whitmire KH (1982) J. Am. Chem. Soc., Chem. Commun., 1081
10 Collins MA, Johnson BFG, Lewis J, Mace JM, Morris J, McPartlin M, Nelson WJH, Puga J, Raithby PR (1983) J. Chem. Soc., Chem. Commun., 689
11 Attard JP, Johnson BFG, Lewis J, Mace JM, Raithby PR (1985) J. Chem. Soc., Chem. Commun., 1526
12 Bradley JS, Hill EW, Ansell GB, Modrick MA (1982) Organometallics, 1:1634
13 Johnson BFG, Kaner DA, Lewis J, Raithby PR, Rosales M (1982) J. Organomet. Chem. 231:C59
14 Johnson BFG, Lewis J, Nicholls JN, Puga J, Raithby PR, Rosales MJ, McPartlin M, Clegg W (1983) J. Am. Chem. Soc., Dalton Trans., 277
15 Johnson BFG, Lewis J, Nicholls JN, Puga J, Raithby PR, Whitmire KH (1983) J. Am. Chem. Soc., Dalton Trans., 787
16 Johnson BFG. Lewis J, Wong K, McPartlin M (1980) J. Organomet. Chem. 185:C17
17 Johnson BFG, Lewis J, Nelson WJH, Nicholls JN, Puga J, Raithby PR, Rosales MJ, Schröder M, Vargas MD (1983) J. Am. Chem. Soc., Dalton Trans., 2447

Table 3.2. (continued)

18 Johnson BFG, Lewis J, Raithby PR, Rosales MJ, Welch DA (1986) J. Am. Chem. Soc., Dalton Trans., 453
19 Jackson PF, Johnson BFG, Lewis J, Nicholls JN (1980) J. Chem. Soc., Chem. Commun., 564
20 Jackson PF, Johnson BFG, Lewis J, Nelson WJH, McPartlin MJ (1982) J. Am. Chem. Soc., Dalton Trans., 2099
21 Braye EH, Dahl LF, Hübel W, Wampler DL (1962) J. Am. Chem. Soc. 84:4633
22 Gourdon A, Jeannin Y (1985) J. Organomet. Chem. 290:199
23 Lopatin VE, Gubin SP, Mikova NM, Tsybenov M Ts., Slovokhotov Yu L, Struchov Yu T (1985) J. Organomet. Chem. 292:275
24 Ciani G, D'Alfonso G, Romiti P, Sironi A, Freni M (1983) J. Organomet. Chem. 244:C27
25 Ciani G, D'Alfonso G, Freni M, Romiti P, Sironi A (1982) J. Am. Chem. Soc., Chem. Commun., 339
26 Churchill MR, Wormald J, Knight J, Mays MJ (1971) J. Am. Chem. Soc. 93:3073
27 Gourdon A, Jeannin Y (1985) J. Organomet. Chem. 282:C39
28 Sirigu A, Bianchi M, Benedetti E (1969) J. Chem. Soc., Chem. Commun., 596
29 Bradley JS, Ansell GB, Hill EW (1980) J. Organomet. Chem. 184:C33
30 Brown SC, Evans J, Webster M (1981) J. Am. Chem. Soc., Dalton Trans., 2263
31 Jackson PF, Johnson BFG, Lewis J, Raithby PR, Will GJ, McPartlin MJ, Nelson WJH (1980) J. Chem. Soc., Chem. Commun., 1190
32 Jackson PF, Johnson BFG, Lewis J, McPartlin M, Nelson WJH (1980) J. Chem. Soc., Chem. Commun., 224
33 Braga D, Henrick K, Johnson BFG, Lewis J, McPartlin, Nelson WJH, Puga J (1982) J. Chem. Soc., Chem. Commun., 1083
34 Farrar DH, Jackson PG, Johnson BFG, Lewis J, Nelson WJH, Vargas MD, McPartlin M (1981) J. Chem. Soc., Chem. Commun., 1009
35 Albano VG, Braga D, Martinengo S (1986) J. Am. Chem. Soc. Dalton Trans., 981
36 Albano VG, Chini P, Ciani G, Sansoni M, Martinengo S (1980) J. Am. Chem. Soc. Dalton Trans., 163
37 Albano VG, Chini P, Ciani G, Martinengo S, Sansoni M, Strumolo D (1974) J. Chem. Soc., Chem. Commun., 299
38 Martinengo S, Strumolo D, Chini P, Albano VG, Braga D (1985) J. Am. Chem. Soc. Dalton Trans., 35
39 Martinengo S, Ciani G, Sironi A, Heaton BT, Mason J (1979) J. Am. Chem. Soc. 101:7095
40 Albano VG, Sansoni M, Chini ,P, Martinengo S (1973) J. Am. Chem. Soc. Dalton Trans., 651
41 Bonfichi R, Ciani G, Sironi A,Martinengo S (1983) J. Am. Chem. Soc. Dalton Trans., 253
42 Albano VG, Braga D, Fumagalli A, Martinengo S (1985) J. Am. Chem. Soc. Dalton Trans., 1137
43 Albano VG, Braga D, Chini P, Strumolo D, Martinengo S (1983) J. Am. Chem. Soc. Dalton Trans., 249

Nickel carbide clusters are also known. From the reaction of the hexanickel anion $[Ni_6(CO)_{12}]^{2-}$ with carbon tetrachloride it is possible to obtain a series of high nuclearity carbide clusters. Among them $[Ni_8(CO)_{16}C]^{2-}$ and $[Ni_9(CO)_{17}C]^{2-}$ with structures based on a square antiprismatic array of nickel atoms.

3.2.4 Compounds with (μ_3-M)E$_3$ Units

There are two types of compounds with stoichiometries $(\mu_3\text{-M})E_3$ and $(\mu_3\text{-M})_2E_3$ in which the main group atoms are phosphorus or arsenic and that can be therefore considered as metal derivatives of the molecules P_4 or As_4.

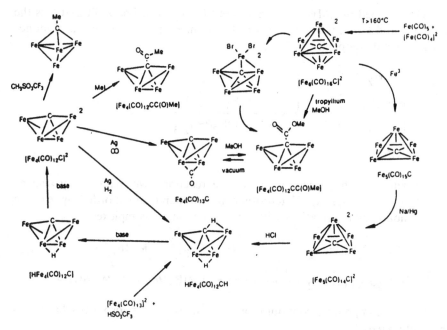

Fig. 3.11. Selected reactions from the chemistry of iron carbides. [Reproduced with permission from Whitmire KH (1988) J. Coord. Chem. 17: 95]

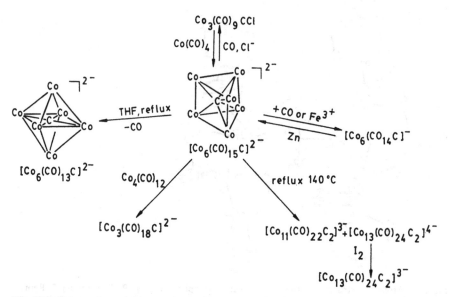

Fig. 3.12. Preparation and interconversion reaction of cobalt carbide clusters. [Whitmire KH (1988) J. Coord. Chem. 17: 95]

The tetrahedral LMP_3 clusters where $M = Co$, Rh, Pd, or Pt and L is the tripod ligand $CH_3C(CH_2PPh_2)_3$ can be obtained from the reaction of mononuclear metal complexes with P_4.

$$\left.\begin{array}{c} [Co(H_2O)_6]^{2+} \\ [Rh(CO)_2Cl]_2 \\ [Rh(C_2H_4)_2Cl]_2 \\ Ir(PPh_3)_2(CO)Cl \end{array}\right\} \xrightarrow[L=CH_3C(CH_2PPh_2)_3]{P_4} L-M-P_3$$

By adapting the stoichiometry of the reaction, two metals may also be attached to the P_3-group. Hetero-dimetal compounds are afforded by reaction of the monometal derivatives with appropriate metal complexes.

$$[M(H_2O)_6][BF_4]_2 \xrightarrow{P_4, L} [(LM)_2P_3]^{2+} \quad M = Ni, Co$$

$$[M(H_2O)_6][BF_4]_2 \xrightarrow{LCoP_3, L} LM(P_3)M'L \quad M, M' = Ni, Co$$

Similar compounds containing arsenic in place of phosphorus have also been prepared.

$$Co_2(CO)_8 \xrightarrow[CO, 200\,°C, 100\,atm]{(AsMe)_5, Hexane} (CO)_3CoAs_3$$

$$[Co(H_2O)_6][BF_4]_2 \xrightarrow{As_4, L} [(LCo)_2As_3]^{2+}$$

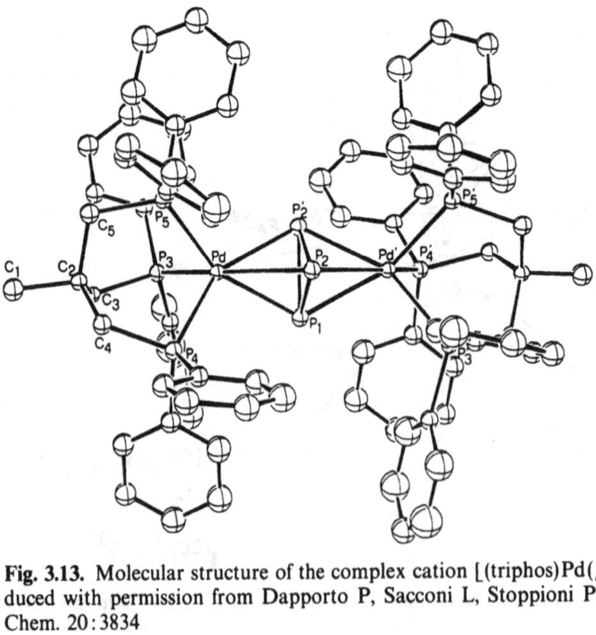

Fig. 3.13. Molecular structure of the complex cation $[(triphos)Pd(\mu, \eta^3-P_3)Pd(triphos)]^+$. Reproduced with permission from Dapporto P, Sacconi L, Stoppioni P and Zanobini F (1981) Inorg. Chem. 20:3834

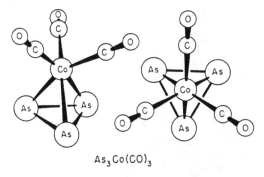

As$_3$Co(CO)$_3$

Fig. 3.14. Two views of the molecular structure of (CO)$_3$CoAs$_3$. Reproduced with permission from Foust AS, Foster HS, Dahl LF (1969) J. Am. Chem. Soc. 91:5631

Typical structures of this kind of phosphor and arsenic compounds are illustrated in Figs. 3.13 and 3.14.

3.2.5 Other Main Group Element-Transition Metal Mixed Clusters

The structures of the compounds illustrated in Figs. 3.15 to 3.17 could be considered as main group element clusters or rings stabilized by the interaction with metal atoms. Formally they could be also considered as derivatives of the type of compounds known as Zintl anions which are part of the clusters of post-transition elements to be discussed in the next Chapter. Since the properties of these mixed cluster compounds are strongly dependent on the nature of main group elements, any systematization of their chemistry is practically impossible. The reactions normally used for obtaining the compounds mentioned above are an example of such a diversity of behavior.

$$[BiFe_3(CO)_{10}]^-$$
$$\text{or} \xrightarrow{\text{CO}} [Bi_4Fe_4(CO)_{13}]^{2-}$$
$$[Bi_2Fe_4(CO)_{13}]^{2-}$$

$$(\pi\text{-Cp})_2Mo(CO)_4 \xrightarrow[\text{(AsMe)}_5]{190\,°C} (\pi\text{-CpMo})_2As_5$$

$$(\pi\text{-Cp*})_2Mo(CO)_4 \xrightarrow{P_4} [(\pi\text{-Cp*})Mo]_2P_6$$

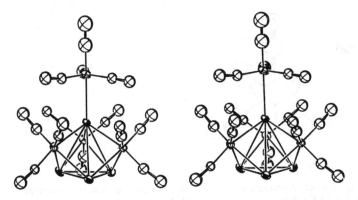

Fig. 3.15. Molecular structure of the anion $[Bi_4Fe_4(CO)_{13}]^{2-}$. Reproduced with permission from Whitmire KH, Albright TA, Kang SK, Churchill MR, Fettinger JC (1986) Inorg. Chem. 25:2799

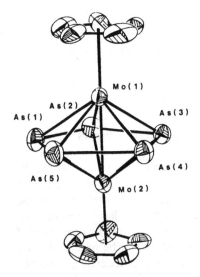

Fig. 3.16. Molecular structure of $(CpMo)_2(\mu, \eta^4\text{-}As_5)$. Reproduced with permission from Reingold AL, Foley MJ, Sullivan PJ (1982) J. Am. Chem. Soc. 104:4727

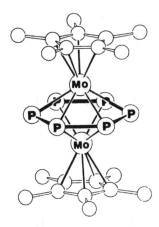

Fig. 3.17. Molecular structure of $[(\pi\text{-CpMo})_2(\mu, \eta^6\text{-P}_6)]$. Reproduced with permission from Meier H, Lauer W, Scholter FU (1985) Angew. Chem. Int. Ed. Engl. 24:350

3.2.6 Bonding in Main Group Element-Transition Metal Mixed Compounds

In Chapter 2, a series of concepts and models for describing bonding in borane and metal cluster were analyzed. The most versatile of such descriptions appears to be those rules related with the Effective Atomic Number (EAN) and with the Skeleton Electron Pairs (SEP). As assumed in some of the rules described in Table 2.16, frequently main group element fragments are actually involved in cluster bonding so that they obey the same electron counting rules.

The number of electrons with which a given main group fragment contributes varies depending on the presence of substitutes on the main group atoms. As analyzed with some detail above for clusters with $\mu_3\text{-EM}_3$ units, the main group element always has an external electron pair which is not part of the cluster itself. These external electrons may be present as a naked lone pair; electron pair coordinated to an external metal center; or a pair formed by one electron from E and another one from an atom, group or metal fragment. As shown schematically in Table 2.13 naked atoms of the groups 13, 14, 15, and 16 donate to the cluster electron count one, two, three and four electrons respectively. The coordination of the external electron pair to a metal acceptor, usually 16-electron fragments, does not change the electron contribution of the main group element to the cluster. However, the link of the main group atom to external groups as halide, alkyl, aryl, or 17 electron metal fragments increases the donation to cluster count by one electron. In the case of interstitial main group atoms, they contribute with all their valence electrons to the cluster skeleton.

Table 3.3. Cluster valence electron counting in main group-transition metal mixed clusters

Compound	Cluster valence electrons	Skeletal electron pairs	Structure[a]	Ref.
$Co_3(CO)_9\{\mu_3\text{-SiCo}(CO)_4\}$	48	6	Tetrahedron	1
$FeCo_2(CO)_9\{\mu_3\text{-NH}\}$	48	6	Tetrahedron	2
$As_2Fe_3(CO)_9$	48	6	c	3
$[Bi_2Fe_2(CO)_6\{\mu\text{-Co}(CO)_4\}]^-$	52	6[b]	e	4
$Fe_4(CO)_{13}C$	62[c]	7	A	5
$H_3Ru_4(CO)_{11}N$	62	7	A	6
$[Fe_4Co(CO)_{14}C]^-$	74	7	C	7
$Ru_5(CO)_{13}(PPh_3)_2C$	74	7	C	8
$Ru_5(CO)_{15}(AuPPh_3)(Cl)C$	88	8	B	9
$Os_5(CO)_{16}C$	76	8	B	10
$Fe_6(CO)_{11}(NO)_4C$	86	7	D	11
$Ru_6(CO)_{17}C$	86	7	D	12
$Co_6(CO)_{12}(S)_2C$	90	9	E	13
$[Rh_6(CO)_{15}N]^-$	90	9	E	14

[a] Structures referred to Fig. 3.18 (a–e) and Table 3.2 (A–E).
[b] $\mu_2\text{-Co}(CO)_4$ is considered as a two-electron donor group.
[c] Carbide and nitride are considered most often interstitial atoms donating all of its electrons to the cluster valence electron count.

References

1 Schmid G, Bätzel V, Etzrodt G (1976) J. Organomet. Chem. 112:345
2 Fjare DE, Keyes DG, Gladfelter WL (1983) J. Organomet. Chem. 250:383
3 Delbaere LTJ, Kruczynski LJ, McBride DW (1973) J. Chem. Soc., Dalton Trans., 307
4 Whitmire KH, Raghuveer KS, Churchill MR, Fettinger JC, See RF (1986) J. Am. Chem. Soc. 108:2778
5 Bradley JS, Ansell GB, Leonowicz ME, Hiel EW (1981) J. Am. Chem. Soc. 103:4868
6 Collins MA, Johnson BFG, Lewis J, Mace JM, Morris J, McPartlin M, Nelson WJH, Puga J, Raithby PR (1983) J. Chem. Soc., Chem. Commun., 689
7 Hriljac JA, Swepston PN, Shriver DF (1985) Organometallics 4:158
8 Cook SL, Evans J, Gray LR, Webster M (1986) J. Chem. Soc., Dalton Trans., 2149
9 Johnson BFG, Lewis J, Nicholls JN, Puga J, Whitmire KH (1983) J. Chem. Soc., Dalton Trans., 787
10 Johnson BGF, Lewis J, Nelson WJH, Nicholls JN, Puga J, Raithby PR, Rosales MJ, Schröder M, Vargas MD (1983) J. Chem. Soc., Dalton Trans., 2447
11 Gourdon A, Jeannin Y (1985) J. Organomet. Chem. 282:C39
12 Sirigu A, Bianchi M, Benedetti E (1969) J. Chem. Soc., Chem. Commun., 596
13 Arrigoni A, Cerriotti A, Pergola RD, Longoni G, Manassero M, Masciocchi N, Sansoni M (1984) Angew. Chem. Int. Ed. 23:322
14 Bonfichi R, Ciani G, Sironi A, Martinengo S (1983) J. Chem. Soc., Dalton Trans., 253

Tetranuclear clusters M_4, M_3E, M_2E_2, ME_3, and E_4 are generally electron-precise polyhedra with six skeletal electron pairs, i.e. one two-center/two-electron bond per edge in a tetrahedral structure. Characteristic cluster valence electron counts are shown in Table 3.3. Analogously to normal metal clusters, the increase in the number of valence electrons in mixed main group-transition metal clusters normally leads to bond disruption. In the series of compounds with stoichiometries M_3E_2 schematized in Fig. 3.18 in which the electron count changes from 48 to 52 electrons the observed structures may be rationalized considering first the disruption of one metal-metal bond thus arising a *nido*-cluster with one vertex missing from the parent octahedron. The structure of the compounds with 52 electrons which shows strong interactions between both main group elements in the molecule may be understood only if the new structure is considered as an electron-precise tetrahedron with an extra bridging metal. Such is the case of the compound $[Bi_2Fe_2(CO)_6\{\mu\text{-}Co(CO)_4\}]^-$ whose structure is illustrated in Fig. 3.19. Encapsuled atoms in the species with stoichiometries M_5E and M_6E may be considered to donate all of its valence electrons.

Cluster of the series M_5E often adopt a square-based pyramidal geometry with the main group element in the center of the square base. Such compounds often have seven skeletal electron pairs and are considered to be *nido*-octahedral molecules. The main group element often acquires a position slightly below the square of metal atoms. The position of the carbon atom in carbides apparently depends upon the charge of the complex. Thus in the compound $Fe_5(CO)_{15}C$ the carbon atom lies 0.09 Å below the square base meanwhile in the anion $[Fe_5(CO)_{14}C]^{2-}$ such deviation grows up to 0.18 Å. In the series of compounds $Ru_5(CO)_{15-x}(PPh_3)_xC$ these distances are 0.11, 0.19, and 0.23 Å for $x = 0$, 1, and 2 respectively.

Simple clusters with stoichiometry M_6E in general adopt either an octahedral or a prismatic trigonal structure. Clusters with seven skeleton electron pairs adopt the octahedral geometry. Trigonal prismatic arrangements require nine skeleton electron pairs for localized edge-bonding. As illustrated schematically in Fig. 3.12, for the M_6E cobalt-carbide, the configuration of these clusters may be modulated by addition or elimination of ligands.

Bonding in species with the unit $(\mu_3\text{-}M)E_3$, e.g. those derived from P_4 and As_4, can be regarded for the purpose of electron-counting either as complexes in which the triphosphorous cycle P_3 behaves as a tri-*hapto* three-electron donor towards the transition metal moiety or as a tetrahedral cluster in which one phosphor in P_4 has been substituted by a three-electron donor. Counting 30 valence electrons corresponds to an electron-precise tetrahedron with six two-electron/two center bonds. In such kind of compound with stoichiometries ME_3 and M_2E_3 (see Figs. 3.13 and 3.14) the E–E distances are shorter than those in the pure element E_4. This has been attributed to electron delocalization toward

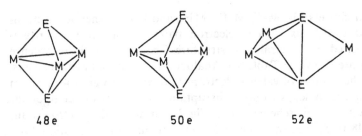

48 e   50 e   52 e

Fig. 3.18. Electron count in M_3E_3 compounds

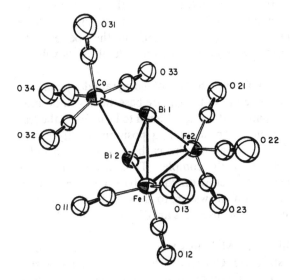

Fig. 3.19. Molecular structure of the anion $[Bi_2Fe_2(CO)_6\{\mu\text{-}Co(CO)_4\}]^-$. Reproduced with permission from Whitmire KH, Raghuveer KS, Churchill MR, Fettinger JC, See RF (1986) J. Am. Chem. Soc. 108:2778

the metal atom reducing thus the E–E antibonding interactions. Although for an electron-saturated tetrahedron with six two-electron/two center bonds corresponds a counting of 30 valence electrons, species with electron counting between 30 and 33 may also be stabilized as illustrated in the voltammogram reproduced in Fig. 3.20.

Somewhat different from that of P_4 and As_4 is the nature of the tetrabismuth compound $[Bi_4Fe_4(CO)_{13}]^{2-}$ described in Fig. 3.15. Three of the tetrahedron faces are capped by $Fe(CO)_3$ groups, but the apical bismuth atom is also coordinated to an external $Fe(CO)_4$ moiety. Actually this species is more related to the cage compound P_4S_3 than to a normal cluster. Indeed the apical bismuth is 3.45 Å from those in the triangular base i.e. both types of bismuth are practically non bonded directly.

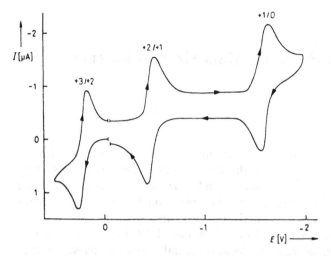

Fig. 3.20. Cyclic voltammogram of [(triphos)Co-μ-(h³-P₃)Co(triphos)]BF₄. From the original cluster species with 31 valence electrons others with 30, 32 and 33 valence electrons are produced. Reproduced with permission from Fabrizzi L, Sacconi L (1979) Inorg. Chim. Acta 36:L407

The arsenic and phosphor compounds shown in Figs. 3.16 and 3.17 respectively can be considered formally as a transition between the normal transition metal cluster and the polynuclear main group derivatives known as Zintl ions which are discussed in the next Chapter. Although it is clear that in these cases it is not possible to consider main group elements as ligands, it is also difficult to find a good description in terms of the electron counting discussed for other mixed compounds. The chemistry of such compounds is reminiscent of that of the main group element rings.

Chapter 4

Cluster Compounds of the Main Group Elements

The arrangement of the elements considering their properties as a function of their atomic number established in the Periodic Table permits us to classify them into two large groups: The Main Group Elements, i.e. the alkaline, alkaline earth, and post-transition elements, and the Transition Metals. Table 4.1 summarizes selected properties of Main Group Elements.

The elements of groups 1 and 2 of the Periodic Table form, in accordance with their low number of valence electrons, metallic solids. The stability of element-element covalent interactions inferred from the existence of gaseous diatomic molecules and the interatomic distances within them appears to be rather low being significative only for the lighter alkali metals.

As observed in Table 4.1 the connectivity of the elements in their solids increases toward the left and downwards in the Periodic Table. The tendency of the elements to form species with a few specific instead of a number of unspecific interactions obeys, to a considerable extent, the availability of both electrons and orbitals in the valence shell. As inferred from the correlation between electronegativity and the number of neighbors in the solid observed inside a Group (see, for instance groups 15 and 16), the connectivity of the elements in their compounds is also affected by the effective nuclear charge felt by valence electrons. The greater the charge, the higher the stabilization produced by formation of discrete covalent bonds.

Data in Table 4.1 also reflect the well known tendency of the most electronegative elements of the second Period to form multiple bonds.

4.1 Clusters and Cages of Main Group Elements

Most main group elements build solids that are either infinite arrangements or species of low dimensionality – chains or rings – packed in molecular solids. Exceptions to this behavior are carbon and phosphorus which lead to allotropic forms made up of cluster species. The best known of them is the elementary white phosphorus P_4 which has a cluster structure itself and retains its shape in solution and in the gas phase. In the case of carbon a new allotropic form consisting of molecular clusters C_n, the fullerenes, has been discovered recently. These species are giant carbon cages with spherically shaped structures which lead to molecular solids and that are also stable in solution.

The formation of homonuclear molecular clusters from a given Main Group Element can be formally thought of as the process of taking apart a small piece of the element stabilizing it thereafter by either replacing the bonds to the original matrix by specific ligands or placing it in an appropriate medium that allows it to exist as an isolated species. From connectivity data in Table 3.1 the ability of the elements to bear chains, rings, or three-dimensional structures can be inferred. From this point of view all the elements with connectivity equal or greater than three, i.e. the elements of the groups 1 to 15, are in principle able to form the type of homonuclear cages or clusters in which we are interested in.

The fundamental difference between transition metal elements and the Main Group Elements is the configuration of the valence electron shells. Main Group Elements have, if any, only closed d-shells.

Although alkaline and alkaline earth-elements like the transition elements have metallic properties, the chemistry of both groups shows significative differences. The relatively low nuclear effective charge undergoing valence d-electrons in transitions metals permits the presence of a number of low-lying and often partially occupied energy levels. As shown in Chapter 2 these orbitals are very important both in the formation of the cluster backbone and in the bonding with ligands stabilizing the conglomerate. Certainly, the stabilization of pre-transition metal clusters by cluster-ligand bonding will be poor, if any, because of the lack of d valence orbitals.

The chemistry of the electron-deficient boranes and carboranes which has indeed been employed as a useful model in the description of bonding in transition metal clusters appears to be a singular case among main group elements. From the point of view of the cluster chemistry, the properties of boron appear to meet with those of the transition metals.

Because of the increase of element effective nuclear charge on going to the right in the Periodic Table, post-transition elements in general show a greater tendency to have a reduced nuclearity as well as to form two center-two electron covalent bonds than the transition elements. This feature, in addition to the relatively low number of low-laying orbitals proper of the s-p block elements, leads to the formation of conglomerates which are mainly electron-saturated species. In them, each atom forms three two-center/two-electron skeleton bonds in addition to either a lone pair or one bond to a ligand. In most of these cases it seems more proper to refer to cages than to clusters. Indeed, as it will be seen in the next sections, many three-dimensional post-transition element aggregates display geometries that remind us more of the polycyclic organic compounds than of the deltahedral structures in the transition metal clusters. Nonetheless there are some cases such as the post-transition metals in which deltahedral structures like those for electron-deficient elements are observed.

In this Chapter, some of the different possibilities that Main Group Elements offer for building atomic conglomerates will be analyzed. In such a context examples of electron-deficient cluster species such as the clusters formed by the alkali metals and some complementary aspects of the chemistry of boranes and

Table 4.1. Main-Group Elements: 1. Electron Configurations. 2. Electronegativity (after Allred and Rochow). 3. Type of Structure Solid Phase[a] and Interatomic Distances[b] (in Parenthesis, Number of Nearest Neighbors). 4. Interatomic Distances[c] in Diatomic Molecule in Gaseous State. 5. Van der Waals Distances[c]

	1	2	13	14	15	16	17
	Li	Be	B	C	N	O	F
1.	$[He]\,2s$	$[He]\,2s^2$	$[He]\,2s^2 2p$	$[He]\,2s^2 2p^2$	$[He]\,2s^2 2p^3$	$[He]\,2s^2 2p^4$	$[He]\,2s^2 2p^5$
2.	0.97	1.47	2.01	2.50	3.07	3.50	4.10
3.	A2 3.039(8); A1 3.114(12)	A3 2.226(6); A2 2.286(6)	rh. B_{12} 1.73–1.79(5); 1.71(1/2); 2.03(1/2)	A4 = diamond 1.554(4); graphite 1.418(3) [3.348(1)]	alfa-N_2 1.15(1)	1.208(1) (gas)	1.49(1) [> 2.82]
4.	2.672	—	1.589	1.243	1.098	1.207	1.417
5.	—	—	—	3.4	3.0	2.8	2.7

	1	2	13	14	15	16	17
	Na	Mg	Al	Si	P	S	Cl
1.	$[Ne]\,3s$	$[Ne]\,3s^2$	$[Ne]\,3s^2 3p$	$[Ne]\,3s^2 3p^2$	$[Ne]\,3s^2 3p^3$	$[Ne]\,3s^2 3p^4$	$[Ne]\,3s^2 3p^5$
2.	1.01	1.23	1.47	1.74	2.06	2.44	2.83
3.	A2 3.716(8); A3 3.767(6) 3.768(6)	A3 3.197(6); A1 3.209(6)	A1 2.863(12)	A4 2.352(4)	black 2.22(2) 2.24(1) violet ⟨2.22⟩(3) white 2.21(3)	α-S_8 2.05(2) [> 3.69] S_∞ 2.07(2) [> 3.5]	2.02(1) [> 3.34]
4.	3.078	—	—	2.246	1.894	1.889	1.998
5.	—	—	—	—	3.8	3.7	3.6

	1	2	13	14	15	16	17
	K	Ca	Ga	Ge	As	Se	Br
1.	$[Ar]\,4s$	$[Ar]\,4s^2$	$[Ar]\,4s^2 4p$	$[Ar]\,4s^2 4p^2$	$[Ar]\,4s^2 4p^3$	$[Ar]\,4s^2 4p^4$	$[Ar]\,4s^2 4p^5$
2.	0.91	1.04	1.82	2.02	2.20	2.48	2.74
3.	A2(78°K) 4.554(8)	A1 3.943(12)	A1 2.442(1) 2.712(2)	A4 2.450	metallic 2.51(3)	Se_∞ 2.32(2)	2.27(1) [> 3.30]

	Rb	Sr	In	Cs	Ba	Tl	Sn	Pb	Sb	Bi	Te	I
(top)	3.923	A3 3.940(6) 3.955(6)	2.742(2) 2.801(2)						[3.15(3)]		[3.46(4)]	
	—	—	—						As_4(gas) 2.44		α-Se_8 2.34(2) [> 3.53]	2.281
									2.288		2.166	3.9
									4.0		4.0	
1.	$[Kr]5s$	$[Kr]5s^2$	$[Kr]5s^2 5p$	$[Xe]6s$	$[Xe]6s^2$	$[Xe]4f^{14}5d^{10}6s^2 6p$	$[Kr]5s^2 5p^2$	$[Xe]\ldots 6p^2$	$[Kr]5s^2 5p^3$	$[Xe]\ldots 6p^3$	$[Kr]5s^2 5p^4$	$[Kr]5s^2 5p^5$
2.	0.89	0.99	1.49	0.86	0.97	1.44	1.72	1.55	1.82	1.67	2.01	2.21
3.	A2(78°K) 4.854(8)	A1 4.303(12); A3(> 215°C) 4.31(6) 4.31(6)	A1 3.244(4) 3.370(8)	A2(78°K) 5.254(8)	A2 4.352(8)	A3 3.408(6) 3.456(6); A1 3.42(12)	A4 = gray 2.810(4); white 3.023(4) 3.182(2)	A1 3.501(12)	$2.91(3)^9$ [3.36(3)]	3.07(3) [3.53(3)]	2.86(2) [3.74(4)]	2.70(1) [> 3.54]
4.	—	—	—	—	—	—	—	—	2.8–2.9	—	2.557	2.667
5.	—	—	—	—	—	—	—	—	4.4	—	4.4	4.3

a A1 = Cubic close packed. A2 = Cubic body centered. A3 = Hexagonal close packed
b Distances in Å

References

Fluck, Heumann (1988) Periodic System der Element. VCH Verlaggesellschaft 1988.
Allamann R (1977) in: Rheingold AL (ed) Homoatomic rings, chains and macromolecules of Main-Group elements

carboranes are discussed. In order to visualize the transition, sometimes imprecise and subtle, from cage compounds to proper cluster species observed for post-transition element derivatives, homonuclear species of the groups 14 and 15 as well as some germane aspects of the chemistry of the naked post-transition metal clusters will be analyzed.

4.2 Cluster Species of Alkali Metals

Alkali metal cluster formation presupposes that the element is in intermediate oxidation states between zero-valent and normal valence ones which are characteristic of metallic and salt-like species respectively. The chemistry of this class of compounds is therefore strongly associated with the stabilization of such states.

The low number of valence orbitals available for cluster-ligand bonding commented on above in addition to the high reduction potentials of alkali metals cause alkali metal clusters to be rather unstable. Disproportionation into the metal and other products with the element in higher oxidation states is often a thermodynamic favorable process.

$$M_n^{m+} \rightarrow (n-m)M + mM^+$$

The presence of any electron donor able to stabilize the cation M^+ will therefore favor cluster disruption. Such limiting conditions make it very difficult to find an adequate medium for the stabilization of such kinds of clusters. Solvent as well as counterions must be highly inert indeed, both as oxidation agent and as a Lewis base. Although the severity of such limiting conditions increases with increasing atomic weight of the metals, it has been possible to achieve the conditions under which clusters of the elements rubidium and cesium can exist, namely as the suboxide to be described in this Section.

As mentioned above, the tendency of the elements to form covalent bonds reducing their connectivity increases with increasing electronegativity of the element. In the case of the alkali elements, the formation of molecular clusters – in which the metal displays a relatively low coordination number – should be possible only for the lighter group elements. A number of lithium clusters are indeed known. Selected examples of them will be discussed below.

4.2.1 Lithium Clusters

Lithium as the first element in the Periodic Table is expected to exhibit special properties. It is the most electronegative of the alkali elements but simultaneously the most electropositive of the Period. That is apparent in lithium chemistry, since there is, for instance, a widely used organometallic chemistry and a rich structural chemistry which is not common for the other group elements. On the other hand, high polar species and lithium bonds with great ionic character are almost a constant in lithium chemistry. The chemical behavior of lithium shows a considerable tendency to aggregation in solution as well as in the solid state. Some aspects of this chemistry being germane to the subject of this book will be discussed in this section.

Structural Properties. Although lithium structural chemistry has been developed only recently, there are examples enough to appreciate that organolithium compounds show an appreciable tendency to aggregation. Although the structures of the products are often determined by subtle medium effects, it is known that the steric size of the organic rest and the coordinating properties of reagents are important for the extent of the association and the geometry around the lithium atom. Dimers, tetramers, and other oligomers are very common in lithium chemistry. The most favorable coordination numbers for lithium are 5, 6 and 7.

Table 4.2. Selected examples of monomeric and oligometric organolithiuim compounds

Compound[a]	Structure	Ref.
$(C_2H_5)_6Li_6$	Li_6C_6 skeleton. Ethyl groups on triangular faces of octahedron Li_6	1
$(C_6H_{11})_6Li_6 \cdot 2C_6H_6$	Li_6C_6 skeleton. See Fig. 4.3	2
$(CH_3)_4Li_4$	Li_4C_4 skeleton. See Fig. 4.1	3
$(CH_3)_4Li_4 \cdot 2TMEDA$	Li_4C_4 skeleton. See Fig. 4.2	4
t-Bu_4Li_4	Li_4C_4 skeleton analagon to $(CH_3Li)_4$	5
$(PhC \equiv C)_4Li_4 \cdot 2TMHDA$	Li_4C_4 skeleton. Acetylide moiety occuping faces of Li_4 tetrahedron	6
$Ph_4Li_4 \cdot 4Et_2O$	Li_4C_4 skeleton. Phenyl moiety occupying triangular faces of Li_4.	7
$(PhLi \cdot Et_2O)_3LiBr$	Li_4C_3Br skeleton. Phenyl groups and Br atom are occupying the triangular faces of Li_4 tetrahedron.	7
$Ph_2Li_2 \cdot 2TMEDA$	Li_2 with bridging phenyl groups and one chelating diamine at each Li atom.	8
$(PhC \equiv C)_2Li_2 \cdot TMPDA$	Li_2 with phenyl-ethynyl bridges.	9
$PhLi \cdot PMDTA$	Monomer	10
$Ph_3CLi \cdot TMEDA$	Monomer	11
$(Fluorenyl) Li \{N(C_2H_4)_3CH\}_2$	Monomer	12

[a] TMEDA = Tetramethylethylendiamine; TMHDA = tetramethylhexamethylendiamine; PMDTA = pentamethyldiethylentriamine.

References

1 Brown TL, Gerties RL, Batus DA, Ladd JA (1964) J. Am. Chem. Soc. 86:2135
2 Zerger R, Rhine W, Stucky GD (1974) J. Am. Chem. Soc. 96:6548
3 Weiss E, Henchen G (1970) J. Organomet. Chem. 21:265
4 Köster H, Thoennes D, Weiss E (1978) J. Organomet. Chem. 160:1
5 McKeever LD, Waak R (1969) J. Chem. Soc. D 1969:750
6 Schubert B, Weiss E (1983) Angew. Chem. Int. Ed. 22:496
7 Hope H, Power PP (1983) J. Am. Chem. Soc. 105:5320
8 Thoennes D, Weiss E (1978) Chem. Ber. 111:3157
9 Shubert B, Weiss E (1983) Chem. Ber. 116:3212
10 Schuemann V, Kopt J, Weiss E (1985) Angew. Chem. 97:222
11 Brooks JJ, Stucky DG (1972) J. Angew. Chem. 94:7333
12 Brooks JJ, Rhrine W, Stucky GD (1972) J. Am. Chem. Soc., 94:7339

In Table 4.2, some selected examples of lithium aggregates with structures determined by crystallographic studies are described.

From the point of view of metal aggregates some simple alkyl lithium compounds are specially interesting. As illustrated in Fig. 4.1, the structure of the methyl lithium may be described as a tetrahedral array of lithium atoms with three face-bridging methyl groups. Li_4 forms a perfect tetrahedron with interatomic distances Li–Li of 2.68 Å. The Li–C bonds within the cluster are 2.41 Å, only somewhat shorter than those to the next cluster, 2.36 Å. This feature reveals relatively strong inter-cluster interactions agreeing with the relatively low solubility of methyl lithium. The effect of the inter-cluster interactions can be counteracted by the addition of strong coordinative ligands. Thus for instance, the reaction of methyl lithium with tetramethylethylendiamine, $(CH_3)_2NCH_2$-$CH_2N(CH_3)_2$ (TMEDA), leads to the compound $(CH_3Li)_4 \cdot 2\,TMEDA$ with the structure reproduced in Fig. 4.2. $(CH_3Li)_4$ units are in this case bonded by TMEDA bridges. Li-Li lengths are 2.56–2.57 Å, somewhat shorter than in methyl lithium. The effect of the coordination of the polyamine is also observed on the lithium-carbon distances being also about 0.1 Å shorter than in CH_3Li.

Donor ligands can influence not only the inter-cluster interactions but also the nuclearity of the clusters. An illustrative example of such an effect is observed for phenyl lithium which, depending on the donor ligands, shows nuclearities of one in [PhLi (pmdta)] (pmdta = pentamethyldiethylentriamine), two in [PhLi-(tmeda)]$_2$ and four in the diethyl ether derivative [PhLi(OEt$_2$)]$_4$.

Voluminous organic derivatives also prevent the formation of higher conglomerates. The monomeric compounds [Ph$_3$CLi (tmeda)] and [(Fluorenyl) Li-{N(C$_2$H$_4$)$_3$CH}$_2$] are some instances that illustrate such an effect.

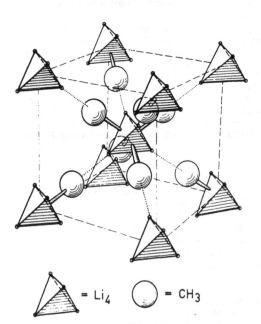

△ = Li$_4$ ◯ = CH$_3$

Fig. 4.1. Molecular structure of methyl lithium. Reproduced with permission from Weiss E, Henchen G (1970) J. Organomet. Chem. 21:265

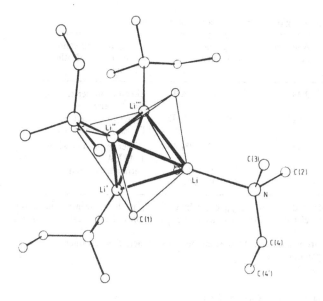

Fig. 4.2. Molecular structure of $(CH_3Li)_4 \cdot 2TMDA$. Reproduced with permission from Köster H, Thoennes D, Weiss E (1978) J. Organomet. Chem. 160:1

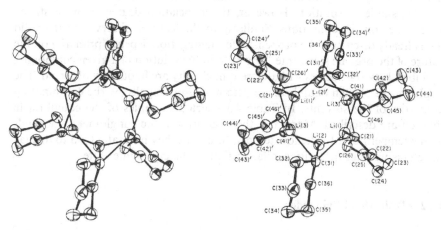

Fig. 4.3. Molecular structure of $[(C_6H_{11}Li)_6 \cdot 2C_6H_6]$. Reproduced with permission from Zerger R., Rhine W, Stucky G. (1974) J. Am. Chem. Soc. 96:6048

Organolithium compounds with a nuclearity greater than 4 are also possible. Thus for instance cyclohexyllithium is, as illustrated in Fig. 4.3, an hexamer with the formula $[(C_6H_{11}Li)_6 \cdot 2C_6H_6]$ in which the lithium atoms are ordered in an octahedral array.

Table 4.3. Equilibria of lithium conglomerates in tetrahydrofuran solutions

Compound	Association degree[a]	Aggregation equilibrium[b]	Aggregation in solid compounds and other solvents
n-Butyllithium	2.38	81% D, 19% T	Hexamer in benzene or cyclohexane
Phenyllithium adduct with Et_2O	1.61	39% M, 61% D	Dimer in adduct with TMEDA[c]. Tetramer in solid
2-Lithium-2-methyl-1,3-dithian	1.18	18% D, 82% M	Dimer in adduct with TMEDA
Lithium-1-cyclopentenolat	2.78	39% T, 61% D	Tetramer in solid
Lithium-diisopropylamid	1.63	63% D, 37% M	

[a] Association degree (n) is defined as $n = c_{nominal}/c_{exp}$ where $c_{nominal}$ is the concentration of the lithium species in the solution referred to the monomer and c_{exp} the concentration calculated from cryoscopic experiments.
[b] Ideal two component mixture is assumed. M = monomer, D = dimer, T = tetramer.
[c] TMEDA = tetramethylethylendiamine.

Reference

Bauer W, Seebach D (1984) Helv. Chim. Acta 67:1972

Aggregation of Organolithium Compounds in Solution. Organolithium cluster are also stable in solution. However, the association degree shows a strong dependence on the conditions. Similarly to the features observed in the solid state already discussed, the propensity for aggregation depends primarily on the nature of the organic substitute. However, in the solution of a given compound, equilibria between species with different nuclearity are frequently observed. The displacement of such equilibria are determined principally by the coordinating properties of the solvents. In Table 4.3 selected examples of the equilibria in solution are described. Moreover, the nuclearity of the conglomerate depends on the temperature. In the low temperature limit they normally tend to display structures with a nuclearity similar to that in the crystalline state.

4.2.2 Alkali Metal Suboxides

The formation of homonuclear bonding between alkali-metal atoms which characterizes the formation of cluster species implies that these elements are in an intermediate oxidation state between 0 and + 1. This condition appears to be met by the suboxides of rubidium and cesium.

Controlled additions of oxygen to rubidium or cesium lead to colored products with metallic properties when the metal is in excess. With the aid of sophisticated techniques for preparing, manipulating, and investigating these substances, detailed thermal studies of the products from these reactions have been carried out.

As observed in the phase diagram Rb/Rb_2O reproduced in Fig. 4.4, from the step-wise controlled oxidation of rubidium two stable rubidium suboxides Rb_9O_2 and Rb_6O are formed. Rb_9O_2, a compound that can be handled safely at room temperature, forms large platelets which look like metallic copper. Although according to the phase diagram, it is not possible to obtain the phase Rb_6O from the melt, amorphous samples by rapid cooling or single crystals in appropriate temperature gradients can be obtained.

The phase diagram Cs/CsO_2 reproduced in Fig. 4.5 is apparently more complex than that of rubidium. In this case there are three stoichiometric suboxides, Cs_7O, Cs_4O and $Cs_{11}O_3$. The $Cs_{11}O_3$ is a permanganate-colored compound stable at room temperature. The phase Cs_4O which decomposes below room temperature is difficult to obtain as a pure compound. Cs_7O is a bronze-colored liquid at room temperature.

Selected characteristics of the suboxides of rubidium and cesium are summarized in Table 4.4.

The stoichiometries deduced from the phase diagrams do not permit us to establish a simple relationship between the different phases. However, the analysis of single crystal X-ray diffraction determinations of the different phases reveals the metallic atoms appearing in the solid ordered in arrangements that can be interpreted as metal cluster occluding oxygen atoms. Schemes of structures of these phases are reproduced in Figs. 4.6 and 4.7 for the rubidium and cesium compounds respectively. In all cases the metal atoms forming distorted octahedra are coordinated to an oxygen atom located in the center. In the case of the rubidium derivatives two of these octahedra share a face bearing the

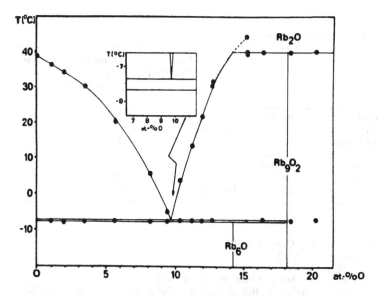

Fig. 4.4. Phase diagram of the system Rb/Rb_2O. Reproduced with permission from Simon A (1973) Z. Anorg. Allg. Chem. 395:301

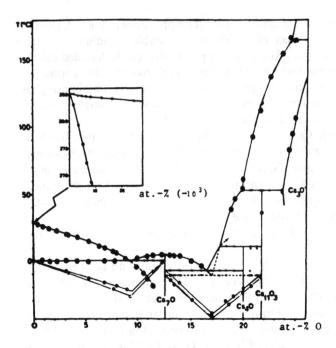

Fig. 4.5. Phase diagram of the system Cs/CsO$_2$. Reproduced with permission from Simon A (1973) Z. Anorg. Allg. Chem. 395:301

Table 4.4. Suboxides of rubidium and cesium

Compound	Structure	Color	Fusion or Disc.Temp. (°C)	Molar Volumes (cm^{-3} mol^{-1}) (extrapolated at 0 K)	Ref.
Rb$_9$O$_2$	Rb$_9$O$_2$-clusters	metallic copper-like platelets	40	387	1, 2
Rb$_6$O	Rb$_9$O$_2$Rb$_3$	brass-colored	− 8	278	1, 3
Cs$_{11}$O$_3$	Cs$_{11}$O$_3$-clusters	permanganat-colored	52	562	1, 3
Cs$_4$O	Cs$_{11}$O$_3$Cs	—	ca. 12	210	1, 4
Cs$_7$O	Cs$_{11}$O$_3$Cs$_{10}$	brass-colored liquid	4	418	1, 5

References

1 Simon A (1973) Z. Anorg. Allg. Chem. 395:301
2 Simon A (1971) Naturwissen. 58:623
3 Simon A, Deiseroth HJ (1976) Rev. Chim. Minér. 13:98
4 Simon A, Westerbeek E (1972) Angew. Chem. Int. Ed. 11:1105
5 Simon A, Deiseroth HJ, Westerbeek E, Hillenkötther B. (1976) Z. Anorg. All. Chem. 423:203
6 Simon A (1971) Naturwissen. 58:622
7 Simon A (1976) Z. Anorg. All. Chem. 422:208

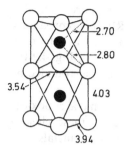

Fig. 4.6. Structure of Rb_9O_2

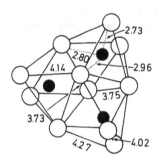

Fig. 4.7. Structure of $Cs_{11}O_3$

cluster Rb_9O_2. In cesium derivatives, trigonal clusters of $Cs_{11}O_3$ formed by three-equivalent, face-sharing octahedra are observed.

As illustrated in Fig. 4.6 in the cluster Rb_9O_2 there are two classes of rubidium atoms. The internuclear distance between the three atoms in the equatorial plane coordinated each to two oxygens is about 3.54 Å. Peripheral atoms instead show internuclear distances of 3.93–3.94 Å. Octahedra of metal atoms are in this structure elongated along the three-fold axis, therefore the distances between equatorial and peripheral atoms are in the range 4.00–4.95 Å. This distortion also results in different Rb–O distances for equatorial and peripheral atoms.

Still stronger is the differentiation of distances in the structure of cluster $Cs_{11}O_3$ (Fig. 4.7). There are three classes of cesium atoms which are coordinated to one, two, and three oxygen atoms respectively. The intermetallic distances in this cluster are between 3.71 Å for three-coordinated atoms and 4.27 Å for the peripheral one-coordinated ones.

In spite of the apparent diversity of the phases only the structures containing oxygen mentioned above are observed. In the structures of the compounds Rb_9O_2 and $Cs_{11}O_3$ the characteristic clusters are close-packed as an arrangement of large spheres. The other phases are formed by a combination of the suboxide with the stoichiometry corresponding to the cluster and the pure metal:

$$2Rb_6O = [Rb_9O_2]Rb_3; \quad 3Cs_7O = [Cs_{11}O_3]Cs_{10};$$
$$3Cs_4O = [Cs_{11}O_3]Cs.$$

In the metal-rich suboxides the metal is concentrated in purely metallic regions. In Fig. 4.8, a scheme of the structure of the phase Rb_6O is reproduced. There the Rb_9O_2 clusters are arranged in layers that alternate with others of metallic rubidium atoms. In the case of the Cs_7O as illustrated in Fig. 4.9 the $Cs_{11}O_3$ clusters are arranged to form columns which are in turn surrounded by purely metallic cesium atoms.

In contrast to transition-metal molecular clusters, the alkali-metal suboxides are stable only in the solid state. As described in Table 4.4, these clusters decompose at temperatures rather below the melting point of the metals. The stability of these species appears to be relatively precarious. It is very probable that the stabilization of this class of extreme electron-deficient compounds is possible only at relatively low temperature and in strong reducing media such as the alkali-metals rubidium or cesium.

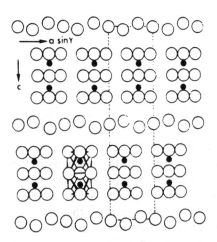

Fig. 4.8. Structure of the phase Rb_6O. Reproduced with permission from Simon A, Deiseroth HJ (1976) Rev. Chim. Minér. 13:98

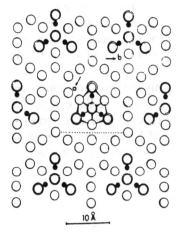

Fig. 4.9. Structure of the phase Cs_7O. Reproduced with permission from Simon A (1976) Z. Anorg. All. Chem. 422:208

Structures analyzed above as metal clusters could be in principle also considered as alkali metals strongly contaminated by oxygen, leading thus to a quasi-infinite crystalline arrangement. Their classification as a cluster species therefore needs further evidence.

Atomic and molecular volumina can be obtained from accurate determinations of the lattice constants of the corresponding solids. In Table 4.4, the molecular volumina determined experimentally for the different phases of the alkali-metal suboxide are also given. Considering that the atomic volumina of rubidium and cesium are 54.7 and 69.4 cm^3 mol^{-1} respectively, it is easy to see that experimental values for the phases Rb$_9$O and Cs$_7$O agree with those calculated by assuming for the phases, the composition [Rb$_9$O$_3$]Rb$_3$ and [Cs$_{11}$O$_3$]Cs$_{10}$ ((559.6 cm^3 mol^{-1} and 1267.4 cm^3 mol^{-1} respectively).

Controlled oxidation of mixtures of both rubidium and cesium metals leads under equilibrium conditions only to the cesium clusters Cs$_{11}$O$_3$ in the form of the compounds [Cs$_{11}$O$_3$]Cs$_{10-x}$Rb$_x$, [Cs$_{11}$O$_3$]Rb$_{7-x}$Cs$_x$, or [Cs$_{11}$O$_3$]Cs$_{1-x}$Rb$_x$. Only after the consumption of all the cesium can the rubidium partially replace the cesium in the clusters. This feature indicates the presence of equilibria between actual chemical species with different relative thermodynamic stabilities. The ionization energy of cesium is lower than that of rubidium. This feature appears to be determinant in the relative stability of the suboxides, higher for Cs$_{11}$O$_3$ than for Rb$_9$O$_2$, deduced from the experiments discussed above.

In Fig. 4.10 the photoelectron spectra of the phases [Cs$_{11}$O$_3$]Cs$_{10}$ and Cs$_{11}$O$_3$ beside that of metallic cesium are reproduced. There are many features

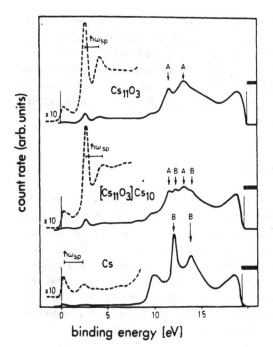

Fig. 4.10. Photoelectron spectra of the phases [Cs$_{11}$O$_3$]Cs$_{10}$ and Cs$_{11}$O$_3$. Reproduced with permission from Simon A (1977) In: Rheingold AL (ed) Homoatomic rings, chains and macromolecules of main-group elements. Amsterdam, Elsevier, p 117

that can be considered to support the chemical description of the cluster above:
1. The spectrum of the phase $[Cs_{11}O_3]Cs_{10}$ corresponds to the superposition of
the spectra of both $Cs_{11}O_3$ and Cs. 2. All samples produce photoelectrons near
the Fermi level as expected for metallic substances. 3. Suboxide spectra always
show a pronounced peak at 2.7 eV corresponding to the $O(2p)$ level. The shape
as well as the position of this band point to the presence of nearly free O^{2-} ions.
4. The spin-orbit split levels $Cs(5p)$ observed in $Cs_{11}O_3$, which are about 0.5 eV
shifted to smaller binding energy respect to those in the pure metal, are also
observed in the spectra of the phase Cs_7O where they coexist with other
doublets similar to that observed for the pure metal.

Chemical Bonding in Alkali-Metal Suboxides. Assuming a very simple model in
which the alkali-metal atoms have a charge $+1$ and the oxygen -2 the species
Rb_9O_2 and $Cs_{11}O_3$ should have a charge $+5$. Such a description agrees with
the ionic character of about 85% estimated for this bonding from the electronegativities according to Pauling. These clusters can be therefore formulated
as $(Rb^+)_9(O^{2-})_2(e^-)_5$ and $(Cs^+)_{11}(O^{2-})_3(e^-)_5$. The five electrons have the role
of compensating the positive charge as well as of providing the metallic bonding
between the clusters and pure metal atoms in the structures. Nonetheless,
calculations of stability of the species $[Rb_9O_2]^{n+}$ show they have an energy
minimum when n is 2.3. For n > 3.2 these species are unstable and the cluster
should lose a rubidium atom. Actually, although a great part of the stability of
the clusters is due to the existence of M–O bonds, the relatively weak M–M
bonds also appear to be essential, especially for neutralizing the excess of
positive charge.

Such a description of the bonding is consistent with the properties of the
alkali-metal suboxide described above. The metallic properties and the concentration of negative charge in the oxygen atom detected in the photoelectron
spectra agree well with the formulas proposed above. From such a point of view
the octahedral M_6O units could be considered as an inverse coordination
compound in which the central atom is holding six ligands in an octohedric
arrangement.

4.2.3 Boron Hydrides and Carboranes

From the point of view of the formation of aggregates, boron constitutes
a singular case among main group elements. With three valence electrons and
a relatively high electronegativity, boron should tend to form three covalent
bonds. However the products would be electron-deficient species. Boron chemistry may be understood by considering the way by which such electron
deficiency is remedied. Mononuclear three-coordinated species are stable only
with π-donor ligands. Boron halides, esters, and amides are examples in which
the electron deficiency is counteracted by π-donation. In such compounds,
boron-ligand bonds therefore have bond orders greater than one.

A second way of stabilization is the formation of Lewis acid-base adducts. Indeed, three-coordinated boron species are typical Lewis acids. Tetracoordinated boron in the adducts has a stable inert-gas configuration.

A third manner of avoiding electron deficiency is the formation of multicenter bonds. Diborane B_2H_6, whose bonding description was discussed in Chapter 1, is the first compound of a big family in which a plethora of cluster compounds are found. The chemistry of boron-hydrogen species initiated at the beginning of the century by Stock has seen great development in the second half of the century reaching a dimension that justifies its study separately as a branch of chemistry. Here only selected aspects of the chemistry of boranes and carboranes related to its contribution to the knowledge of cluster chemistry will be refered to.

Structures of Boranes and Carboranes. As mentioned above, elementary boron occurs in its most stable modification as B_{12} icosahedric units connected to each other by six boron-boron two-center bonds. Boron-boron distances within the icosahedron range from 1.73 to 1.79 Å. From a structural point of view, boranes can be systematized by considering them as derivatives of the elementary icosahedron, but built by B-H units. Some examples of geometrical forms that can be derived from icosahedron were shown in Fig. 2.27.

Although neutral boranes with formula B_nH_n are not known, this stoichiometry is characteristic of anionic species which are thermodynamically as well as kinetically notably stable (vide infra).

The structures of known neutral boranes may be classified in two analogous series of compounds with formulas B_nH_{n+4} and B_nH_{n+6}. Following the systematization mentioned above these two series may be considered to be formed by degradation of the corresponding *closo*-structures B_nH_n. The unsaturation arising from the break down of boron-boron bonds in the close polyhedra is compensated by the formation of B-H-B three-center bonds similar to those described for the diborane as well as additional B-H bonds. These three-center bonds lie flanking the open face of the incomplete polyhedron. According to the extent of the degradation of parent polyhedron, semi-closed arrays called *nido* structures or open arrangements known as *arachno* structures with general formulas B_nH_{n+4} and B_nH_{n+6} respectively are obtained. Figs. 4.11 and 4.12 illustrate schematically some *nido-* and *arachno*-geometries respectively.

Syntheses and Properties of Boranes. Some properties of selected boranes are described in Table 4.5.

Most boranes are prepared by thermal decomposition of diborane under predetermined conditions. B_2H_6 is prepared by reduction of boron halides with metal hydrides as $NaBH_4$, $LiAlH_4$, or LiH.

$$6MH + 8BF_3 \rightarrow 6M^+[BF_4]^- + B_2H_6$$

Boranes are, in general, unstable in strong donor media. The reaction frequently occurs with degradation of the borane clusters through displacement

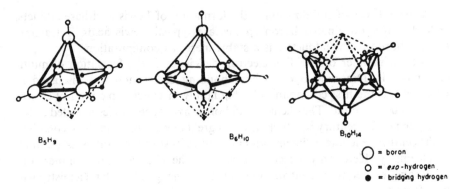

Fig. 4.11. The *nido*-boranes B_5H_9, B_6H_{10} and $B_{10}H_{14}$ showing the fundamental polyhedra. Reproduced with permission from Wade K (1976) Adv. Inorg. Chem. Rodiochem. 18 : 1

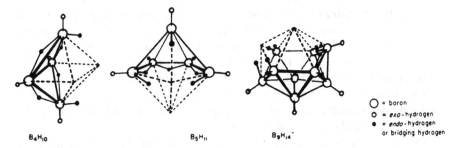

Fig. 4.12. The *arachno* boranes B_4H_{10}, B_5H_{11}, and $[B_9H_{14}]^-$ showing the fundamental polyhdera. Reproduced with permission from Wade K (1976) Adv. Inorg. Chem. Radiochem. 18 : 1

of borane fragments from three-center bonds to form Lewis acid-base adducts. *arachno*-Boranes are specially sensitive to this class of reactions.

$$B_2H_6 + 2NMe_3 \rightarrow 2BH_3NMe_3$$

$$B_4H_{10} + 2NMe_3 \rightarrow 2BH_3NMe_3 + B_3H_7NMe_3$$

$$B_6H_{12} + PMe_3 \rightarrow BH_3PMe_3 + B_5H_9$$

The capacity of boranes for accepting electron density without cluster disruption increases with increasing nuclearity. Thus, *nido*-boranes with more multicenter boron bonds and fewer B-H-B bonds than the corresponding arachno-derivatives are able to form adducts with different Lewis bases.

$$B_5H_9 + 2L \rightarrow B_5H_9L_2 \qquad L = NMe_3, PMe_3, etc..$$

Similar reactions are also known among other for B_6H_{10} and $B_{10}H_{14}$.

In basic media, it is also possible to induce a proton transference with formation of borane anions.

$$B_{10}H_{14} + 2NEt_3 \rightarrow [Et_3NH]_2^+[B_{10}H_{10}]^{2-} + H_2$$

Table 4.5. Properties of selected neutral binary boranes

Compounds	Properties
nido-B_5H_9	Pentaborane (9). Structure: square pyramid (Fig. 4.11). The most stable among lower boranes. Colorless liquid, mp $-46.8\,°C$, bp $60.0\,°C$. Preparation: Pyrolysis of B_2H_6 and H_2 at $200-250\,°C$. It reacts slowly with water ($H_2BO_3 + H_2$). With bases (NR_3, PR_3) it forms adducts $B_5H_9L_2$.
nido-B_6H_{10}	Hexaborane (10). Structure: pentagonal pyramid (Fig. 4.11). Rather stable when pure. Colourless liquid, mp $-62.3\,°C$, bp $108\,°C$. Preparation: $LiB_5H_8 + B_2H_6$. With alkali metal hydrides or alkyls it yields salts of $B_6H_9^-$. By protonation the cationic species nido-$B_6H_{11}^+$ are obtained.
nido-$B_{10}H_{14}$	Decaborane (14). Polyhedral structure determined by X-ray and neutron diffraction studies (Fig. 4.11). Air stable crystalline solid, mp $99.6\,°C$ (one of the first boranes isolated by A. Stock). Preparation: Thermal decomposition of B_2H_6. Product reacts slowly with water and can be titrated with acids ($pK_a = 3.5$ in acetonitrile; 2.7 in ethanol). Deprotonation by strong bases (OH^-, NH_3) occurs at a bridged site.
arachno-B_4H_{10}	Tetraborane (10). Structure determined X-ray crystallography, gas-phase electron diffraction and microwave spectroscopy shows "butterfly" shape (C_{2v}) (Fig. 4.12). Colourless volatile liquid mp $-120\,°C$, bp $18\,°C$. Thermically rather unstable. It reacts slowly with water. With KH at $78\,°C$ it yields $B_4H_9^-$.

References

Barton L (1982) Top. Current Chem. 100:169
Greenwood NN (1989). In: Roesky HW (ed) Ring clusters and polymers of Main Group and Transition Elements. Elsevier, Amsterdam, Oxford, New York, Tokyo, p. 49

This acidic behavior produced by transformation of a hydrogen bridge bonding into a boron-boron bond with loss of one proton occurs preferentially in nido-compounds.

The correlation between the capacity of accumulating negative charge and the degree of homonuclear bonding in boranes agrees with the existence of the family of boranes anions $[B_nH_n]^{2-}$ (n = 6–12) which are closed polyhedra with BH vertexes. In these clusters the stabilization is produced exclusively by the formation of multicenter boron-boron bonds. Borane anions have, in general, a remarkable stability. Selected examples of these boron clusters are described in Table 4.6.

Carboranes. The accumulation of charge observed in borane anions discussed above may also be afforded by changing some boron atoms by other elements with more valence electrons than boron. The best known examples of this kind of isoelectronic compound are the carboranes. The most important series of carboranes is that with molecular formulas $C_2B_{n-2}H_n$, the dicarba-*closo*-boranes, where n is any integer from 5 to 12. These neutral species are isoelectronic with the borane anions $B_nH_n^{2-}$. The structures of these compounds are closed deltahedra with terminal hydrogen atoms attached to each carbon or

Table 4.6. Properties of selected *closo*-borane anions.

Anion	Structure	Other properties
$B_6H_6^{2-}$	*Closo*-octahedron. Interatomic distances (Å) in $K_2B_6H_6$: B–B = 1.72 Å. B–H = 1.07 Å.	Preparation: Thermal decomposition of NaB_3H_8 at 160 °C. Acidification to pH 5 in water leads to monoanion $B_6H_7^-$.
$B_8H_8^{2-}$	*Closo*-distorted dodecahedron. Interatomic distances (Å) increase with increasing cluster-connectivity: 1.56 between 4-connected though 1.93 between two 5-connected vertices.	Preparation: Air oxidation of *closo*-$B_9H_9^{2-}$ salts in 1,2 dimethoxy ethane at 70 °C. Cs-salt stable up to 600 °C. Colorless compounds when pure. In solution it is oxidized to the burgundy red radical anion $B_8H_8^-$.
$B_9H_9^{2-}$	Trigonal prismatic structure. Interatomic distances depend on connectivity. ^{11}B-NMR (D_{3h} structure) remains unmodified up to 200 °C i.e. high barrier for fluxional cluster rearrangement.	Preparation: Pyrolysis of $Na_2[B_{10}H_{12}]$ at 240 °C. Cs-salt stable up to 600 °C in sealed tube. Solid stable to air up to 575 °C. In solution it undergoes stepwise air oxidation to $B_8H_9^-$, $B_8H_8^{2-}$, $B_7H_7^{2-}$, and $B_6H_6^{2-}$. *Closo*-$B_9H_9^{2-}$ is stable in neutral and alkaline aqueous solution but it is degraded by acid.
$B_{10}H_{10}^{2-}$	Bicapped square antiprismatic structure. B–B distances (Å) range from 1.69 for (B(apix)-B (antiprism)) bonds to 1.81 for B–B bonds in equatorial triangular-faces.	Preparation: $B_{10}H_{14}$ + 2NEt$_3$, reflux in xylene. Colorless or pale-yellow salts. They are amongst the most stable polyhedral borane compounds.
$B_{11}H_{11}^{2-}$	Probable structure, octadecahedron, is not definitively stabilised. Dianion is highly fluxional in solution.	Preparation: Pyrolysis of $Cs_2B_{10}H_{13}$ at 250 °C. It is stable indefinitely at 400 °C but disproportionates to $Cs_2B_{10}H_{10}$ + $Cs_2B_{12}H_{12}$ at 600 °C. More stable to hydrolysis than smaller $B_nH_n^{2-}$ species. It remains unaffected by neutral or aqueous alkali solutions as well as by 2N HCl.
$B_{12}H_{12}^{2-}$	Icosahedron with B–B distances of 1.78 Å	Preparation: $2Et_3NBH_3$ + $B_{10}H_{14}$ at 900 °C or thermolysis of Et_4NBH_4. This ion is the most stable of all *closo*-$B_nH_n^{2-}$. Cs-salt being stable at least up to 600 °C. The hydrated free acid $[H_3O]_2[B_{12}H_{12}]\cdot4H_2O$ is known and it is apparently slightly stronger than H_2SO_4.

References

See Table 4.5.

boron. A series of structures of dicarba-closo-boranes are illustrated in Fig. 4.13. The characteristics of some selected dicarboranes are described in Table 4.7.

Another important series of carboranes are the double dianions $[C_2B_{n-2}H_n]^{2-}$. These species which have n cage atoms and (n + 2) pairs of skeleton electrons adopt *nido*-structures in which one corner of the polyhedron is left vacant as shown in the scheme of the dicarba-*nido*-hexaborane (8) illustrated in Fig. 4.14. Dicarba-*closo*-boranes in general are unreactive in

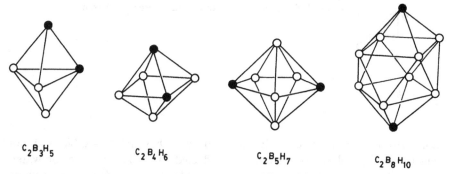

$C_2B_3H_5$ $C_2B_4H_6$ $C_2B_5H_7$ $C_2B_8H_{10}$

Fig. 4.13. Selected examples of dicarba-*closo*-borane skeletons with different nuclearities. Reproduced with permission from Corbett JD (1976) Prog. Inorg. Chem. 21:129

Table 4.7. Properties of dicarboranes

Compound	Properties
$C_2B_3H_5$	1.5-Dicarba-*closo*-pentaborane*. Mp $-126\,°C$; bp $-37\,°C$. Air stable and inert to acetone, water, amines, etc. at room temp.; at high temperature isomerization to the most stable isomer. Isomers may be identified by ^{11}B-NMR. Preparation as a mixture of small carboranes by thermal decomposition of *nido*-2,3-$C_2B_4H_8$ (obtained at 450–460 °C from pentaborane (9) and acetylene).
$C_2B_4H_6$	1,6-dicarba-*closo*-hexaborane*. Mp $-32°$, bp 27.7 °C. Reactivity and preparation are similar to $C_2B_3H_5$.
$C_2B_5H_7$	Pentagonal bipyramidal 2,4-dicarba-*closo*-heptaborane*. ^{11}B-NMR spectrum shows three different types of boron atoms. Reactivity similar to lower dicarba-*closo*-boranes.
$C_2B_6H_8$	1,7-dicarba-*closo*-octaborane*. Dodecahedron. Preparation by thermal decomposition of $C_2B_7H_{13}$ (200 °C, Ph$_2$O; yield ca. 30%).
$C_2B_7H_9$	1,6-dicarba-*closo*-nonaborane*. Tricapped trigonal prism. Preparation similar to $C_2B_6H_8$ with ca. 70% yield.
$C_2B_8H_{10}$	1,10-dicarba-*closo*-dodecaborane*. Bicapped Archimedeam antiprism. Preparation similar to $C_2B_6H_8$ with ca. 30% yield.
$C_2B_9H_{11}$	2,3-dicarba-*closo*-undecaborane*. Octahedron. Preparation from $C_2B_{10}H_{12}$ via the formation of *nido*-$C_2B_9H_{13}$ by nucleophilic degradation followed of thermal decomposition. Structure established by X-ray diffraction analysis of the C,C′-dimethyl derivative. Oxidation of dicarba-undecaborane by $Cr_2O_7^{2-}$ in acetic acid leads to dicarba-*arachno*-nonaborane. $C_2B_7B_{13}$ is often used for the synthesis of lower *closo*-corboranes.
$C_2H_{10}H_{12}$	1,2-dicarba-*closo*-dodecaborane*. Icosahedron. Unpleasant smelling, volatile solid mp 295 °C. It is transformed in the 1,7-$C_2B_{10}H_{12}$ meta-isomer at 470 °C. At 615 °C the *para*-isomer may be obtained. Synthesis: from $B_{10}H_4$ by reaction with Et$_2$S in *n*-propylether to give $B_{10}H_{12}$ (Et$_2$S)$_2$ followed by reaction with acetylene at 85 °C.

* Thermally most stable isomer.

References

Beall H, Onak T (1975) In: Muetterties EL (ed) Boron hydride chemistry. Academic Press, New York
Dunks GB, Hawthorne MF (1973) Acc. Chem. Res. 6:174
Grimes RN (1971) Carboranes. Academic Press, New York

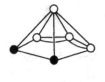

Fig. 4.14. Schematic structure of 1,2 dicarba-*nido*-hexaborane (8)

$C_2B_4H_8$

comparison with the neutral boron hydrides. They are air-stable and not sensitive to hydrolysis, being indeed unaffected by boiling ethanol or water. Strong oxidants such as permanganate, hypochlorite, and hydrogen peroxide as well as reducing agents do not affect the carborane skeleton. They are however susceptible to attack by nucleophiles such as methoxide ion or hydrazine which can abstract a boron atom leading to *nido*-anions.

$[Me_3NH]^+ [C_2B_9H_{13}]^-$

$[Me_3NH]^+Cl^-$

$1,2-C_2B_{10}H_{12}$ $\xrightarrow[\text{MeOH}]{\text{OMe}^-}$ $[(3)-1,2-C_2B_9H_{12}]^-$

NaH/THF

$(Na^+)_2 [C_2B_9H_{12}]^{2-}$

nido-Anions $[C_2B_{n-2}H_n]^{2-}$ have been found to form a wide range of coordination complexes with metal ions in which the metal occupies the position of the lost boron atom regenerating thus a closed polyhedron. Figure 4.15 illustrates the structures of metal complexes of the anion $[C_2B_9H_{11}]^{2-}$ also known as "dicarbollide". The bis(carbollyl) complexes with some transition metal cations are considerably more thermally stable than their bis (cyclopentadienyl) analogues. Although in these complexes metal-ligand interactions are similar to those in cyclopentadienyl complexes, it is not essential that the element completing the polyhedron is a transition metal. That can be appreciated in Fig. 4.16 which shows similar complexes of the anion $[C_2B_9H_{11}]^{2-}$ with main group metal cations. It is worth noting that, in the beryllium derivative, the cation coordination sphere is completed by a donor ligand while in the group 14 derivatives this position is occupied by a non-bonding pair of electrons which practically does not interact with the carborane cluster. This feature is also observed in the naked clusters of post-transition metals that are discussed in other Section of this Chapter.

Bonding in Boranes and Carboranes. The relationship between valence electrons and valence orbitals in boron is precisely the most suitable for the formation of

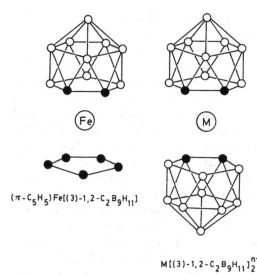

Fig. 4.15. Metal complexes of the amion "dicarbollide" $[C_2B_9H_{11}]^{2-}$. The vacant corner is regarded as number 3

(Fe)

(M)

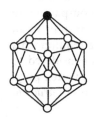

$(\pi\text{-}C_5H_5)Fe[(3)\text{-}1,2\text{-}C_2B_9H_{11}]$

$M[(3)\text{-}1,2\text{-}C_2B_9H_{11}]_2^{n-}$

$M = Fe(II), Co(III)$

$Ni(III), Ni(IV)$

$C_2B_9H_{11}M$

$M = Ge, Sn, Pb;$
or $M = Be \leftarrow NMe_3$

Fig. 4.16. Main group metal complexes of $[C_2B_9H_{11}]^{2-}$. The vacant corner is regarded as number 3

clusters specially in absence of good π-donor ligands. The formation of aggregates is the best way of avoiding electron deficiency and for reaching an inert-gas electron configuration.

A qualitative discussion of the theoretical basis for the description of bonding in boranes and carboranes was carried out in Chapter 2. There are three orbitals per boron atom disposable for the formation of the cluster skeleton. Thus, for a given member of the series of the polyhedra derived from the icosahedron illustrated in Fig. 2.27, the linear combination of these boron orbitals gives raise to a set of $(n + 1)$ low-lying skeleton molecular orbitals, n being the nuclearity of the polyhedron. Since each boron atom in boranes distracts one electron in the formation of one two-center/two electron "exo"-bond with hydrogen, there are only two electrons per boron atom available for skeleton bonding. Among binary boranes, only the *closo* borane anions $[B_nH_n]^{2-}$ therefore have enough electrons for filling the available skeleton-bonding molecular orbitals. Isoelectronic species with these borane dianions as the dicarboranes discussed in the preceding Section also fulfill the electron requirements of closed polyhedra. Further examples of isoelectronic relationships are provided by the formation of complexes of the *nido* dianions $[C_2B_{n-2}H_n]^{2-}$ with both transition and main group metals as those illustrated above in Figs. 4.15 and 4.16. Selected examples of correlations between electron counting and polyhedron geometries observed for *closo*-boranes and carboranes are displayed in Table 4.8. Chemical properties of this class of compounds with *closo*-structures, which, as described above, are much more stable than equivalent derivatives with open structures, can be understood by their closed electron configurations. HOMOs and LUMOs in such structures are located inside a closed polyhedron; moreover, on the outside only relatively inert B-H and C-H groups with two-electron/two-center bonds are found. There, bonding molecular orbitals lie lower and antibonding molecular orbitals higher than those arising from skeletal bonding (see Chapter 2). These features agree well with the relatively high inertness of *closo*-species.

Open structures, such as *nido* and *arachno* compounds, may also be described by the same bonding concept. *nido*-Compounds B_nH_{n+4}, or in the case of the dicarboranes the anions $[C_2B_{n-2}H_n]^{2-}$, have one electron-pair more than required for filling the $(n + 1)$ low-lying skeletal molecular orbitals in an n-vertex polyhedron. However this number of electrons meets the requirements of a polyhedron with $(n + 1)$ vertices. *nido*-Species may be then considered to consist of an atomic arrangement with the geometry of a $(n + 1)$-vertex polyhedron in which one corner is left vacant. Analogously the structures of *arachno*

Table 4.8. Electron counting in *closo*-boranes and carboranes

Compound	No. of skeletal bond electron pairs	Fundamental polyhedron
$C_2B_3H_5$	6	Trigonal bipyramid
$C_6H_6^{2-}$ $C_2B_4H_6$	7	Octahedron
$B_7H_7^{2-}$ $C_2B_5H_7$	8	Pentagonal bipyramid
$B_8H_8^{2-}$ $C_2B_6H_8$	9	Dodecahedron
$B_9H_9^{2-}$ $C_2B_7H_9$	10	Tricapped trigonal prism
$B_{10}H_{10}^{2-}$ $CB_9H_{10}^-$ $C_2B_8H_{10}$	11	Bicapped Archimedean antiprism
$B_{11}H_{11}^{2-}$ $CB_{10}H_{11}^-$	12	Octadecahedron
$CB_{11}H_{12}^-$ $C_2B_{10}H_{12}$	13	Icosahedron

Table 4.9. Electron counting in *nido*-boranes and carboranes

Compounds	No. of skeletal bond electron pairs	Fundamental polyhedron[a]
$B_4H_7^-$	6	Trigonal pyramid
B_5H_9 $B_5H_8^-$ $C_2B_2H_7$	7	Octahedron
B_6H_{10} $B_6H_9^-$ CB_5H_9 $C_2B_4H_8$ $C_3B_3H_7$ $C_4B_2H_6$	8	Pentagonal bipyramid
B_8H_{12} $C_2B_6H_{10}$	10	Tricapped trigonal prism
$B_9H_{12}^-$ $C_2B_7H_{11}$	11	Bicapped Archimedean antiprism
$B_{10}H_{14}$ $B_{10}H_{13}^-$ CB_9H_{13} $C_2B_8H_{12}$	12	Octadecahedron
$B_{11}H_{15}$ $B_{11}H_{14}^-$ $CB_{10}H_{13}^-$ $C_2B_9H_{11}^{2-}$ $C_4B_7H_{11}$	13	Icosahedron

[a] n-Skeletal atoms in *nido*-species define all but one of the (n + 1) vertices of the corresponding fundamental polyhedron. See Fig. 4.11.

species correspond to those of the (n + 2)-vertex polyhedra in which two corners are vacant. Tables 4.9 and 4.10 describe selected electron-counting examples in *nido* and *arachno*-species respectively.

The reactivity of species with open structures is certainly relatively much higher than that of the equivalent closed derivatives since open polyhedron faces make the interaction of other chemical agents with cluster HOMOs and LUMOs much more probable. That is directly related to the formation of Lewis acid-base *nido*-carborane complexes described above. Furthermore the reactivity of open-cluster species is also enhanced by the presence of three-center B-H-B bonds which originate occupied and unoccupied molecular orbitals with relatively high and low energy respectively. This class of compounds is, in general, therefore an excellent intermediate for synthesis.

4.3 Cage Compounds of Non-Metal Elements

Elements of the 14 and further groups have valence electrons enough for reaching an inert-gas electron configuration by building only two-electrons/

Table 4.10. Electron counting in *arachno*-boranes and carboranes

Compounds	No. of skeletal bond electron pairs	Fundamental polyhedra[a]
$B_3H_8^-$	6	Trigonal bipyramid
B_4H_{10} $B_4H_9^-$	7	Octahedron
B_5H_{11} $B_5H_{10}^-$ $B_5H_9^{2-}$ $C_2B_3H_7^{2-}$	8	Pentagonal bipyramid
B_6H_{12} $B_6H_{11}^-$	9	Dodecahedron
$B_7H_{12}^-$	10	Tricapped trigonal prism
B_8H_{14} $C_2B_6H_8^{4-}$	11	Bicapped Archimedean antiprism
B_9H_{15} $B_9H_{14}^-$ $B_9H_{13}^{2-}$ $C_2B_7H_{13}$	12	Octadecahedron
$B_{10}H_{15}^-$ $B_{10}H_{14}^{2-}$ $C_2B_8H_{10}^{4-}$	13	Icosahedron

[a] n-Skeletal atoms in *arachno*-species define all but two of the (n + 2) vertices of the corresponding fundamental polyhedron. See Fig. 4.12.

two-center bonds. The driving force for forming multicenter bonds, by which a relatively low number of electrons can saturate the valence orbitals of the components, does not exist in the case of the electron-rich non-metal elements. The formation of three-dimensional structures occurs therefore mostly by two-electron/two-center bonds leading to electron-precise species. This makes a big difference between these compounds and the cluster species formed by electron-deficient elements analyzed above.

Cage compounds are well known in organic chemistry. There are a great number of homo and heteronuclear compounds with the most diverse structures and peculiarities. However the extension of this chemistry exceeds by far the scope of this book. Therefore the discussion here will be limited to some fundamental aspects of the chemistry of the third allotropic form of carbon, the fullerenes, recently discovered.

4.3.1 Molecular Carbon Clusters. The Fullerenes

The chemical species with spherically shaped structures are found only seldom in the chemistry of main group elements. Some examples of this type of compounds are the boranes and carboranes described above and some post-transitional deltahedral clusters to be described later in this Chapter.

In this section, some aspects of the chemistry of a family of carbon cluster species describing large polycyclic structures with nearly spherical symmetries called fullerenes are discussed.

A precocious example of cage structures resembling a soccerball is that of the heteronuclear anion $[P_{12}S_{12}N_{14}]^{6-}$ illustrated schematically in Fig. 4.17. This polycyclic P–N species shows a relatively high stability due, at least partially, to the enhanced electron delocalization associated with the formation of the cage. This kind of conglomerate, consisting of a polycyclic carbon network with relatively large interior void spaces, can be designated as a cluster but whose nature is certainly different from that of the transition metal clusters described in Chapter 2.

The Discovery of the Fullerenes. In experiments carried out at Rice University in 1985 associated with the presence of carbon in stars and space [Kroto HW, Heath JR, O'Brien SC, Curl RF, Smalley RE (1985) Nature 318:162], the C_{60} molecule was discovered. This species is one of the many carbon clusters which can be generated when a plasma of carbon vapor produced in the surface of graphite by laser irradiation is cooled by an inert gas jet. The detection of this type of carbon aggregate which appears to constitute a third allotropic and the first molecular form of carbon by mass spectroscopy is illustrated in Fig. 4.18.

Clusters produced by vaporization of graphite are indeed a big family. Thus, some mass spectrometric determinations have shown the presence of clusters with mass peaks up to 720. The experiments reveal, however, that the size distribution is not statistic and that only ions with even number of carbon atoms are detected. Moreover, as illustrated in Fig. 4.18, the intensity of the C_{60} peak is always the highest; its relative intensity depends however upon the experiment conditions. As shown in Fig. 4.19, in particular cases, conditions are found for

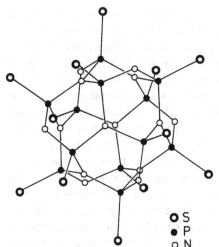

o S
● P
o N

Fig. 4.17. Molecular structure of the anion $[P_{12}S_{12}N_{14}]^{6-}$ in the compound $K_6[P_{12}S_{12}N_{14}] \cdot 8H_2O$ after Fluck E, Lang M, Horn F, Hädicke E, Sheldrick GE (1976) Z. Naturfosch. 31b:419. The anion has a spherical shape with a smaller diameter of ca. 4.56 Å

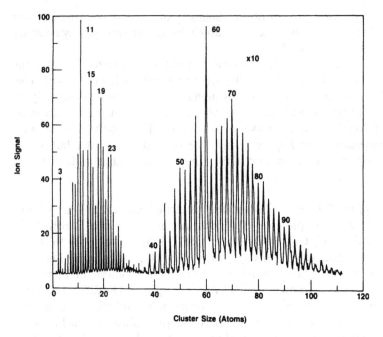

Fig. 4.18. Mass spectrum of carbon clusters produced by laser vaporization of graphite. Reprinted with permission from Rohlfing EA, Cox DM, Kaldor A (1984) J. Chem. Phys. 81:3322. These experiments in which clusters with 30–100 atoms were detected at the first time showed that only even-numbered clusters are stable

which the mass spectrum is totally dominated by the C_{60} peak. From these features it can be concluded that this species is specially stable to further nucleation.

The common nomenclature used for this class of carbon species arises from considerations about the very special structure that can be formed by a 60-carbon atom cluster. The arrangement of atoms defining the vertex of a truncated isohedron bears to the structure shown in Fig. 4.20 that reminds one of the seam pattern on a soccerball. This arrangement which satisfies the valence requirements of all the atoms can be considered as a hexagonal graphitic sheet which by incorporating pentagons is able to curl into a ball. Two factors are considered to be specially important for the stability of this structure, namely the high degree of electron delocalization reached by a perfect alternation of C–C double bonds and the uniform distribution of the atoms in the surface of a sphere. Such a geodesic atom array implies a symmetrical distribution of the strain related to the curvature of a normally planar, graphitic conjugated bond system. Considering the resemblance of this atom arrangement with the geodesic domes invented by the architect Buckminster Fuller, the compound C_{60} was named *buckminsterfullerene*. Since the IUPAC alternative for identifying these species is very long and difficult to derive, as observed in Fig. 4.21, the

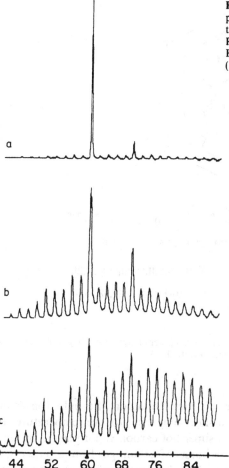

Fig. 4.19. Mass spectrum of carbon clusters produced by laser vaporization under conditions of increasing extent of clustering (a to c). Reproduced with permission from Kroto HW, Heath JR, O'Brien SC, Curl RF, Smalley RE (1985) Nature 318:162

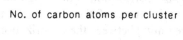

No. of carbon atoms per cluster

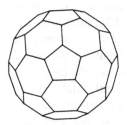

Fig. 4.20. Schema of the truncated icosahedral structure expected for the carbon backbone in the cluster C_{60}

name *fullerenes* has been adopted for the whole family of closed carbon cages with 12 pentagons and N ($N > 1$) hexagons in an sp^2 network.

Synthesis of Fullerenes. As mentioned above, the experiments leading to the first observation of the fullerenes involved the formation of carbon vapor plasma by

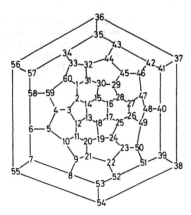

Hontriacontacyclo $\left[29.29.0.0^{2.14}.0^{3.12}.0^{4.59}.0^{5.10}.0^{6.58}.0^{7.55}.0^{8.53}.0^{9.21}.0^{11.20}.0^{13.19} \right.$

$.0^{15.30}.0^{16.28}.0^{17.25}.0^{19.24}.0^{22.52}.0^{23.50}.0^{26.49}.0^{27.47}.0^{29.45}.0^{32.44}.0^{33.60}.0^{34.57}.0^{35.43}$

$\left. .0^{36.56}.0^{37.41}.0^{38.54}.0^{39.51}.0^{40.48}.0^{42.46} \right]$ hexaconta-1,3,5(10),6,8,11,13(15),14,16,19,21,

23,25,27,29(45),30,32(44),33,35(43)36,38(54),39(51),40(45),41,46,49,52,55,57-

triacontaene

Fig. 4.21. IUPAC nomenclature for C_{60} derived from computer analysis according to P. Rose cited in Kroto HW, Allaf AW, Balm SP (1991) Chem. Rev. 91:1213

laser irradiation of a graphite surface. In fact the action of a pulsed laser on a spot on a rotating graphite disk in a short time can produce temperatures of more than 10 000 °C which generate a super hot carbon vapor plasma on that surface. By cooling such a plasma with bursts of helium gas from a pulsed gas nozzle the clustering reaction takes place. Size distribution of the conglomerates may be modulated by regulating the cooking time of the elemental carbon species on the surface adjusting the relative timing between the vaporization and carrier-gas pulses. In general, the clustering degree increases with increasing residence time of the cluster in the source. As shown in Fig. 4.22, conditions can be found for which the formation of clusters with 300 to 350 atoms may be optimized. This fact apparently contradicts the cluster distributions illustrated in Fig. 4.19 in which increasing chemical "cooking" time favored the formation of C_{60}. However it has been demonstrated that its apparently high relative concentration is due to the fact that this species is the sole survivor in a clustering process leading to a large number of giant clusters so big that they practically cannot be detected. As it will be discussed further below, C_{60} is indeed not always the principal product in the clustering of carbon vapor but is the most inert to further rearrangements.

Laser evaporation has been a very valuable method not only for discovering the fullerenes but also, as it will be treated with some more detail in a following section, for studying its formation mechanism. However, the method does not

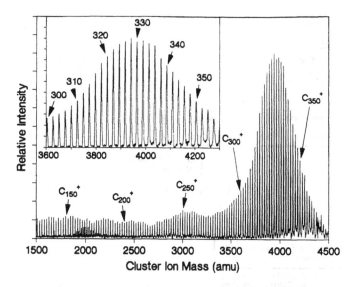

Fig. 4.22. Mass spectrum of carbon clusters produced by laser vaporization of graphite and detected by Fourier transform ion cyclotron resonance mass spectrometer. The experiment has been optimized to trap positive clusters in the C_{300}^{+} to C_{350}^{+} mass range. Reproduced with permission from Maruyama S, Anderson LR, Smalley RE (1990) Rev. Sci. Instrum. 61:3686

permit us to obtain the products in amount appropriate for conventional chemical studies. A qualitative improvement in this direction represents the method for the synthesis of some fullerenes which was found by Krätschmer W, Lamb LD, Fostiropoulos K and Huffman R (1990) Nature 347:354.

The method is based on the extraction of the fullerenes contained by the soot obtained by resistive heating (in an arc lamp) of pure graphite in an inert atmosphere. The soot containing a few per cent of C_{60} molecules is dispersed in benzene. The fullerene dissolves giving rise to wine-red to brown colored solutions. After separation from the soot, solvent evaporation, and mild drying, dark brown to black crystalline material is obtained. Alternatively, soot with a higher content of fullerene may be obtained by sublimation. According to analysis by mass spectroscopy as well as absorption spectra (vide infra), the product contains primarily buckmisterfullerene.

Development of this method leads to various separation protocols which permit us to obtain macroscopic quantities of buckminsterfullerene and the separation of other fullerenes and their isomers as well. Figure 4.23 illustrates an effective protocol for isolation and separation of fullerene. Typical chromatographic profiles for such separations can be observed in Fig. 4.24.

Mechanism of the Self-Assembly of Fullerenes. The spontaneous formation of fullerenes under the extreme high energetic conditions imposed by the carbon vapor plasma implies a considerable entropy loss which should be considered in

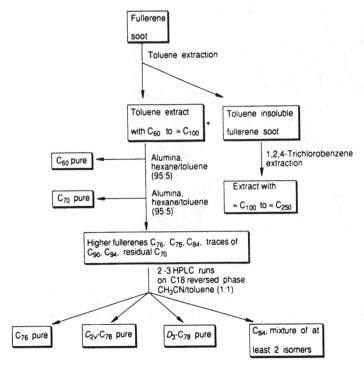

Fig. 4.23. A protocol for the isolation and separation of fullerenes. Reproduced with permission from Diederich F, Whetten RL (1992) Acc. Chem. Res. 25:119

any mechanistic explanation of the formation of this allotropic variety of carbon.

A plausible nucleation mechanism could be the gradual formation of carbon structures: For instance by formation of linear carbon chains which via the addition of other linear chains and the addition of small carbon radicals can grow to bear graphite-like polycyclic networks.

Open aromatic carbon networks constitute, however, relatively energy-rich, highly reactive species because of the great number of peripheral dangling bonds they have. Thus, for instance, a flat graphite-like sheet of 60 atoms would have at least 20 dangling bonds. Acquiring a certain curvature in order to diminish the number of dangling bonds by forming closed cages could be a way of stabilizing such a systems. Energy costs associated with the lower delocalization degree implied by loss of planarity should be compensated by the energy released produced by the elimination of the edge dangling bonds. This process is schematically illustrated by the diagram reproduced in Fig. 4.25 that illustrates a hypothetical carbon vapor nucleation scheme proposed for explaining the formation of concentric shell graphite microparticles observed experimentally by electron microscopy.

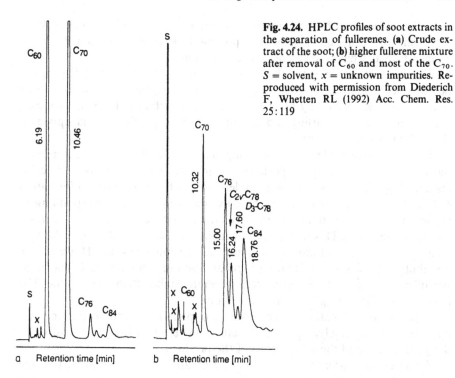

Fig. 4.24. HPLC profiles of soot extracts in the separation of fullerenes. (**a**) Crude extract of the soot; (**b**) higher fullerene mixture after removal of C_{60} and most of the C_{70}. S = solvent, x = unknown impurities. Reproduced with permission from Diederich F, Whetten RL (1992) Acc. Chem. Res. 25:119

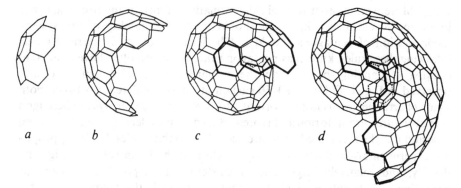

Fig. 4.25. Diagram respresenting a hypothetical carbon vapor nucleation scheme leading to concentric shells as those observed in graphite microparticles. Fullerene formation may occur by statistical closure at stage (c). Reproduced with permision from Kroto HW, McKay KG (1988) Nature 331:328

The reactivity of the intermediates in the nucleation process depends fundamentally on the number of nucleation sites available for binding other fragments. From this point of view, a closed structure is specially inert. The formation of a given aggregate from the great variety of them produced originally always requires the occurrence of a series of intra and internetwork rearrangements. Consequently such a process is strongly determined by the temperature at which the synthesis is carried out.

Experiments performed with mixtures of pure ^{12}C and pure ^{13}C graphite powders give rise to isotope distributions which clearly indicate that the formation of fullerenes occurs by aggregation of small carbon radicals as proposed above rather than via curling and closing performed graphite sheets ripped off the carbon starting material.

Some of the special features appearing in the species distribution observed in the experiments on laser evaporation of graphite mentioned above (see for instance Figs. 4.18 and 4.22) can be understood by the mechanism of formation of fullerene already discussed. In the mass spectrum region corresponding to clusters with more than 30 atoms, only ions with an even number of carbon atoms are observed. This feature can be interpreted as a result of the relatively high kinetic stability of cluster structures as the fullerenes are. However the distribution of such species is rather peculiar. Some of them, especially buckminsterfullerene, appear to be considerably more stable than the others. The fullerene-70 follows C_{60} in such distributions.

A plausible explanation of the existence of preferred carbon conglomerates may be found by analyzing their structures in terms of both the number of dangling bonds and the aromaticity of the products.

An accepted approach to the formation of fullerenes is to consider that the open graphitic sheets formed originally solely by hexagons may rearrange for incorporating pentagons in its structure, thus permitting the sheet to curl more easily. Moreover, in order to avoid the formation of locally antiaromatic cycles in the polycyclic network, which should lower the stability of the system, incorporation of pentagons should not be located adjacently. Arrangements following these rules, known as pentagon rules, do indeed often have a different number of dangling bonds than the graphitic arrays. That can be clearly observed in Fig. 4.26. Up to 20–30 carbon atoms, the difference between both kinds of arrangement is not important. However for bigger arrays the pentagon rule structures are notoriously favored. Buckminsterfullere is indeed a special case and it is the first closed structure according to these rules. Under appropriate experimental conditions–namely a relatively high temperature, high gas density, and relatively large time of clustering–it is possible to reach by annealing a distribution of clusters that agrees with the predictions of the pentagon rule. The general failure to produce macroscopic amounts of fullerenes by laser vaporization of graphite can be explained by considering experimental conditions. The carbon plasma generated on the graphite surface having an extremely high density expands rapidly to the surrounding gas carrier which normally is at room temperature. That results, on the one hand, in a very rapid growth rate and, on the other hand, in a rapid cooling that prevents the system from annealing and thus reaching more stable structures following the pentagon rules. Under such circumstances most clusters grow up well beyond 60 carbon atoms giving rise to a broad distribution of giant fullerenes.

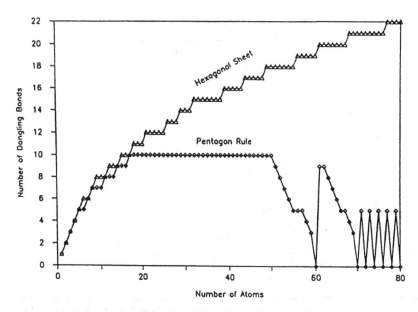

Fig. 4.26. Comparison of the number of dangling bonds in structures formed by graphite sheets containing only hexagons which obey the pentagon rules as a function of the number of carbon atoms in the sheet. Reproduced with permission from Haufler RE, Chai Y, Chibante LPF, Conceicao J, Jin C, Wang LS, Maruyama S, Smalley RE (1991) Res. Soc. Symp. Proc. 206:627

In the synthesis of fullerenes by resistive heating of graphite it is possible, as commented above, to obtain macroscopic amounts of fullerenes. In contrast to the laser vaporization already discussed, in this new way for preparing fullerenes both carbon vapor on the graphite surface and the rate of carbon clusters growth are rather low. Moreover, by adjusting the pressure of the helium buffer gas around the evaporating rod it is possible to control the rate of cluster growth and thus to optimize the cluster distribution.

Molecular Structure of Fullerenes. Obtaining macroscopical amounts of fullerenes has permitted us to study the properties of the solid. Thus, in the same work where the C_{60}-synthesis commented above is described, both the X-ray and the electron diffraction patterns reproduced in Fig. 4.27 are reported. Single-crystal X-ray diffraction analysis has not been possible mainly because large crystals do not exhibit long-range periodicity in all directions. However results as those described in Fig. 4.27 permit us to establish unequivocally that this new form of solid carbon is a molecular solid consisting of a compact array of round balls of about 10 Å in diameter, thus in agreement with the dimensions expected for Buckminsterfullerene. Estimated intermolecular distances as well as compressibility measurements reveal that intermolecular bonding is dominated by van der Waals interactions. As can be observed in the isothermal

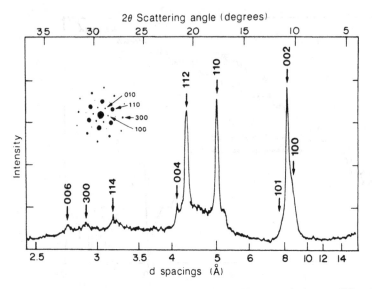

Fig. 4.27. Powder X-ray diffraction pattern and single-crystal electron diffraction pattern (*inset upper left*) of buckminsterfullerene. Reproduced with permission from Krätschmer W, Lamb LD, Fostiropoulos K and Huffman DR (1990) Nature 347: 354

volume compressibilities displayed in Table 4.11, C_{60} is clearly the softest among known carbon solid forms.

In the characterization of the molecular structure as well as in the synthesis of the fullerenes, the spectroscopic properties of this carbon species have been very important.

In Fig. 4.28, the IR spectrum of buckminsterfullerene which shows only four strong bands is reproduced. They are indeed the bands expected for the free, truncated icosahedral molecule depicted in Fig. 4.20.

Table 4.11. Comparison of the isothermal compressibilities and van der Waals diameters of different forms of carbon

Property	Diamond	Graphite	C_{60}
Carbon van der Waals Diameter Å	—	3.35	2.94
One axis compressibility $-d(\ln x)/dp$ m^{2}N$^{-1} \times 10^{11}$	—	2.3	2.3
Volume compressibility $(-1/V)(dV/dP)$ m^{2}N$^{-1} \times 10^{11}$	0.18	2.7	6.9

Reference

Fischer JE, Heiney PA, McGhie AR, Romanow WJ, Denestein AM, McCauley Jr. JP, Smith III AB (1991) Science 252: 1288

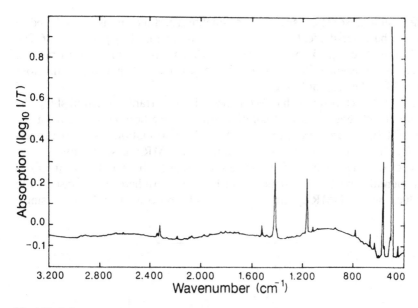

Fig. 4.28. Infrarot spectrum of the fullerene C_{60}. Reproduced with permission from Krätschmer W, Lamb LD, Fostiropoulos K and Huffman DR (1990) Nature 347:354

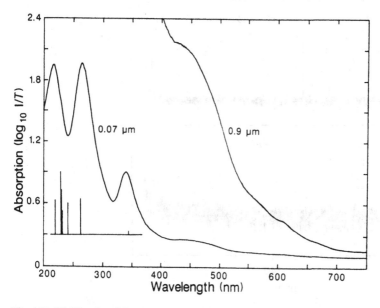

Fig. 4.29. Visible-ultraviolet spectrum of fullerene C_{60}. Theoretical spectrum based on calculations of the allowed transitions of buckminsterfullerene (Larsson S, Volosov A, Rosen A (1987) Chem. Phys. Lett. 137:501). Reproduced with permission from Krätschmer W, Lamb LD, Fostiropoulos K and Huffman DR (1990) Nature 347:354

The visible absorption spectrum of buckminsterfullerene is reproduced in Fig. 4.29. Characteristic features of this spectrum are the absorptions at 216, 264 and 339 nm. These peaks which are somewhat narrower in solution may be assigned, as consigned in the same figure, to a set of allowed transitions calculated for the C_{60}-molecule.

^{13}C-NMR spectroscopy has been specially important for establishing the structure of fullerenes. Relevant solution studies have been carried out for the two most stable fullerenes C_{60} and C_{70}. Valuable information about the properties of C_{60} has also been obtained by solid-state NMR measurements.

The exact equivalence of all 60 atoms implied by the I_h symmetry is a fundamental aspect of the structure of buckminsterfullerene. As observed in Fig. 4.30a, the ^{13}C-NMR spectrum of a C_{60} solution consists indeed of a single

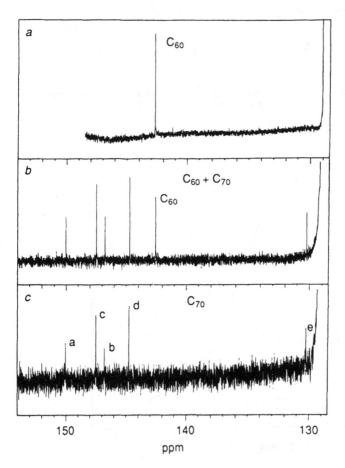

Fig. 4.30. ^{13}C-NMR spectrum of the fullerenes C_{60} and C_{70} in benzene solutions without ^{1}H-decoupling. (a) Pure C_{60}, (b) mixture of C_{60} and C_{70}, (c) pure C_{70}. Signal in spectrum of C_{70} are labeled according the asignment shown in Fig. 4–31. Reproduced with permission from Taylor R, Hare JP, Abdul-Sada AK, Kroto HW (1990) J. Chem. Soc., Chem. Commun., 1423

sharp line. The chemical shift of 142.7 ppm agrees with the resonances of carbon atoms in positions similar to that in buckminsterfullerene e.g. in quaternary carbons in azulene, 140.2 ppm, and fluorene, 141.6 and 143.2 ppm. Although ^{13}C-NMR spectra of organic substances must be normally acquired with ^{1}H decoupling for removing short and long-range ^{13}C–^{1}H coupling, spectra of fullerenes do not need this technique, due to the special nature of this class of chemical species.

A substantial confirmation of the structure of fullerenes has been obtained from the NMR study of purified samples of C_{70}. The ^{13}C-NMR spectrum of a solution of C_{70} shown in Fig. 4.30b and c consists of a set of five lines in the ratio 10:20:10:20:10 with chemical shifts 150.07, 147.52, 146.82, 144.77, and 130.28 ppm. This spectrum corresponds to the D_{5h} structure shown in Fig. 4.31b. These features not only confirm the 5–6 ring topology of the fullerenes but also disprove the possibility that the carbon may be fluxional. Two-dimensional NMR experiments, such as that illustrated in Fig. 4.30 which correlates the ^{13}C-NMR signal of a carbon with that of his bonded neighbor, permit to establish the connectivity within the molecule. Such results agree with the assignment of the carbon atoms in C-70 discussed above (see scheme in Fig. 4.31b).

Solid-state ^{13}C-NMR spectroscopy has proven to be also a valuable tool for studying the properties of buckminsterfullerene. In Fig. 4.33 a series of spectra of a solid sample of C_{60} at different temperatures is reproduced. The single, relatively narrow signal observed at room temperature indicates that under these conditions the molecules reorient rapidly and isotropically in the solid. However, as the temperature goes down, a broad pattern characteristic for the chemical shift anisotropy of powder samples is observed. The evolution of the spectrum described in Fig. 4.33, corresponding to the transformation of a situation of rapid molecular rotation into one characterized by static molecules, points to a phase equilibrium between a mobile phase called the "rotator phase"

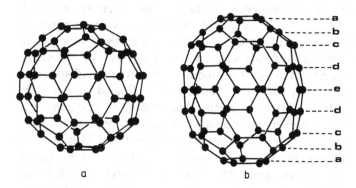

Fig. 4.31. Schematic diagrams comparing the magnetic equivalence of carbon atoms in fullerenes C_{60} and C_{70}. In (a) all sixty atoms are equivalent while in (b) there are five different types of carbon atom

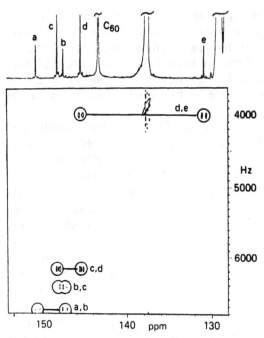

Fig. 4.32. Two-dimensional ^{13}C-NMR spectrum of C_{70} showing the carbon connectivity pattern within the molecule labeled according to the corresponding diagram in Fig. 4.31. Reproduced with permision from Johnson RD, Bethune DS, Yannoni CS (1992) Acc. Chem. Res. 25:169

and a more rigid phase the "ratchet" phase. The activation energies estimated for these phase are consistently 1.4 and 4.5 kcal/mol respectively.

Solid-state ^{13}C-NMR spectrum at temperatures which are low enough for preventing dynamic disorder commented above may be a powerful tool for structural research in fullerenes. This technique has indeed provided the first bond-length measurements for C_{60}. Thus, internuclear distances between neighbor atoms in this molecule were obtained from the ^{13}C–^{13}C magnetic dipolar coupling in ^{13}C enriched samples. Figure 4.34 illustrates a ^{13}C-NMR spectrum of a powder sample of such enriched substance. This spectrum, in which broadening effects due to ^{13}C chemical shift anisotropy have been removed, consists fundamentally in a very strong center line flanked by a weak pair of doublets. The center line corresponds mainly to ^{13}C nuclei with no ^{13}C neighbors (omitted in the spectrum in Fig. 4.34) and the doublets to an AB spectrum belonging to two different kinds of C–C bond in C_{60}. Considering that the splitting of the doublets is inversely proportional to the cube of the internuclear distances, it is possible to obtain a measurement of the C–C bond length from these features. Spectra simulations as those illustrated in Fig. 4.34 lead to an icosahedral structure with 60 longer pentagon edges (1.45 Å) and 30 shorter (1.40 Å) links between the pentagons. The molecular diameter of 7.10 Å calculated from these figures agrees with the structural data obtained for the backminsterfullerene from gas phase electron diffraction, solid state neutron diffraction, and X-ray diffraction studies.

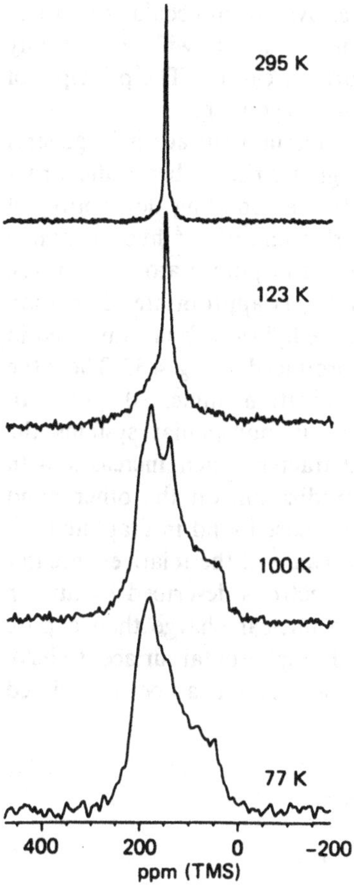

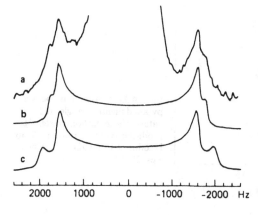

295 K

123 K

100 K

77 K

400   200   0   −200
ppm (TMS)

Fig. 4.33. Temperature dependance of the ^{13}C-NMR spectrum of a solid buckminsterfulleredce sample. Reproduced with permission from Yannoni CS, Johnson RD, Meijer G, Bethume DS, Salem JR (1991) J. Phys. Chem. 95:9

a

b

c

2000   1000   0   −1000   −2000   Hz

Fig. 4.34. Comparison of a part of the experimental ^{13}C NMR spectra of a C_{60} powder sample showing the signals arising from dipolar coupling of directly bonded carbon atoms (a) with simulated spectra calculated assuming bond lengths of 1.45 and 1.40 Å (b) and 1.451 and 1.345 Å (c). Reproduced with permission from Yannoni CS, Bernier PP, Bethume DS, Meijer G, Salem JR (1991) J. Am. Chem. Soc. 113:3190

Electronic Structure of Fullerenes. As discussed above, the molecular structures of fullerenes show some very conspicuous features which will be certainly determinant in the electronic structure of this form of carbon. The principal of such structural features is certainly their spheroidal geometry.

The best arrangement of trivalent carbon atoms in a surface is in general achieved by the formation of six-member rings giving rise to benzoidic structures as occurring in graphite. However, the typical bonding description of graphite, i.e. a network of sigma bonds by overlapping of sp^2 hybrid atomic orbitals and a π-system formed by the overlapping of pure p atomic orbitals, cannot be directly applied to spheroidal surfaces. A very appropriate description of the effect of the curvature on the character of the hybrid orbitals involved in the π-system for fullerene-type surfaces is that reproduced in Fig. 4.35. There the curvature degree is expressed by the pyramidalization angle, $90 - \Theta_{\sigma\pi}$. In contrast to the orbital distribution of graphite, in non-planar systems the π molecular orbital has a certain s-orbital character which increases with increasing curvature of the surface. The σ-hybridization on the other hand displays features which are intermediate between those found in graphite (sp^2) and diamond (sp^3). As discussed in Chapter 1, because of the relatively greater penetration effect of the s atomic orbitals, the electrons described by an s–p hybrid orbital will experience a higher effective nuclear charge than a pure p-orbital. This feature causes the orbitals involved in spheroidal surfaces to have a higher electron affinity than those in a graphite sheet. As it can be appreciated

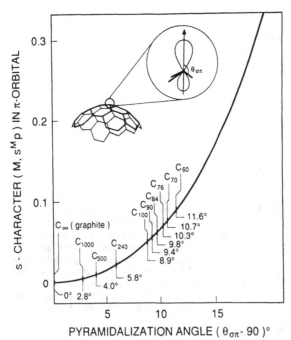

Fig. 4.35. Correlation between pyramidalization angle and curvature of the carbon cluster. Reproduced with permision from Haddon RC (1992) Acc. Chem. Res. 25:127

in Fig. 4.35, the effect of the rehybridization on the chemical properties of the fullerenes will increase with decreasing conglomerate size.

In addition to the hybridation changes already discussed, the existence of spheroidal surfaces implies that the structures cannot contain just six-member but also five-member rings. As discussed above, in the case of the fullerene, 12 five-carbon rings of the type found in the cyclopentadiene are indeed required. Considering the remarkable tendency of cyclopentadiene to gain one electron giving rise to the anion $C_5H_5^-$ it is expected that the inclusion of five-member rings in the spheroidal structures of the fullerene should be a factor that upgrades their ability to accept electron density. Similarly to the influence of the rehybridization on the electron affinity discussed above, the effect of the presence of the 12 five-member rings in the fullerenes increases with decreasing nuclearity of the clusters.

A practical result of the enhancement of the electron affinity induced by the formation of fullerenes is the relatively easy formation of negative species as the anion C_{60}^- which will also be discussed later in this chapter.

Qualitative answers to questions about the stability of fullerenes may be indeed found using the rough bonding description outlined above. In spite of the rehybridization, the system may be still described by the superposition of a set of σ-bonds formed by the surface carbon orbitals. That gives rise to a network of trivalent carbon atoms, and to a delocalized π-system over the surface in such a way that an aromatic character appears as highly likely. These features constitute a real example of tridimensional aromaticity. The degree of aromaticity which could be exemplified by considering the number of resonance structures is often used as an argument for the stability of aromatic species. For the fullerene C_{60} with a closed-shell electronic structure, a total of 12 500 Kekulé structures have been indeed calculated. That is certainly a factor that, in addition to the lack of dangling bonds (vide supra), should be considered as determinant in the stability of the fullerenes.

There are a number of more quantitative calculations of the electronic structure of fullerenes. Some of them were indeed carried out prior to the discovery of buckminsterfullerene. However, the results obtained by different approaches, going from relatively simple calculations with a calculator to rather complex algorithms resolved by powerful computational tools, are qualitatively similar. In Fig. 4.36 an orbital energy-level diagram for C_{60} obtained by Hückel calculations can be observed. The nature of the frontier orbitals can be indeed experimentally confirmed by ultraviolet photoelectron spectroscopy. The signals observed in the spectrum of the species C_{60}^- illustrated in Fig. 4.37 are nicely in accord with the expectations for the buckminsterfullerene structure, i.e. a highly symmetric truncated icosahedral geometry. The signal near 3 eV corresponds to a removal of the most weakly bound electron. This extra electron is forced to occupy the threefold degenerate t_{1u}, the lowest unoccupied molecular orbital (LUMO) in the molecular orbital arrangement of C_{60}, because the HOMO of the neutral molecule is already filled with 10 equivalent electrons. These electrons arising from the h_u symmetry HOMO appear to produce the

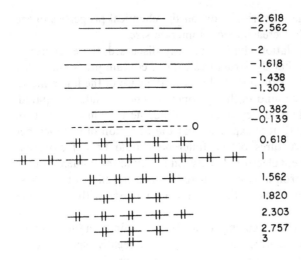

−2.618
−2.562

−2

−1.618
−1.438
−1.303

−0.382
−0.139

0

0.618

1

1.562

1.820

2.303

2.757
3

Fig. 4.36. Hückel molecular orbital calculations for C_{60}. The orbital energy levels are given in units of beta. Reproduced with permission from Davinson RA (1981) Theor Chim Acta 58:193

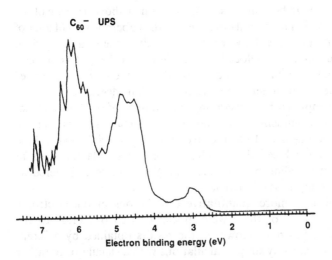

Fig. 4.37. Ultraviolet photoelectron spectrum (UPS) of the anion C_{60}^-. Reproduced with permission from Curl RF, Smalley RE (1988) Science 242:1017

features observed in the UV photoelectron spectrum in the 4 to 5 eV range. Larger signals in the 5.5 to 7 eV are assigned to the group of 18 electrons in the nearly degenerate g_g and h_g orbitals.

Chemical Properties of Fullerenes. One of the most relevant facts in the chemical behavior of fullerenes appears to be their relatively high electron affinity. This property does indeed give rise to a series of reactions thus resulting in the modification and functionalization of the clusters. Reduction, nucleophilic and oxidative additions of low-valent transition metals constitute important ways for obtaining fullerene derivatives.

The key in many of known reactions of fullerenes appears to be the "pyra-cyclene" moieties existing in this type of clusters. Figure 4.38 illustrates a representation of backminsterfullerene emphasizing the presence of such a moiety. The fact that in this structure each double bond is surrounded by four other electron-withdrawing groups is possibly related with the high electron affinity of the fullerenes already mentioned.

Reduction of Fullerenes. In agreement with the existence of low-lying un-occupied molecular orbitals described above, fullerenes are mild oxidation agents. Reduction potentials for buckminsterfullerene can be appreciated in the cyclic voltammetry studies illustrated in Fig. 4.39 in which the electrochemical behavior of C_{60} with the diphylderivative Ph_2C_{61} is compared (vide infra). Interestingly C_{70} shows a similar electrochemical behavior.

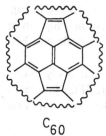

C_{60}

Fig. 4.38. "Pyracyclene" moiety in buckminsterfullerene. Each double bond is surrounded by electron withdrawing groups

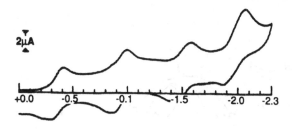

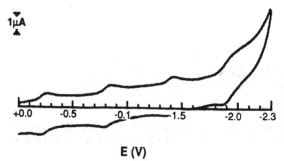

E (V)

Fig. 4.39. Cyclic voltammogram of C_{60}^{-} (above) and Ph_2C_{61} (below) in tetrahydrofuran solutions with tetra-butylammonium fluoroborate as supporting electrolyte. Potentials are refered to the couple Ag/Ag^+. Reproduced with permission from Suzuki T, Li Q, Khemani KC, Wudl F, Almarsson Ö (1991) Science 254:1186

The electrochemical reduction of backminsterfullerene constituting the anode in a cell containing Ph_4PCl as supporting electrolyte gives rise to the formation of crystals of a charge-transfer adduct on the surface of the electrode:

$$C_{60} \xrightarrow[C_6H_4Cl_2]{Ph_4PCl} C_{60}^- \cdot Ph_4P^+ \cdot (Ph_4PCl)_2$$

A series of anionic derivatives of backminsterfullerene have been also obtained as phases of type $A_x C_{60}$ (x = 1 − 6) by exposing the fullerene films to alkali-metal vapor. Since some of these phases lead to conducting and superconducting solids, they are therefore promising as new materials and they will be discussed separately later in this section.

Functionalizing the Fullerenes. The Fulleroids. Interesting results in the functionalizing of fullerenes have been obtained by means of dipolar additions of diazoalcanes to C_{60}. As shown in the reaction scheme in Fig. 4.40, the addition of diazomethane to a C_{60}-double bond gives rise to a 1-pyrazoline intermediate which loses nitrogen spontaneously at room temperature producing a ring-opened structure. Other examples of this class of compounds are shown in Fig. 4.41. Curiously, the UV-visible spectrum as well as the cyclic voltammogram of such monoadducts are essentially identical to that of buckminsterfullerene. For this reason such compounds have been called fulleroids.

Apparently the formation of the fulleroids, as shown schematically above (Fig. 4.41) implies only subtle changes in the electronic structure of the fullerenes which are not enough for substantially changing their electronic spectrum and their electrochemistry.

Important evidence about the formation of species with an open *trans* annular bond has been obtained from the synthesis of the compound $PhHC_{61}$

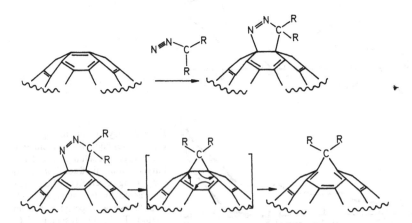

Fig. 4.40. Example of functionalization of fullerenes. Preparation of "fulleroids" by reaction of fullerenes with diazoalkane [Suzuki T, Li Q, Khemani KC, Wudl F, Almarsson Ö (1991) Science 254:1186]

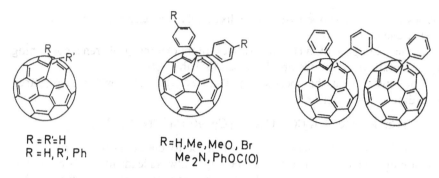

R = R' = H
R = H, R', Ph

R = H, Me, MeO, Br
Me₂N, PhOC(O)

Fig. 4.41. Selected examples of mono and bifulleroids [Wudl F (1992) Acc. Chem. Res. 25:157]

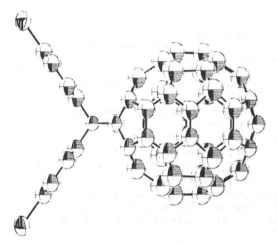

Fig. 4.42. Molecular structure of bis (bromophenyl) fulleroid. Reproduced with permission from Wudl F (1992) Acc. Chem. Res. 25:157

with a highly ^{13}C-enriched methine carbon. The chemical shift of this carbon atom (140 ppm) as well as its C–H coupling constant (140 Hz) agrees with those for an annulene. Further evidence is provided by the single-crystal molecular structure obtained for the compound (p-BrC₆H₄)₂C₆₁ illustrated in Fig. 4.42. The bond length found for the *trans* annular bond was 1.84 Å, considerably longer than a fullerene bond (1.37–1.47 Å).

Metal Complexes of Buckminsterfullerene. Although the fullerenes may be described as spheres covered by benzene rings, their properties differ from those of the normal aromatic rings. The characteristic long/short bond alternation of the six-member rings in C₆₀ discussed above suggests a greater localization degree of the double bonds than in aromatic species. That is also apparent in the organometallic chemistry of C₆₀. Indeed, as it will be discussed below, fullerenes behave in the formation of metal compounds more like an electron-deficient, species as the tetracyanoethylene, than like a relatively electron-rich species such as ethylene or benzene.

Some examples of metal derivatives of buckminsterfullerene containing direct metal-carbon bonds are described in Table 4.12.

The existence of the species $[Cp^*Ru(MeCN)_2C_{60}]^{3+}$ which is prepared according to the reaction

$$[Cp^*Ru(MeCN)_3]X + C_{60} \rightarrow [Cp^*Ru(MeCN)_2C_{60}]X_3$$

stresses the different behavior of C_{60} respect to planar aromatic rings. The reaction of the latter with $[Cp^*Ru(MeCN)_3]^+$ always leads to the formation of sandwich compounds displacing the three coordinated acetonitrile from ruthenium. The reaction above is similar however to that with electron deficient alkenes which lead to products of type $[Cp^*Ru(MeCN)_2 (\eta^2 \text{ alkene}]^+$.

Table 4.12. Selected examples of fullerene metal complexes

Compound	Remarks	Ref.
$(Ph_3P)_2Pt(\eta^2\text{-}C_{60})$	$(Ph_3P)_2Pt(\eta^2\text{-}C_2H_4) + C_{60}$, 75% yield. Structure see Fig. 4.43.	1
$(Et_3P)_2M(\eta^2\text{-}C_{60})$	M = Ni, Pd, Pt; $M(PEt_3)_4 + C_{60}$; spectroscopic characterization.	2
$(Et_3P)_6M(\eta^2\text{-}C_{60})$	M = Ni, Pd, Pt; $M(PEt_3)_4 + C_{60}$ (excess); Spectroscopic and crystallographic characterization	3
$(\eta^5\text{-}C_9H_7)Ir(CO)(\eta^2\text{-}C_{60})$	$(\eta^5\text{-}C_9H_7)Ir(CO)(\eta^2\text{-}C_8H_{14}) + C_{60}$, 58%. Spectroscopic characterization.	4
$(PPh_3)_2Ir(CO)Cl(\eta^2\text{-}C_{70})$	$(PPh_3)_2Ir(CO)Cl + C_{70})$. X-ray crystallography show structure similar to Pt compound in Fig. 4.43 in which Ir is bound to an a–b edge of C_{70} (see Fig. 4.31b)	5
$\{[Cp^*Ru(CH_3CN)_2]_3C_{60}\}^{3+}(X^-)_3$	$[Cp^*Ru(CH_3CN)_3]^+(X^- + C_{60}$ (excess).	1
$(TPP)Cr(THF)_3(C_{60})$	TPP = Tetraphenylporphinato. Cr(II)TPP in THF + C_{60}. Compound is a Cr(III)-species containing the $C_{60}\cdot^-$ radical.	6
$La@C_{82}$	Arc vaporization of composite rod made of graphite and La_2O_3. Soot contains ca. 3% of the product. Characterization by mass, EPR and photoelectron spectroscopies.	7, 8

References

1 Fagan PJ, Calabrese JG, Malone B (1991) Science 252:1160
2 Fagan PJ, Calabrese JC, Malone B (1992) Acc. Chem. Res. 25:134
3 Fagan PJ, Calabrese JG, Malone B (1991) J. Am. Chem. Soc. 113:9408
4 Koefod RS, Hudgens MF, Shapley JR (1991) J. Am. Chem. Soc. 113:8957
5 Balch AL, Catalano VJ, Lee JW, Olmstead MM, Parkin SR (1991) J. Am. Chem. Soc. 113:8953
6 Penicaud A, Hsu J, Reed CA (1991) 113:6698
7 Heath JR, O'Brien SC, Zhang Q, Liu Y, Curl RF, Kroto HW, Tittel FK, Smalley RE (1985) J. Am. Chem. Soc. 107:7779
8 Weaver JH, Chai Y, Kroll GH, Jin C, Ohno TR, Haufler RE, Guo T, Alford JM, Conceicao J, Chibante LPF, Jain A, Palmer G, Smalley RE (1992) Chem. Phys. Lett. 190:460

The reaction of a magenta toluene solution of C_{60} with the complex $(Ph_3P)_2Pt(\eta^2\text{-}C_2H_4)$ produces a dark emerald green solution from which a black crystalline compound is obtained:

$$(Ph_3P)_2Pt(H_2C=CH_2) + C_{60} \xrightarrow[\text{2 h, 25 °C}]{\text{toluene}} (Ph_3P)_2Pt\text{-}C_{60} + H_2C=CH_2$$

Figure 4.43 illustrates the structure of the product obtained by X-ray crystallography. The platinum atom is bound to just two carbon atoms of the C_{60} cluster. Analogously to the osmium derivative $(t\text{-}BuC_5H_5)_2OsO_4(C_{60})$ as well as to the iridium complexes, $(Ph_3P)_2(CO)ClIr(C_{60})$ and $(\eta^5\text{-indenyl})(CO)Ir(C_{60})$, the addition occurs to a bond at the fusion of two six-member rings. The higher reactivity of this type of bond respect to that of the linkages at the fusion of a five-member and a six-member ring appears to be due to the higher double character of the former reflected in its comparatively shorter internuclear lengths (vide supra).

The nature of the metal-fullerene bond, which is expected to be greatly influenced by the electron affinity of the fullerene discussed above, is also reflected in the changes in the coordination sphere of the metal atom induced by complex formation. A general feature in the coordination of alkenes to metal atoms is that the four groups attached to the alkene bend back away from the

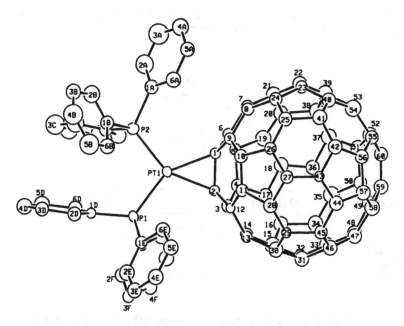

Fig. 4.43. Schematic molecular structure of $(Ph_3P)_2Pt(\eta^2C_{60})$ obtained from single crystal X-ray diffraction analysis of the complex $(Ph_3P)_2Pt(\eta^2\text{-}C_{60})\cdot C_4H_8O$ [Fagan PJ, Calabrese JG, Malone B (1991) Science 252:1160]

metal. Probably that results as a consequence of the metal d-orbital back-bonding into the alkene π^* orbital. A measurement of such an influence is the angle Θ that these groups form with the alkene plane as described in the scheme on the heading of Table 4.13. From data in this Table it is apparent that the neighborhood of the C–C double bond in the fullerene presents a deviation from planarity which, although it is already appropriate for metal bonding, increases slightly with complex formation. The two carbon atoms interacting directly with the metal in the fullerene are indeed pulled away from the cluster surface. Both, the curvature of the fullerene and their electron affinity, favor the η^2 binding manner and the $d\pi$–$p\pi$ back-bonding process. In this respect it is interesting to note that in the iridium complex of the fullerene C_{70}, $(Ph_3P)_2(CO)ClIr(C_{70})$, the metal atom binds to the oblong extreme of the cluster where the curvature is at its greatest.

Fullerene complexes of low-valent transition metal complexes may also be obtained by substitution of phosphine groups in the tetraphosphine derivative of the group 10 metals:

$$(Et_3P)_4M + C_{60} \rightarrow (Et_3P)_2M(\eta^2\text{-}C_{60}) + 2Et_3P \qquad \text{M: Ni, Pd, Pt}$$

All these products show properties similar to those of the well-characterized platinum derivatives described above.

Table 4.13. Effect of the coordination of metals to alkenes. Deviation Θ of the groups attached to the carbon–carbon double bond from the original alkene plane

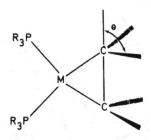

Type of compound	Θ (degree)		Reference
	Ligand	Complex	
Pt-alkene complexes	0	22–35	1, 2
Pt-C_{60} complex	31	41	3

References

1 Kumar A, Lichtenhan JD, Critchlow SC, Eichinger BE, Worden WT (1990) J. Am. Chem. Soc. 112:5633
2 Morokuma K, Borden WT (1991) J. Am. Chem. Soc. 113:1912
3 Fagan PJ, Calabrese JG, Malone B (1991) Science 252:1160

Fullerene can attach more than one metal atom. Although the isolation of pure products from the reaction mixtures is generally very difficult, it has been possible to separate and characterize a series of hexametal derivatives:

$$6(Et_3P)_4M + C_{60} \rightarrow [(Et_3P)_2M]_6(\eta^2\text{-}C_{60}) + 12\,Et_3P \qquad M: Ni, Pd, Pt$$

Spectroscopic studies show that the products exist as single structural isomers. They are all indeed isostructural and appear to have structures as that of the palladium derivative illustrated in the scheme in Fig. 4.44. The metals are in an octahedral array surrounding the exterior of the C_{60} core. Analogously to $(Ph_3P)_2Pt(\eta^2\text{-}C_2H_4)$, each metal atom is bound across a six-member ring junction. Carbon-carbon bond lengths in the complexes $[(Et_3P)_2M]_6(\eta^2\text{-}C_{60})$ for M=Pt and Pd are shown in Fig. 4.45 using an scheme that represents the most favorable electronic resonance structure based on these internuclear distances. Each metal atom pulls out two carbon atoms from the C_{60} framework, the internuclear distances C in this scheme being significantly shorter than in C_{60}. Also the carbon-carbon bonds at which each metal atom is attached (A) are lengthened significantly. The short/long bond difference of ca. 0.06 Å is much less in the metal complexes, namely 0.036 Å in the platinum compound. This feature may be interpreted as an increase of the degree of aromaticity in the metal derivatives.

The relatively high stability of the C_{60} hexametal derivatives is probably due to two main factors. One is the fact that an octahedral array of metal atoms around C_{60} is sterically the most favorable situation; there is no room for a seventh metal fragment. The other factor could be the electronic stabilization of the system due to the higher aromaticity of the metal derivatives already mentioned.

Complex formation also produces some alterations in the electronic structure of fullerenes. In general, the reduction potentials of the metal derivatives are

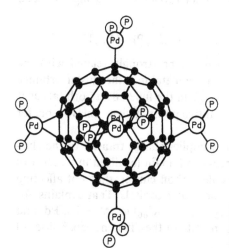

Fig. 4.44. Schematical structure of $[(Et_3P)_2Pd]_6C_{60}$. The ethyl groups have been removed for clarity. [Fagan PJ, Calabrese JC, Malone B (1992) Acc. Chem. Res. 25:134]

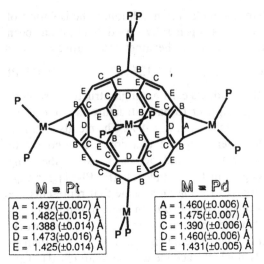

M = Pt
A = 1.497(±0.007) Å
B = 1.482(±0.015) Å
C = 1.388 (±0.014) Å
D = 1.473(±0.016) Å
E = 1.425(±0.014) Å

M = Pd
A = 1.460(±0.006) Å
B = 1.475(±0.007) Å
C = 1.390 (±0.006) Å
D = 1.460(±0.006) Å
E = 1.431(±0.005) Å

Fig. 4.45. Electronic resonance structures and corresponding bond distances in the platinum and paladium compounds $[(Et_3P)_2M]_6C_{60}$. Reproduced with permission from Fagan PJ, Calabrese JC, Malone B (1992) Acc. Chem. Res. 25:134

higher than for buckminsterfullerene itself. Compounds $(Et_3P)_2MC_{60}$ and $[(Et_3P)_2M]_6(C_{60})$ (M = Ni, Pd, Pt) discussed above are indeed about 0.30 V and 2.7–3.0 V harder to reduce than C_{60} respectively. This effect could be due to a rise in the energy of the lowest unoccupied molecular orbital produced by the virtual removal of one double bond from the conjugated fullerene system. Such an electron rearrangement appears to induce a significant decrease in the electron affinity of the system.

The stability of the metal compounds also decreases upon reduction. Thus, the rate of decomposition of mono platinum derivatives for releasing the metal moiety into solution follows the order

$$[(Et_3P)_2MC_{60}]^{3-} > [(Et_3P)_2MC_{60}]^{2-} > [(Et_3P)_2MC_{60}]^{-}$$

The stability of metal complexes appears to be strongly related with the possibility of π-electron back donation from the metal into the cluster orbitals. Therefore any effect which increases the charge in the fullerene π^* orbitals as it occurs in the reduction process will be, from this point of view, unfavorable for the complex formation.

As mentioned, metal-cluster interaction implies charge transfer to the clusters. However such a charge density appears to remain localized in a sector of the sphere and not to extend into the whole carbon cluster, thus not affecting directly the energy of the cluster frontier molecular orbitals. That explains, for instance, the fact that the compounds $(Et_3P_2)_2M(\eta^2-C_{60})$ for M = Ni, Pd, and Pt have all the same reduction potential in spite of the different capabilities of these compounds for π-back bonding. The influence of the metal coordination observed in the successive metallations – in which the tendency of further metal atoms to add decreases dramatically with increasing metallation – is mainly due to the changes in the cluster electron delocalization produced by removing a double bond as described above.

Fullerenes have proved to be rather reactive. However, it is in general difficult to generate pure distinct products. One of the first reactions producing discrete stable products able to be studied crystallographically was an osmilation reaction.

Osmilation processes of polycyclic aromatic hydrocarbons with osmium tetroxide in the presence of pyridine, such as that illustrated for anthracene in the scheme in Fig. 4.46, may be indeed applied to fullerenes. Figure 4.47 describes the osmilation of buckminsterfullerene to give the 1:1 and 1:2 adducts. As observed in such a scheme, these osmilations are processes that can be reversed thermically. This feature points out products in which the fullerene structure remains fundamentally intact. The osmilation degree depends not only on the stoichiometry of the reagents but also of the solubility of the products. By selecting the ligand on osmium, it is possible to obtain a product with a given stoichiometry to precipitate and protect it from further reaction. Scheme in Fig. 4.48 describes some instances of the ligands adequated for obtaining products with different stoichiometries.

The osmilation of fullerenes has been important not only for investigating the reactivity of these clusters but also for obtaining for the first time an X-ray crystallographic analysis at atomic resolution level of the C_{60} framework. As mentioned above, orientational disorder in the fullerene crystals – arising fundamentally from their nearly spherical symmetry which permits these ball-like molecules to rotate freely – systematically prevents us from obtaining the atomic

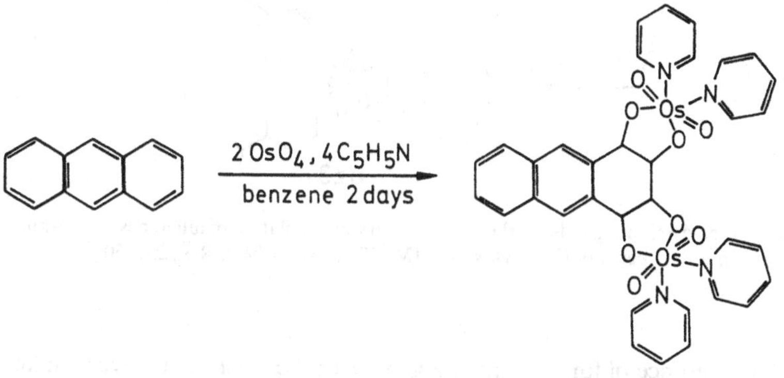

Fig. 4.46. Reaction of anthracene with osmium tetraoxide in the presence of pyridine [Wallis JM, Kochi JK (1988) J. Am. Chem. Soc. 110:8207]

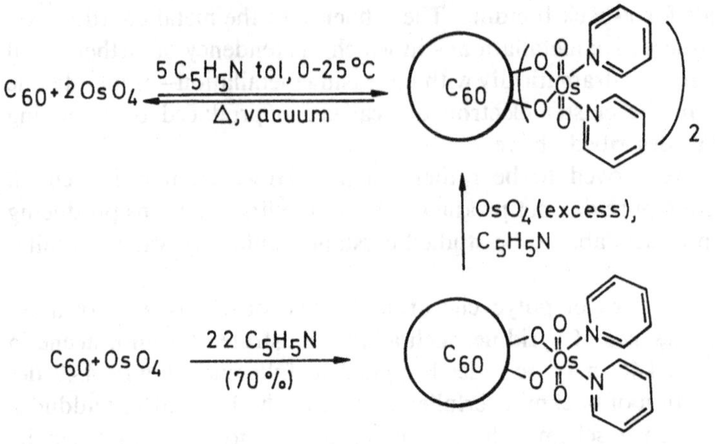

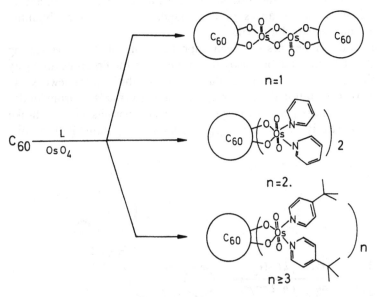

Fig. 4.47. Scheme of the osmilation reactions of buckminsterfullerene [Hawkins JM, Lewis TA, Loren SD, Meyer A, Heath JR, Shibato Y. Saykally RJ (1990) J. Org. Chem. 55:6250 and Hawkins JM (1992) Acc. Chem. Res. 25:150]

Fig. 4.48. Effect of the osmium ligands on the stoichiometry of osmilation of fullerenes. n = osmilation degree osmium atoms per mol C_{60}. [Hawkins JM (1992) Acc. Chem. Res. 25:150]

positions. The presence of functionalizing groups in the cluster, however, makes it possible for the carbon framework to be organized relative to the attached group, thus giving rise to crystals in which the carbon clusters are uniformly oriented.

As shown by the molecular structure of the compound $C_{60}(OsO_4)$-(4-*tert*-butylpyridine)$_2$ reproduced in Fig. 4.49, the fulleride part of the molecule actually corresponds to a soccerball-like arrangement of carbon atoms. There, 20 six-member rings are fused with 12 five-member rings to give a carbon cluster with 32 faces. These features totally agree with the structure of the buckminsterfullerene discussed above.

Fullerene Endohedral Complexes. The possibility of trapping atoms or molecules inside carbon frameworks is a very peculiar property of fullerenes. Today there is indeed strong evidence that metal atoms can actually be trapped inside fullerene cages. A nomenclature for this peculiar kind of compounds has been even suggested: $(M@C_n)$ for a metal atom M located inside a C_n fullerene.

Endohedral metal fullerenes can be detected in relatively small amounts in the mass spectra in the laser vaporization cluster beams (vide supra). However, macroscopic quantities of these compounds may be produced rather readily either by vaporization in a laser furnace apparatus or by arc-burning of a composite rod of graphite and the corresponding metal oxide. In Fig. 4.50, a mass spectrum which illustrates the formation of a series of fullerene endohedral yttrium complexes obtained by laser vaporization of a composite graphite/Y_2O_3 rod at 1200 °C is reproduced. Among these species there is also one, $Y_2@C_{82}$, which corresponds to the inclusion of a metal cluster in the fullerene ball.

Interesting advances in investigating the nature of this new class of compounds have been obtained by using electron paramagnetic resonance spectroscopy.

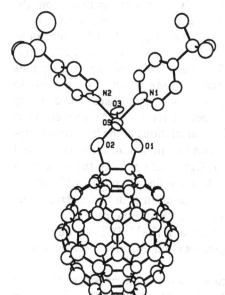

Fig. 4.49. Molecular structure of the compound $C_{60}(OsO_4)$(4-*tert*-butylpyridine)$_2$. Reproduced with permission from Hawkins JM, Meyer A, Lewis TA, Loren S, Hollander FJ (1991) Science 252:312

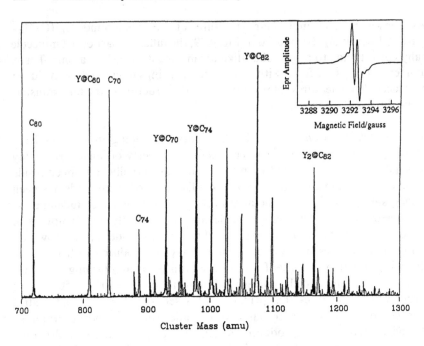

Fig. 4.50. Mass spectrum of fullerene containing ytrium prepared by laser vaporization of a composite graphite/Y_2O_3 rod at 1200 °C. Reproduced with permission from Smalley RE (1992) Acc. Chem. Res. 25:98

Bulk samples and toluene extracts of (La@C_{82}) can be obtained from the toluene extract of the soot produced by arcburning a graphite/La_2O_3 composite, which contains mainly C_{60}, C_{70}, and (La@C_{82}). The EPR spectra at room temperature of (La@C_{82}) solid and of a solution in 1,1,2,2, tetra-chloroethane are shown in Fig. 4.51. The eight-line spectrum corresponds indeed to that expected for the hyperfine coupling of an electron spin to a nuclear magnetic moment with spin 7/2 as in the ^{139}La nucleus. The coupling constant of ca. 1.25 G is smaller than expected for the La^{2+} species. Thus experimental evidence as well as theoretical calculations of the hyperfine coupling indicates that the preferred oxidation state for lanthanum in this compound is +3. As observed in Fig. 4.52, between the eight peaks of ^{139}La there are other hyperfine lines which can be assigned to the hyperfine coupling to ^{13}C nuclei thus reflecting the association of the lanthanum atom with the fullerene. Detailed analysis of the features observed in these EPR spectra leads to a description of this endohedral complex as consisting of a (La^{3+}@C_{82}^{3-}) ground state doublet in which the paramagnetism originates in an unpaired electron spin in the π-system of the fullerene framework.

As expected, the reactivity of the encapsuled metal atoms is low. It is well known that yttrium ions react with N_2O bearing YO^+. This reaction also occurs

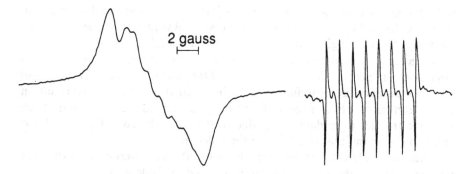

Fig. 4.51. EPR spectrum of an extract resulting from arc-burning of a composite graphite/La₂O₃ rod. (*a*) Solid sample. (*b*) solution in 1,1,2,2,-tetrachloroethane. Reproduced with permission from Johnson RD, de Vries MS, Salem J, Bethume DS, Yannoni CS (1992) Nature 355:239

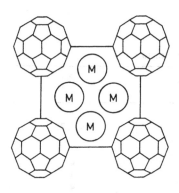

Fig. 4.52. Schematic representation of a face normal to z in the body-centered cubic structure of M_6C_{60} (M = K, Rb, Cs). Lattice constants are 11.39, 11, 54, and 11.79 Å respectively [Zhou O, Fischer JE, Coustel N, Kycia S, Zhu Q, McGhie AR, Romanow WJ, McCauley Jr. JP, Smith III AB, Cox DE (1991) Nature 351:462]

with yttrium complex species in which the metal ion is bound externally to the fullerene, as those generated by gas phase association.

$$Y^+ + C_{60} \longrightarrow Y(C_{60})^+ \xrightarrow{N_2O} YO^+ + C_{60} + N_2$$

However, yttrium species generated by direct laser vaporization of mixtures of graphite and yttrium oxide are unreactive with N_2O.

Potentiality of Fullerenes as New Materials. Fullerene-derived solids are interesting not only from a fundamental but also from a technological point of view. Reactivity of fullerenes has given birth to products with very interesting properties as materials.

As discussed above, fullerenes produce molecular van der Waals-bonded solids in which the cluster molecules are ordered in a fcc lattice. Such solids may indeed originate guest-host compounds by intercalating atoms and molecules in the spaces formed by the tetrahedral and octahedral voids in the fullerene lattice, retaining thus the original identity of the pristine solid.

In general intercalation compounds often show interesting properties as electronic conductors. That is also the case of the products of the intercalation of alkali metals in buckminsterfullerene. As discussed above, this process results in an electron transfer from guest to host thus giving rise to fulleride anions. Such a charge transfer is fundamentally due to the large electron affinity and low reduction potential that characterize the fullerenes.

There are different methods for obtaining $M_x C_{60}$ materials with different x values. Some of these materials are mentioned in Table 4.14.

Structural evidence for the intercalation of K, Rb, and Cs in the saturated phase $M_6 C_{60}$ indicates that all these solids have non-close-packed body-centered cubic structures as that shown schematically in Fig. 4.52, in which each C_{60} is surrounded by 24 alkali metal atoms and each metal atom is in a distorted tetrahedral hole formed by four C_{60} molecules.

Table 4.14. Critical temperatures of metal doped C_{60} superconductors

Compound	Critical temperature (K)	Ref.
$K_3 C_{60}$	18	1
$Rb_3 C_{60}$	29.8, 28	2, 3
$Cs_3 C_{60}$	29.4	4
$Cs_x C_{60}$ x = 1.2–3	30	5
$K_2 RbC_{60}$	21.8	4
$K_{1.5} Rb_{1.5} C_{60}$	22.15	4
$KRb_2 C_{60}$	26.4	4
$Rb_2 CsC_{60}$	31.3	4
$K_3 Tl_{4.5} C_{60}$	23.2	5
$RbTl_{2.0} C_{60}$	48.0	5
$Rb_{2.0} Tl_{2.5} C_{60}$	39	5
$Rb_{2.7} Tl_{2.2} C_{60}$	45	5

References

1 Hebard AF, Rosseinsky MJ, Haddon RC, Murphy DW, Glarum SH, Paistra TTM, Ramirez AP, Kortan AR (1991) Nature 350:600

2 McCauley Jr. JP, Zhu Q, Coustel N, Zhou O, Vaughan G, Idziak SHJ, Fischer JE, Tozer SW, Groski DM, Bykovetz N, Lin CL, McGhie AR, Allen BH, Romanow WJ, Denenstein AM, Smith III AB (1991) J. Am. Chem. Soc. 113:8537

3 Rosseinsky MJ, Ramirez AP, Glarum SH, Murphy DW, Haddon RC, Hebard AF, Paistra TTM, Kortan AR, Zahurak SM, Makhija AV (1991) Phys. Rev. Lett. 66:2830

4 Fleming RM, Ramirez AP, Rosseinsky MJ, Murphy DW, Haddon RC, Zahurak SM, Makhija AV (1991) Nature 352:787

5 Kelty SP, Chen CC, Lieber CM (1991) Nature 352:223

6 Iqbal Z, Baughman RH, Ramakrishna, Khare S, Murthy NS, Bornemann HJ, Morris DE (1991) Science 254:826

Theoretical as well experimental studies show that although there are some differences in the electron structure of pure and alkali metal-doped fullerenes, the band structures of both type of compounds are fundamentally similar. Band width as well as the energy gap between the bands remain qualitatively unchanged after the fullerene doping. In other words, the molecular features of C_{60} dominates the electronic structure of the solid phases.

Thus, the electronic structures of fullerene anionic species can be obtained by filling the 3-fold degenerate LUMOs of the cluster. Accordingly, the products of the intercalation of alkali metals in fullerenes are expected to show, depending on the intercalation degree, an enhanced conductivity. In some cases superconductivity has been also observed (vide infra).

According to Hückel molecular orbitals of the isolated molecule discussed above (see Fig. 4.36), the optimum metallic behavior of these solids should be obtained by adding three electrons to the pseudo aromatic electronic system. Such a situation is achieved by intercalation of three metal atoms in C_{60} if a complete charge transfer of one electron per alkali metal atom is assumed. Similarly, it is clear that the compounds M_6C_{60} will be insulators, since the corresponding derived band (t_{1u}) is filled at this doping level. Although for the M_4C_{60} derivatives a metallic character could also be expected, they appear to have an insulator ground state. Such behaviour can be explained by assuming that the C_{60}^{4-} undergoes a static Jahn-Teller distortion leading to a gap in the band structure. The M_3C_{60} phases are however not expected to be subject to Jahn-Teller distortions and the metallic ground state survives.

Figure 4.53 gives a schematical representation of the morphology of a K_xC_{60} thin film giving a picture of the different stages of potassium incorporation in a buckminsterfullerene film together with the positions of the HOMO and LUMO-derived bands relative to the Fermi level (E_F).

The alkali metal doping first produces a dilute solution. The interaction of K with the surrounding C_{60} molecules leads to new energy levels derived mainly from the LUMO levels. Phase separation occurs when the alkalimetal concentration exceeds the solubility limit. Theoretical calculations indicate that the inclusion of K atoms gives rise to an energy gain per atom of about 0.5 eV for the occupancy of the octahedral sites in K_1C_{60}; of ca. 0.8 eV for tetrahedral holes in K_2C_{60}; and of about 1 eV for the octahedral and tetrahedral sites in K_3C_{60}.

Further K incorporation beyond the K_3C_{60} phase results in the formation of the K_4C_{60} and the K_6C_{60} phases with ions in the tetrahedral sites of the C_{60} lattice.

As can be appreciated in the examples illustrated in Table 4.14, a number of superconductors based on the inclusion of alkali metals and thalium have been prepared. Superconducting transition temperatures in these compounds are significantly higher than those observed for the correponding graphite intercalates. Apparently, in the case of the fullerenes there is a near linear relationship between the superconducting transition temperature, T_c, and the alkali metal size. Such a relationship, deduced from the correlation between T_c and the

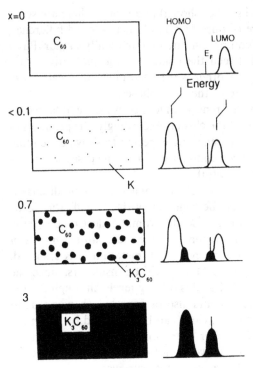

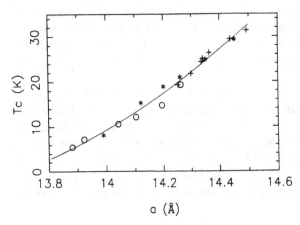

Fig. 4.53. Schematic representation of the inclusion of K in K_xC_{60} thin films indicating the relative positions of the HOMO and LUMO-derived bands and Fermi energy level, E_F. Reproduced with permission from Weaver JH (1992) Acc. Chem. Res. 25:143

lattice constants for a series of compounds at constant pressure, suggests that this transition temperature depends more on the overlap between near-neighbor C_{60} molecules than on the nature of the intercalate. Figure 4.54 illustrates the dependence of T_c on the lattice parameters over a large range of "a" obtained from studies of the influence of pressure on the lattice.

Fig. 4.54. Dependence of the superconducting transition temperature on the lattice parameter a including data obtained for a number of $M_{3-x}M'_xC_{60}$ compounds at 300 K and 1 bar (x) as well as data converted from pressure dependence for K_3C_{60} (○) and Rb_3C_{60} (*). Reproduced with permission from Fischer JE, Heiney PA, Smith III AB (1992) Acc. Chem. Res. 25:112

4.3.2 Phosphorus Cage Compounds

Tetraphosphorus Clusters. As mentioned at the beginning of this Chapter, phosphorus is a unique element displaying a modification formed by discrete molecular units. White phosphorus is a molecular solid made up of P_4 tetrahedral clusters which are stable in solution as well as in the gas phase.

Theoretical investigations of the molecule P_4 show that the valence shell electron density in this species should be concentrated inside and on the faces of the tetrahedron taking part in multicenter molecular bonding. Although after such a description little or no nucleophilic properties of P_4 can be expected, there are some metallic compounds in which the unit participates as a whole. Examples of some transition metal complexes containing the unit P_4 are shown in Table 4.15.

As observed in Table 4.11, the metallic fragments bonded to the P_4 molecules are, in general, soft Lewis acids that can both accept and back-donate electron density to the ligand. The π-acceptor capability characteristic of three-coordinated phosphor species appears thus to play also an important role in the case of P_4. Figure 4.55 reproduces a drawing of the structure of the compound $(np)_3 Ni(P_4)$ determinated by X-ray diffraction analysis. The coordination of

Table 4.15. Metal complexes of the tetraphosphorus molecule

Compound	Remarks	Ref.
$RhL_2X(\eta^2\text{-}P_4)$	$L = PPh_3$; $X = Cl$, Br, I; $L = P(m\text{-tol})_3$, $P(p\text{-tol})_3$, $AsPh_3$; $X = Cl$. Synthesis by reaction of L_3RhCl with P_4 at low temperature. Products are yellow solids, stable under inert atmosphere. Evidence from coordination of P_4 from ^{31}P and ^1H-NMR spectroscopy and X-ray diffraction analysis.	1, 2
$Ir(PPh_3)_2Cl(\eta^2\text{-}P_4)$	Orange solids with a chemistry similar to $RhL_2X(\eta^2\text{-}P_4)$	1
$Fe_3(CO)_{12}(P_4)$	Diamagnetic complex obtained from the reaction of $Fe_2(CO)_9$ with P_4 and characterized by IR and Mössbauer spectroscopies. ^{31}P-NMR spectrum exhibits only a simple line at room temperature. That has been interpreted in terms of ligand fluxionality.	3
$M(nP_3)(\eta^1\text{-}P_4)$	$M = Ni$, Pd; $nP_3 = tris(2\text{-diphenylphosphino-ethyl})amine$. Diamagnetic compounds prepared by reaction of the complexes $M(nP_3)$ with P_4 in THF at ca. 0 °C. Slightly air sensitive in the solid state. Structures determined by X-ray diffraction analysis. See Fig. 4.17.	4, 5

References

1   6 Ginsberg AP, Lindsell WE (1971) J. Am. Chem. Soc. 93:2082
2   8 Lindsell WE (1982) J. Chem. Soc., Chem. Commun., 1422
3   11 Schmid G, Kempny HP (1977) Z. Anorg. Allg. Chem. 432:160
4   12 Dapporto P, Midollini S, Sacconi L (1979) Angew. Chem. Int. Edn. 18:469
5   13 Dapporto P, Sacconi L, Stoppioni P, Zanobini F (1981) Inorg. Chem. 20:3834

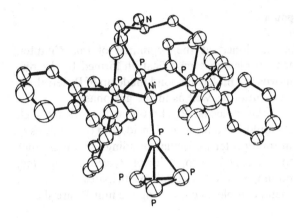

Fig. 4.55. Molecular structure of $(nP_3)Ni(P_4)$, $nP_3 = tris[2-di-(phenylphosphino) ethyl]amin. Reproduced with permission from Dapporto P, Millodini S, Sacconi L (1979) Angew. Chem. Int. Engl. 18:469

P_4 occurring by one apical P-atom appears to be common to this type of complex. The relatively short Ni-P(P_4) distance, about 0.25 Å shorter than the other four distances Ni-P(phosphine), agrees with the back-donation interaction mentioned above.

Of special interest in this class of compounds is the paramagnetic complex $P_4[Fe(CO)_4]_3$. Infra-red as well as Mössbauer spectroscopic investigations clearly indicate that there are two distinct $Fe(CO)_3$ groups, one with penta and two with hexa-coordinated iron atoms. The structure illustrated in Fig. 4.56 with two bridging and one apex-bond $Fe(CO)_4$ groups interprets such results. However ^{31}P-NMR of the compound at room temperature shows only one resonance signal for the P_4 fragment. These results can be conciliated with the structure in Fig. 4.56 by considering that the P_4 group is rotating, thus preventing the observation of magnetically non-equivalent phosphor atoms in the time scale of the NMR experiments.

Metal Phosphides. Elementary phosphorus reacts directly with electropositive metals under suitable conditions to give a series of phosphides with different phosphorus-metal ratios ranging from 0.3 in $M_3^I P$ to 16 in $M^I P_{16}$. Although the structural analysis of such metal phosphides reveals that many of them are

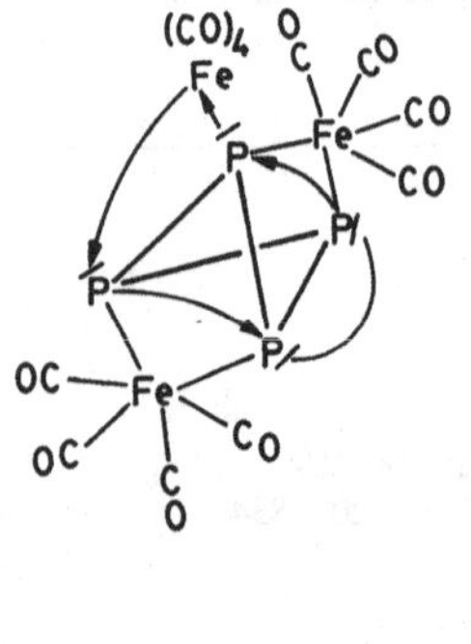

Fig. 4.56. Bonding and fluxionality in $[Fe(CO)_4]_3P_4$ [Schruid G, Kempny HP (1977) Z. Anorg Allg. 432-160]

polymeric species surrounded by metal atoms, there are a few that are made up of discrete polynuclear units isolated in the metal matrix. Examples of this last type of metal phosphide are described in Table 4.16. Figures 4.57 and 4.58 illustrate the structures of the polynuclear anions P_{16}^{2-} and P_{11}^{3-} determined in the compounds $(PPh_4)_2P_{16}$ and Na_3P_{11} respectively.

In general, the sensitivity towards hydrolysis of metal phosphides depends on the metal content. Metal-rich compounds undergo relatively straightforward hydrolysis while the behavior of those with higher phosphorus content resembles more that of the elementary phosphorus, specially with respect to their inertness toward water. As observed in Table 4.16, many polynuclear metal phosphide undergo hydrolysis giving rise to corresponding phosphor hydrides or phosphanes.

$$M_3P_7 + 3H_2O \rightarrow P_7H_3 + 3MOH$$

$$M_3P_{11} + 3H_2O \rightarrow P_{11}H_3 + 3MOH$$

Table 4.16. Selected metal Phosphides with cluster structures

Compound	Remarks	Ref.
$M_3^IP_7$	M = Li, Na, K, Rb, Cs. Bright yellow, extremely sensitive solids. Product of hydrolysis, P_7H_3, as well as physical characteristics indicate the presence of P_7^{3-} units. See text and Fig. 4.59.	1
$M_3^{II}P_{14}$	M = Sr, Ba. Red solids formed by quasi-molecular P_7^{3-} units which are hydrolyzed to P_7H_3.	2, 3
$M_3^IP_{11}$	M = Na. Orange solid formed by P_{11}^{3-} groups which are easily hydrolyzed to $P_{11}H_3$. See Fig. 4.58.	4
M^IP_7	M = Li, Na, Rb, Cs. Red solids, formed by P_7^- units stable to hydrolysis.	1
$M_2^IP_{16}$	M = Ph$_4$P. Red, remarkably stable solid that contains the polycyclic anion P_{16}^{2-}. (See Fig. 4.57).	5

References

1 Schnering HG, Wichelhaus W (1972) Naturwissenschaften 59:78
2 Dalhman W, Schnering HG (1972) Naturwissenschaften 59:420
3 Dalhman W, Schnering HG (1973) Naturwissenschaften 60:429
4 Wichelhaus W, Schnering HG (1973) Naturwissenschaften 60:104
5 Schnering HG, Manriquez V, Hönle W (1981) Angew. Chem. Int. Edn. 20:594

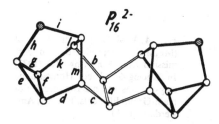

Fig. 4.57. Structure of the anion P_{16}^{2-} in $(PPh_4)_2P_{16}$. Reproduced with permission from Schnering HG, Manriquez V, Hönle W (1981) Angew. Chem. 93:606

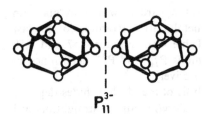

P_{11}^{3-}

Fig. 4.58. Structure of the amion P_{11}^{3-} in Na_3P_{11}. Reproduced with permission from Wichelhaus W, Schnering HG, (1973) Naturwissenschaften 60:104

Specially interesting is the anion P_7^{3-} which appears as the most stable of the phosphides. The molecular unit P_7 is found not only in M_3P_7 but also in other phosphides such as MP_7, MP_{11} and in M_4P_{21} as part of polymeric structures. Lithium salts of the anion P_7^{3-} may also be prepared in solution from either white phosphor or diphosphin by different dis- and comproportionation reactions.

$$3P_4 + 6LiPH_2 \longrightarrow 2Li_3P_7 + 4PH_3$$

$$9P_2H_4 + 3n\text{-BuLi} \xrightarrow[-20\,°C]{THF} Li_3P_7 + 11PH_3 + 3n\text{-BuH}$$

$$9P_2H_4 + 3LiPH_2 \xrightarrow[-20\,°C]{Monoglyme} Li_3P_7 + 14PH_3$$

Crystallographic studies of the compounds Sr_3P_{14} and Ba_2P_7Cl reveal that the anion P_7^{3-} is structurally similar to P_4S_3. It is worth noting that this phosphorsulfide which is isoelectronic with the anion P_7^{3-} is a specially stable and inert phosphor compound. The structure of the anion is essentially the same as determined for the molecular compound $(Me_3Si)_3P_7$ illustrated in Fig. 4.59.

The solubility of the salt Li_3P_7 in some organic solvents permits us to analyze the structure of the anion by ^{31}P-NMR. As illustrated by the spectra reproduced in Fig. 4.60, the number of the signals as well as their chemical shifts

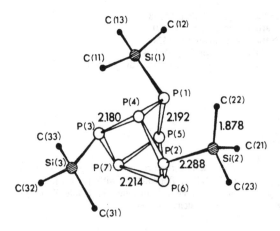

Fig. 4.59. Molecular structure of $(Me_3Si)_3P_7$. Reproduced with permission from Schnering HG (1977) In: Rheingold Al (ed) Homoatomic rings, chains and macromolecules of main-Group Elements. Elsevier, Amsterdam, p 330

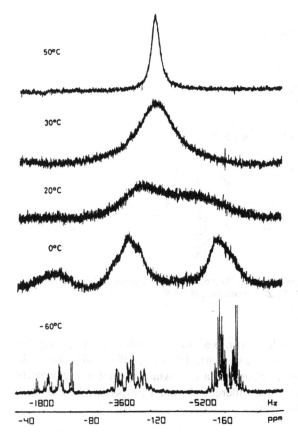

Fig. 4.60. ^{31}P-NMR spectrum of Li$_3$P$_7$ in d$_8$-tetrahydrofuran. Reproduced with permission from (1982) Baudler M Angew. Chem. 94:520

depends on the temperature. At low temperatures it is possible to distinguish clearly three groups of signals centered at $\delta = -57$, 103 and 162 ppm (vs. H$_3$PO$_4$ 85%) with intensity ratios 1:3:3. The high-field fine-structured doublet (162 ppm) is assigned to the phosphorous atom in the basal triangle; the downfield quartet corresponds to the apical phosphor atom while the pseudo triplet at the center of the spectrum belongs to the phosphor atoms linked to lithium atoms. At higher temperatures the signal coalesces pointing to a dynamic process. This process that is considered to be caused by valence tautomery should occur according to the scheme described in Fig. 4.61. The activation enthalpy associated to this process is estimated to be about 59.1 kJ mol^{-1}.

By comparing the behavior of these polycyclic species with that of the transition metal clusters, it is apparent that both situations are not totally equivalent. In metal clusters, intermetallic interactions are in general strong and normally it is impossible to distinguish the behavior of individual atoms. As discussed in Chapter 2, the assignation of oxidation numbers to individual metal atoms frequently does not make sense in cluster chemistry. Chemical and

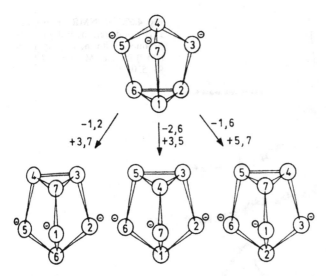

Fig. 4.61. Valence tautomery in the P_7^{3-} cage. Reproduced with permission from Baudler M, Ternberger H, Faber W, Hahn J (1979) Z. Naturfosch. 346:1690

electrochemical reactions often lead to mixed valence states. In the polycyclic phosphide anion P_7^{3-} described above there are, however, individual atoms such as those bridging base and apical phosphor atoms which are not only magnetically non-equivalent but also show different chemical properties. In this case there is a trapped-valence situation.

4.3.3 Arsenic and Antimony Derivatives

In spite of the more metallic properties of arsenic and antimony compared to those of phosphor, known homonuclear compounds of these elements do not differ very much from equivalent phosphor derivatives. The most stable arsenic and antimony cage compounds are indeed those isoelectronic with the anion P_7^{3-} described above.

The polyarsenic compound Ba_3As_{14} is prepared from the reaction of the elements at high temperature.

$$3\,Ba + 14\,As \xrightarrow[1000-1100\ K]{} Ba_3As_{14}$$

Single crystal X-ray diffraction analysis of the product shows that the structure of a heptaarsenic tetracyclene is qualitatively similar to that of the anion P_7^{3-}. A scheme of this structure as well as its principal molecular parameters are reproduced in Fig. 4.62.

The addition of an excess of ethylenediamine (en) to Na–Sb alloy in the presence of ligand 2,2,2-crypt (vide infra), a strong complexing agent of alkali

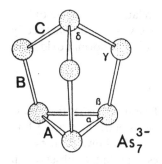

Fig. 4.62. Structure of the polycyclic anion As_7^{3-}. A = 249.8, B = 239.9, and C = 243.2 pm, $\alpha = 60$, $\beta = 105.2$, $\gamma = 99.0$, and $\delta = 101.2°$. Reproduced with permission from Schmettow W, Schnering HG (1977) Angew. Chem. 89:895

metal ions, leads to the formation of a crystalline product.

$$SbNa \xrightarrow[\text{[2,2,2-Cript]}]{\text{en}} [(2,2,2\text{-cript})Na^+]_3[Sb_7^{3-}]$$

The structure of this compound (see Fig. 4.67) shows that the species Sb_7^{3-} is isostructural with the species P_7^{3-} and As_7^{3-} described above.

Although all the anions E_7^{3-} (E = P, As, Sb) commented in this section may be considered formally as Zintl anions (vide infra), their properties do not agree properly with the anions with equivalent stoichiometries formed by the post-transition metals which will be discussed in the next Section.

4.4 Homo and Heteropolyatomic Anions of the Post-Transition Elements. Zintl Anions

Metallic lead as well as many other post-transition metals react with sodium dissolved in liquid ammonia changing the characteristic blue color of the solution, in the case of lead into an intense green. The electrolysis of the solutions shows that this green color is associated with anionic lead species that are plated out on the anode at about 2.25 Pb per Faraday. Accordingly, about the same amount of lead per sodium atom may be dissolved in a solution of the alkali metal in NH_3. These features correspond to the formation of the polyatomic anion Pb_9^{4-}.

$$9Pb + 4Na(NH_3) \xrightarrow{\text{liq. } NH_3} [Pb_9]^{4-} + 4Na^+$$

Although the first evidence of this type of reaction was reported in the last decade of the nineteenth century [Johannis A (1891) C.R. Hebd. Acad. Sci. 113:795], extensive systematic studies on this chemistry were reported by Zintl and co-workers in the 1930s (see references in Table 4.13).

Relevant information about polyatomic species formed in these reactions was originally obtained by potentiometric titrations. Although the reactions of the metals with sodium in liquid ammonia are often relatively slow, the oxidation of sodium in such experiments may be better followed potentiometrically by titrating the sodium solution with a salt of the metal in a normal oxidation state.

$$(4 + 2x)Na + xPbI_2 \longrightarrow Na_4Pb_x + 2xNaI$$

Figure 4.63, reproduced from an original article of Zintl, describes the course of two potentiometric and one conductometric titrations performed according to the equation above. The experiment labelled Nr. 6 clearly shows the presence of two ions Pb_7^{4-} and Pb_9^{4-}. The first of these species is no longer observed in experiment Nr. 7 carried out at a higher sodium concentration.

Relatively analogous results are obtained with other post-transition metals. The most representative polyanions obtained by Zintl and co-workers in liquid ammonia are listed in Table 4.17.

Although the presence of Zintl polyanions appears unquestionable in liquid ammonia their separation from this medium as solid products is not possible. The ammonia-free solids obtained by evaporation of metal saturated solutions at low temperatures followed by vacuum treatment are in general binary intermetallic phases and or simple mixtures of the components.

$$[Na(NH_3)_n^+]_4[Pb_9]^{4-} \longrightarrow 4NaPb_{2.25} + 4nNH_3$$

The isolation of solid compounds of some Zintl ions is however possible by carefully selecting both solvents and, specially, cations. Thus for instance, the compound $(Na)_4(en)_7Sn_9$ can be obtained by dissolving the alloy $NaSn_{2.4-2.6}$

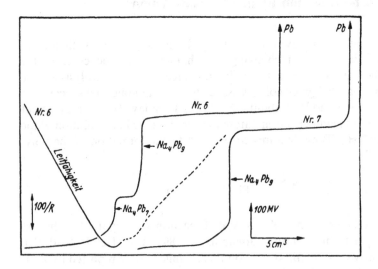

Fig. 4.63. Titrations of sodium in liquid ammonia with lead iodide. Sodium concentration in Experiment Nr. 7 is higher than that in experiment Nr. 6. Reproduced with permission from Zintl E, Gobeau J, Dullenkopt J (1931) Z. Phys. Chem. Abt. A. 154:1

Table 4.17. Examples of polyanions of post-transition metals obtained by E. Zintl in Liquid Ammonia[a]

Polyanion	Color of NH_3-solutions
Sn_9^{4-}	blood red
Pb_7^{4-}	green
Pb_9^{4-}	green to violet
As_3^{3-}	yellow
As_5^{3-}	dark red-brown
As_7^{3-}	dark red-brown
Sb_3^{3-}	deep red
Sb_7^{3-}	red-brown
Bi_3^{3-}	deep violet
Bi_5^{3-}	red-brown
Te_3^{2-}	violet
Te_4^{2-}	dark red

[a] Polyselenides and polysulfides described by the same authors have been omitted.

References

1 Zintl E, Goubeau J, Dullenkopf W (1931) Z. Phys. Chem., Abt. A 154:1
2 Zintl E, Harder A (1931) Z. Phys. Chem., Abt. A 154:47
3 Zintl E, Dullenkopf W (1932) Z. Phys. Chem., Abt. B 16:183
4 Zintl E, Kaiser H (1933) Z. Anorg. Allg. Chem. 211:113

in ethylenediamine (en) followed by the addition of monoglyme or tetra-hydrofuran (THF).

$$NaSn_{2.25} \xrightarrow{\text{en}} [Sn_9]^{4-} + 4Na^+(en)_n \xrightarrow[\text{or THF}]{\text{monoglyme}} Na_4en_7Sn_9$$

Specially important for obtaining solid products retaining the Zintl anions is the coordination of the cation. The high reduction power of these polyanions frequently produces, as mentioned above for the case of ammoniates, the formation of intermetallic species by cation reduction.

In order to stabilize the solids, large counter-ions as inert as possible toward reduction are therefore needed. Quaternary ammonium cations appear to be not inert enough. The use of sequestering agents for the counterions has shown itself to be a good method for stabilizing Zintl polyanions in solid compounds. Excellent results have been obtained with cryptands. Thus, for example, by using the compound 2,2,2-crypt (4,7,13,16,21,24-hexaona-1,10-diazabicyclo[8.8.8] hexacosane) with the structure illustrated in Fig. 4.64, it is possible to isolate a compound containing the species Pb_5^{2-} not detected in solution.

$$2KPb_{2.5} + 2(2,2,2\text{-cript}) \xrightarrow[50\,°C]{\text{en}} (2,2,2\text{-cript-K}^+)_2[Pb_5]^{2-}$$

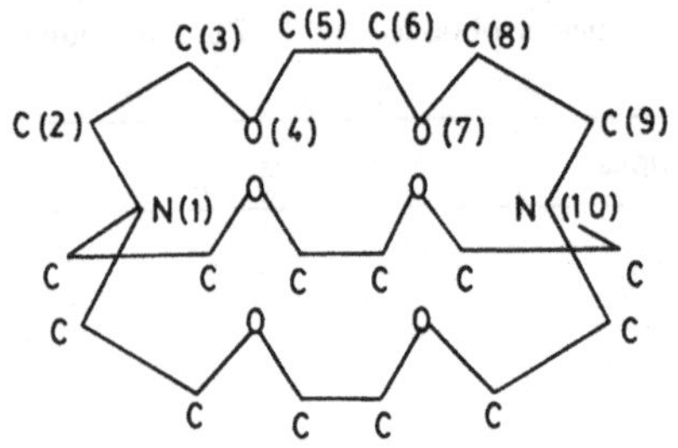

Fig. 4.64. 2,2,2-Crypt

Table 4.18 lists a series of homonuclear Zintl-anions derivatives with structures determined by single crystal X-ray diffraction most of which have been stabilized by this method.

The comparison of Tables 4.17 and 4.18 shows some significative differences in the species described in them. As exemplified with the species Pb_5^{2-} mentioned above, there are species that appear to be stable only in either solution or in the solid state. In the case of many polyanions with relatively high charge as the anions described by Zintl M_3^{3-} (M = As, Sb, or Bi) and M_5^{3-} (M = As or Bi), the demands in solvent dielectric constants and the stability to reduction of both solvent and counterions appear to be higher than those found until now. Although polyanion Pb_9^{4-} was the first to be discovered its stabilization in the solid state is still not possible. All experiments lead to the pentanuclear species Pb_5^{2-} mentioned above, probably due to the comproportionation reaction.

$$Pb + Pb_9^{4-} \longrightarrow 2Pb_5^{2-}$$

Stabilization with criptands also permits the synthesis of compounds containing heteropolyatomic cluster species similar to those described by Zintl. As can be seen in part B of Table 4.14 many of them are conformed by elements of neighbouring groups in the Periodic Table. The polarization of the system and the lost of electron delocalization normally associated with heteroatomic species appear not to be particularly important in this case. Possibly because in many of the structures of this class of clusters (vide infra), there are nonequivalent positions which can accomodate heteroatoms better.

Heteropolyatomic Zintl anions like the polyhomoatomic ones may also exist in solution. Actually, many heteronuclear compounds, which are not obtained as crystalline species, possibly due to the same causes discussed above for the homonuclear species, may be studied in solution. Multinuclear NMR spectroscopy has proved to be an excellent technique for such studies. Mixed polyanions containing atoms of the same group can apparently be prepared in solution practically in any proportion within the normal nuclearity range of the homonuclear species by dissolving the appropriate ternary alloys. Thus for instance, from sodium-thin-lead alloys in ethylenediamine solution the following series of cluster anions $[Sn_xPb_{9-x}]^{4-}$ may be stabilized.

$$4NaSn_{0.25x}Pb_{2.25-0.25x} \xrightarrow{\text{en}} [Sn_xPb_{9-x}]^{4-} + 4Na^+(en)$$

Table 4.18. Synthesis and structure of compounds containing zintl anions

Polyanion	Symmetry	Compounds	Synthesis and characteristics	Ref.
Homo polyatomic anions				
Ge_9^{2-} Ge_9^{4-}	$C_{2v}(D_{3h})$ C_{4v}	$(Crypt-K^+)_6Ge_9^{2-}Ge_9^{4-}\cdot2.5en$	KGe + 2,2,2-crypt in ethylene-diamine (en). Deep red rods.	1
Sn_9^{4-}	C_{4v}	$(Crypt-Na^+)_4Sn_9^{4-}$	$NaSn_{2.25}$ + 2,2,2-crypt in en. Dark-red to black rods.	2
Sn_9^{3-}	D_{3h}	$(Crypt-K^+)_3Sn_9^{3-}\cdot1.5en$	KSn_2 + 2,2,2-crypt in en.	3
Sn_5^{2-}	D_{3h}	$(Crypt-Na^+)_2Sn_5^{2-}$	NaSn to $Na_{1.7}Sn$ + 2,2,2 crypt in en. Orange-brown plates.	4
Pb_5^{2-}	D_{3h}	$(Crypt-M^+)_2Pb_5^{2-}$ M = Na, K	$NaPb_{2.25}$ + 2,2,2-crypt in en. Ruby red prisms.	4
Sb_4^{2-}	D_{4h}	$(Crypt-K^+)_2Sb_4^{2-}$	KGeSb + 2,2,2-crypt in en. Dark red wedges.	5
Bi_4^{2-}	D_{4h}	$(Crypt-K^+)_2Bi_4^{2-}$	K_5Bi_4 or K_3Bi_2 + 2,2,2-crypt in en. Dark green prisms or plates.	6
As_{11}^{3-}	D_3	$(Crypt-K^+)_3As_{11}^{3-}$	KAs_2 + 2,2,2-crypt in en. Deep red rods.	7
Sb_7^{3-}	C_{3v}	$(Crypt-Na^+)_3Sb_7^{3-}$	Na–Sb alloys + 2,2,2-crypt in en. Brown prisms or rods.	5, 8
Hetero polyatomic anions				
$Sn_2Bi_2^{2-}$	T_d	$(Crypt-K^+)_2Sn_2Bi_2^{2-}$	KSn_2 + K_2Bi_2 or KSnBi + 2,2,2-crypt in en. Black crystals.	9
$Pb_2Sb_2^{2-}$	T_d	$(Crypt-K^+)_2Pb_2Sb_2^{2-}$	KPbSb in en. Black plates.	10
$Tl_2Te_2^{2-}$	C_{2v}	$(Crypt-K^+)_2Tl_2Te_2^{2-}\cdot en$	KTlTe + 2,2,2-crypt in en. Dark brown crystals.	11
$TlSn_8^{3-}$ $TlSn_9^{3-}$	$C_{2v}(D_{3h})$ $C_{4v}(D_{4d})$	$(Crypt-K^+)_3(TlSn_9^{3-}$ $\cdot TlSn_8^{3-})_{1/2}\cdot en$	KTlSn + 2,2,2-crypt in en.	12
$As_2Se_6^{3-}$	C_{2h}	$(Crypt-Na^+)_2As_2Se_6^{2-}$	As + Se + 2,2,2-crypt in en. Orange crystals.	13

References

1 Belin CHE, Corbett JD, Cisar A (1977), J. Am. Chem. Soc. 99:7163
2 Corbett JD, Edwards PA (1977) J. Am. Chem. Soc. 99:3313
3 Critchlow SC, Corbett JD (1983) J. Am. Chem. Soc. 105:5715
4 Edwards PA, Corbett JD (1977) Inorg. Chem. 16:903
5 Critchlow SC, Corbett JD (1984) Inorg. Chem. 23:770
6 Cisar A, Corbett JD (1977) Inorg. Chem. 16:2482
7 Belin CHE (1980) J. Am. Chem. Soc. 102:6036
8 Adolphson DG, Corbett JD, Merryman DJ (1976) J. Am. Chem. Soc. 98:7234
9 Critshlow SC, Corbett JD (1982) Inorg. Chem. 21:3286
10 Critshlow SC, Corbett JD (1985) Inorg. Chem. 24:979
11 Burns RC, Corbett JD (1981) J. Am. Chem. Soc. 103:2627
12 Burns RC, Corbett JD (1982) J. Am. Chem. Soc. 104:2804
13 Belin CHE, Charbonnel MM (1982) Inorg. Chem. 21:2504

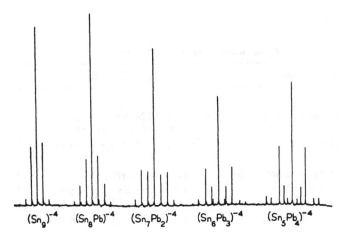

$(Sn_9)^{-4}$ $(Sn_8Pb)^{-4}$ $(Sn_7Pb_2)^{-4}$ $(Sn_6Pb_3)^{-4}$ $(Sn_5Pb_4)^{-4}$

Fig. 4.65. ^{119}Sn-NMR spectra of polyhetero nuclear anions $(Sn_xPb_{9-x})^{4-}$ in ethylediamine. $J(^{119}Sn-^{117}Sn) = 260\,Hz$; $J(^{119}Sn-^{27}Pb) = 560\,Hz$. Reproduced with permission from Rudolp RW, Taylor RC, Young DC (1979) In: Tsutsui M (ed) Fundamental research in homogeneous catalysis. Plenum, New York, p 997.

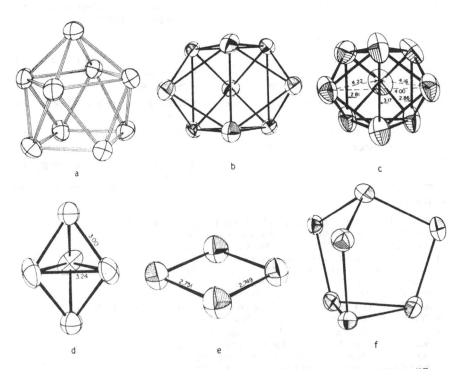

Fig. 4.66. Structures of some homoatomic Zintl anions established by single crystal X-ray diffraction determinations (a) Sn_9^{4-}, (b) Sn_9^{3-}, (c) Ge_9^{2-}, (d) Pb_5^{2-} (e) Bi_4^{2-}, (f) Sb_7^{3-}. Reproduced with permission from Corbett JD (1985) Chem. Rev. 85:383

As illustrated by Fig. 4.65 reproducing parts of a group of ^{119}Sn-NMR spectra corresponding to the reactions above, the complete series of nine compounds may be obtained. The chemical shifts are quite regular and depend linearly on the effective charge per tin atom in the cluster. Fine structure of the NMR spectra is due to ^{119}Sn coupling with both, ^{117}Sn and ^{207}Pb.

The structures of some of the homo and heteronuclear anion clusters mentioned in Table 4.14 are illustrated in Figs. 4.66 and 4.67. Almost all these structures are cluster species fundamentally similar to those of the transition metal clusters discussed in Chapter 2. They belong however to the so-called naked clusters because they do not need to be stabilized by ligands.

Configurations and bonding in naked clusters formed by the post-transition metals follow a pattern similar to that established for boranes and carboranes discussed in Sect 2.3 i.e. that in *closo*-deltahedra with n vertices there are n + 1 low-lying skeletal orbitals so the relationship between polyhedral shape and electron count corresponds to that described in Table 2.11.

Correlations between structure and electron counting for homo and heteronuclear Zintl ions are described in Table 4.19.

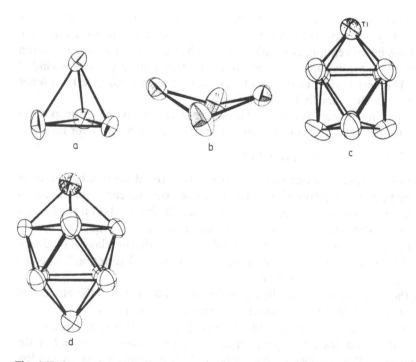

Fig. 4.67. Structures of some heteroatomic Zintl anions established by X-ray crystallography. (a) $Pb_2Sb_2^{2-}$. (b) $Tl_2Te_2^{2-}$, (c) $TlSn_8^{3-}$, (d) $TlSn_9^{3-}$. Reproduced with permision from Corbett JD (1976) Chem. Rev. 85:383.

Table 4.19. Electron counting and structure in homo and heteropolyatomic anions

Anion	Structure (symmetry)	No. of skeletal electreon pairs	Isoelectronic analogues
$Sn_2Bi_2^{2-}$	Tetrahedron (T_d)	6	Sb_4, Bi_4
$Ph_2Sb_2^{2-}$	Tetrahedron (T_d)	6	
Bi_4^{2-}, Sb_4^{2-}	Square planar (D_{4h})	7[a]	Te_4^{2+}, Se_4^{2+}
Pb_5^{2-}, Sn_5^{2-}	Trigonal bipyramid (D_{3h})	6	Bi_5^{3+}, $B_5H_5^{2-}$
Ge_9^{2-}	Tricapped trigonal prism (D_{3h})	10	$B_9H_9^{2-}$
$TlSn_8^{3-}$	Tricapped trigonal prism (D_{3h})	10	$B_9H_9^{2-}$
Sn_9^{3-}	Tricapped trigonal prism (D_{3h})	10.5	
Ge_9^{4-}, Sn_9^{4-}	Unicapped antiprism (C_{4v})	11[b]	Bi_9^{5-}, $B_9H_9^{4-}$
$TlSn_9^{3-}$	Bicapped square antiprism (D_{4h})	11	$B_{10}H_{10}^{2-}$

[a] *arachno*-Species.

[b] *nido*-Species.

4.5 Homopolyatomic Cations of the Post-Transition Elements

Naked polyatomic clusters of post-transition metals can also be obtained as cations. They correspond to intermediate species between the metals and the compounds in conventional oxidation states. The well known dimercury cation Hg_2^{2+}, stable in acidic conditions, may be considered a prototype of this kind of species. The best known examples of this class of compounds are the proper cluster species Bi_9^{5+} and Bi_8^{2+} (vide infra).

The limiting process for the stability of polycations is the disproportion reaction to the metal and compounds in normal oxidation states, for instance,

$$M_n^{m+} \longrightarrow (n-1)M + M^{m+}$$

Homonuclear cations can exist therefore only in the absence of basic groups which might donate electron pairs since the complexing via either solvent anion or added ligand will stabilize preferently species in higher oxidation states. This fact contrasts strongly with the nature of molecular transition metal clusters in which the polyatomic backbone is stabilized by different classes of donor ligands. In this sense, in the case of naked polycations the description "anticoordination chemistry" has been used.

On the other hand, the stability of polycations also requires the absence of oxidizable species that might induce the reduction to the metal.

The reaction conditions required for the preparation of polycationic ions are thus essentially opposed to those used for Zintl ions discussed above. For the stabilization of these species, extremely poor electron donor media are needed. Even the less basic usual molecular solvents often turn out to be not inert enough toward polyatomic cations. They can be better stabilized in strong acidic media. Acidic melt systems such as halometalate salt have proved to be suitable solvents in the synthesis of this class of compounds. Because of the disproportionization equilibrium mentioned above, salts of the same metal M are the most suitable.

The requirements of the counterions are also extreme. Large anions with relatively high inertness toward oxidation and with low basicity through coordination or dissociation to more basic fragments are the more appropriate for stabilizing naked cationic clusters. Best results have been obtained with the tetrachloroaluminate anion $[AlCl_4]^-$ which is normaly generated in situ from the metal halides and aluminum chloride used as solvents.

$$MCl + AlCl_3 \longrightarrow MAlCl_4$$

$NaAlCl_4$ itself serves as an excellent solvent for studies up to 151 °C (m.p.); it has moreover very wide "windows" for optical and spectrochemical studies.

Table 4.20. Homopolyatomic cations of post-transition metals

Polycation	Compound or phase	Ref.
Cd_2^{2+}	$Cd_2(AlCl_4)_2$	1, 2
Hg_2^{2+}	Hg_2Cl_2, $Hg_2(AlCl_4)_2$	3, 4
Hg_3^{2+}	$Hg_3(AlCl_4)_2$	4, 5
Bi_3^+	Metal-BiCl (6–10% Bi)	6, 7
Bi_5^{3+}	$Bi_5(AlCl_4)_3$	8, 9
Bi_8^{2+}	$Bi_8(AlCl_4)_2$	8
Bi_9^{2+a}	$BiCl_{1.167} = Bi_{12}Cl_{14} = Bi_9^{5+}\{(BiCl_5)^{2-}\}_2\{(Bi_2Cl_8)^{2-}\}_{1/2}$ $Bi_9^{5+}Bi^+\{(HfCl_6)^{2-}\}_3$	10–12
Se_8^{2+a}	$Se_8(AlCl_4)_2$	13
Te_4^{2+a}	$Te_4(Al_2Cl_7)_2$, $Te_4(AlCl_4)_2$	14
Te_6^{2+}	$Te_6(AlCl_4)_2$	15
Se_4^{2+}	$Se_4(AlCl_4)_2$, $Se_4(HS_2O_7)_2$	15, 16
Bi_8^{2+}	$Bi_8(AlCl_4)_2$	8

[a] Cation structure determined by X-ray diffraction.

References

1 Corbett JD, Burkhard WJ, Druding LF (1961) J. Am. Chem. Soc. 83:76
2 Potts RA, Barnes RD, Corbett JD (1968) Inorg. Chem. 7:2558
3 Yosim SJ, Mayer SW (1960) J. Phys. Chem. 64:909
4 Torsi G, Fung KW, Begun GM, Mamantov G (1971) Inorg. Chem. 10:2285
5 Ellison RD, Levy HA, Fung KW (1972) Inorg. Chem. 11:833
6 Topol LE, Yosim SJ, Osteryoung RA (1961) J. Phys. Chem. 65:1511
7 Corbett JD, Albers FC, Sallach RA (1968) Inorg. Chim. Acta 2:22
8 Corbett JD (1968) Inorg. Chem. 7:198
9 Corbett JD (1958) J. Phys. Chem. 62:1149
10 Hershaft A, Corbett JD (1963) Inorg. Chem. 2:979
11 Friedman RM, Corbett JD (1973) Inorg. Chim. Acta 7:525
12 Friedman RM, Corbett JD (1973) Inorg. Chem. 12:1134
13 McMullan RK, Prince DJ, Corbett JD (1971) Inorg. Chem. 10:1749
14 Couch TW, Lokken DA, Corbett JD (1972) Inorg. Chem,. 11:357
15 Prince DJ, Corbett JD, Garbisch B (1970) Inorg. Chem. 9:2731
16 Brown ID, Crump DB, Gillespie RJ (1971) Inorg. Chem. 10:2319

In Table 4.20 the properties of most of known polycationic clusters of the post-transition metals are described. In this Table selected examples of dinuclear species as well as some polycationic species of post-transition elements are also mentioned. Although the latter cannot be classified as clusters they have a chemistry similar to the compounds discussed in this section.

The element bismuth appears to have a special tendency to form polycations. As observed in Table 4.20 at least three cluster species, namely Bi_5^{3+}, Bi_8^{2+}, and Bi_9^{5+} have been clearly identified.

The general procedure for the synthesis of polybismuth cations involves the reduction of $BiCl_3$ with metallic bismuth in a medium composed by molten $(BiCl_3 + 3AlCl_3)$ or $(2BiCl_3 + 3HfCl_4)$:

$$4\,Bi + BiCl_3 + 3\,AlCl_3 \xrightarrow{350\,°C} Bi_5(AlCl_4)_3$$

By addition of bismuth and in the presence of an excess (20–33%) of aluminum chloride further reduction of the cluster may be afforded leading to the octanuclear species:

$$2(BiCl_3 + 3\,AlCl_3) + 22\,Bi \xrightarrow[270-320\,°C]{AlCl_3} 3\,Bi_8(AlCl_4)_2$$

The synthesis of the species with nuclearity nine is not formed in the tetrachloroaluminate system but it can be obtained from the mixture of $BiCl_3$ with hafnium tetrachloride:

$$(2BiCl_3 + 3\,HfCl_4) + 8\,Bi \xrightarrow{450\,°C} [(Bi_9)^{5+}\,Bi^+] + [(HfCl_6)^{2-}]_3$$

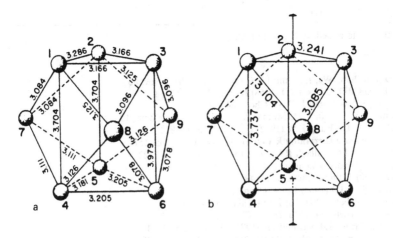

Fig. 4.68. Structures of the Bi_9^{5+} cation. (a) In $BiCl_{2.167}$ and (b) in $Bi_{10}(HfCl_6)_3$. Reproduced with permission from Corbett JD (1976) Prog. Inorg. Chem. 21:129

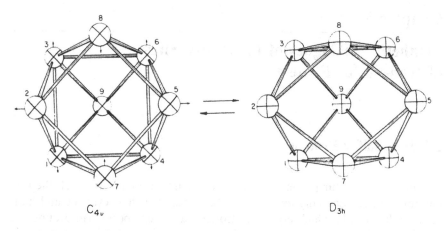

Fig. 4.69. Interconversion between the nine-atom polyhedra unicapped antiprism (C_{4v}) and the tricapped trigonal prism (D_{3h})

Figure 4.68 reproduces the single crystal X-ray diffraction structures determined for the species Bi_9^{5+} obtained from the reaction above (right) and from the solid phase $BiCl_{1.167}$ that corresponds properly to the compound $Bi_{12}Cl_{14}$, actually $Bi_9^{5+}([BiCl_5]^{2-})_2([Bi_2Cl_8]^{2-})_{1/2}$, obtained by reduction of the $BiCl_3$ melt:

$$14\,BiCl_3(\text{melt}) + 22\,Bi \xrightarrow[315-285°]{KCl-BiCl_3} Bi_{12}Cl_{14}$$

Bonding in polycationic clusters of the post-transition metals follows a pattern similar to Zintl ions, boranes, and transition metal clusters. All p-electrons are available for binding the skeleton cluster. According to the rules established from the analysis of boranes and carboranes (see Chapter 2), the cluster Bi_5^{3+} with 12 skeleton electrons being isoelectronic with the Zintl anions Pb_5^{2-} should have a trigonal-bipyramidal structure (D_{3h}). The structure of Bi_8^{2+} with 22 skeletal electrons can be systematized as an *arachno*-structure with n + 6 electrons (n = number of vertices). The structure observed for the species Bi_9^{5+} (Fig. 4.68) with 22 electrons has a symmetry D_{3h} that contrasts with the configuration of C_{4v} for the isoelectronic species Sn_9^{4-} appearing thus as rule-breaking. However as observed in Fig. 4.69, the interconversion between a unicapped antiprismatic structure (C_{4v}) and a tricapped trigonal prismatic structure (D_{3h}) only implies relatively slight changes in angles and interatomic distances. The differences in the energies of valence electrons as well as in the total cluster charge should surely compensate the small energy differences between the molecular orbitals corresponding to both configurations.

Chapter 5

Synthetic Analogues of the Active Sites of Iron-Sulfur Proteins

5.1 Iron-Sulfur Proteins

The name iron-sulfur protein applies to those iron proteins in which the iron environment is essentially formed by sulfur ligands such as cysteine and inorganic sulfur and in which no other strong ligands, e.g. porphyrin groups, are simultaneously present. This class of proteins is widely distributed in nature forming part of a number of organisms at different evolution levels such as in a variety of bacteria, algae, plants and mammals. Iron-sulfur proteins constitute a well-defined group of proteins that contain one, two, four, or eight iron atoms per molecule and whose molecular weights range between 6000 and 20 000. Selected examples of iron-sulfur proteins from different sources are shown in Table 5.1.

In the Table, iron-sulfur proteins are classified into three groups. The Rubredoxins (Rb) containing only lone Fe-atoms held to the protein by four cysteine groups. The Ferredoxins (Fd) containing from two to eight iron and an equal number of inorganic sulfur atoms per molecule. And finally the Molybdo-proteins containing molybdenum besides a number of both iron and inorganic sulfur atoms. Although the biological function of many of these proteins is not known in satisfactory detail, all of them appear to have at least two redox states coupled by one-electron transfer reactions per active site and to function principally as electron carriers. This section refers exclusively to the simple, low molecular weight iron-sulfur proteins known as Ferredoxins.

Among the principal characteristics of the ferredoxins the following can be mentioned: 1. Presence of one or two active centers per molecule containing iron and inorganic sulfur. 2. Relatively low potential electron-transfer reactions. However High-Potential (HP) proteins show exceptionally high potentials (see Table 5.1) so they have been sometimes classified into a special group. 3. Appearance of one signal at $g = 1.94$ in Electron Spin Resonance (ESR or EPR) spectra of frozen solutions at low temperature. 4. Participation as an electron carrier in different types of biological processes such as nitrogen fixation, nitrite reduction, photosynthesis, steroid hydroxilation in mammals etc.

Examples of ferredoxin reactions can be observed in the scheme reproduced in Fig. 5.1. Some of these reactions such as the reduction of NAD to NADH by H_2,

$$H_2 + Fd_{ox} \longrightarrow 2H^+ + Fd_{red}$$

$$Fd_{red} + NAD \longrightarrow Fd_{ox} + NADH$$

Table 5.1. Iron-sulfur proteins

Protein	Source	Mol wt $\times 10^{-3}$
A. Rubredoxins (Rb)		
One center Rb	Anaerobic bacteria (*clostridium pasteurianum*) Aerobe photosynthetic bacteria	6
Two centers Rb	*Pseudomonas oleovorans*	19
B. Ferredoxins (Fd)		
2Fe–2S Ferredoxins		
Plants Ferredoxins (Chloroplast ferredoxins)	Photosynthetic organisms, Spinach; blue-green, yellow, and green algae.	11.5
Putidaredoxin	*Pseudomonas putida.*	12
	Azotobacter vinlandii.	21, 24
Adrenodoxin	Bovine adrenal cortex	13
Mitochondrial non-heme iron-protein	Bovine heart mitochondria	26
4Fe–4S Ferredoxins		
One center Fd		
Bacterial Ferredoxins	*Desulfovibrio gigas*	6
Ferredoxins	*Bacillus polymyxa*	9
	Chromatium (HIPIP)	
	Azotobacter vinlandii	13
Two centers Fd		
Bacterial Ferredoxins	*Chromatium*	6
	Clostridia	6
	Azotobacter	15
C. Molybdoproteins		
Xanthine oxidase		
2Mo, 8Fe–8S	Mammals (milk, liver)	275
Nitrogenase (I)		
2Mo, 24–24(Fe–S)	*Clostridium pasteurianum*	220 (4 Sub-units)
1Mo, 17–18(Fe–S)	*Klebsiella phneumoniae*	218 (4 Sub-units)
1Mo, 33(Fe–S)	*Azotobacter vinlandii*	270 (2 Sub-units)

References

1 Mason R, Zubieta JA (1973) Angew. Chem. 86:390
2 Averill BA (1978) In: Sigel H (ed) Metal ions in biological systems. Marcel Decker, New York, p. 127
3 Lovenberg W (1973) (ed) Iron-Sulfur Proteins, vol 1. Academic Press, New York

or the anaerobic pyruvate oxidation,

$$\text{Pyruvate} + \text{Fd}_{ox} + \text{CoA} \longrightarrow \text{acetyl-CoA} + \text{CO}_2 + \text{Fd}_{red}$$

$$\text{Fd}_{red} + 2\text{H}^+ \longrightarrow \text{Fd}_{ox} + \text{H}_2$$

may be used for testing the activity of a number of Ferredoxins no matter what their origin. In general, Ferredoxins show a low selectivity. This feature is related to the similarity of the active sites in all iron-sulfur proteins (vide infra).

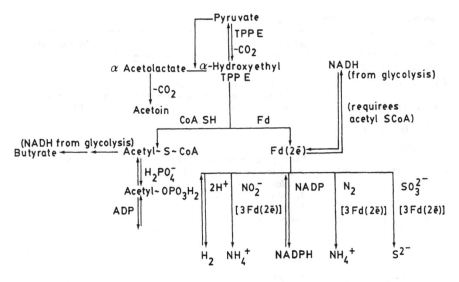

Fig. 5.1. Examples of chemical reactions involving ferredoxins

The structures of the active sites of 4-Fe and 8-Fe Ferredoxins has been univocally established by X-ray diffraction methods. Proteins of the 4-Fe type contain the $[Fe_4S_4(S\text{-}Cys)_4]$ site with a cubane-like structure as that illustrated schematically in Fig. 5.2. The structure of HP from *Chromatium vinosum* has been determined for two oxidation levels, HP_{red} and HP_{ox}. In both cases the active sites are tetranuclear clusters as shown in Fig. 5.2. Characteristic structural parameters for such centers can be observed in Fig. 5.3 where they are reproduced beside those of some synthetic analogues. As illustrated in Fig. 5.4, in the 8-Fe Fd_{ox} from *Peptococcus aerogenes* there are two identical tetranuclear active centers which are separated by about 12 Å. They are dimensionally very similar to the site in HP_{red}.

According to extensive physicochemical information the minimal composition of the active sites in 2Fe-Fd is $[Fe_2S_2(S\text{-}Cys)_4]$ with a structure as that

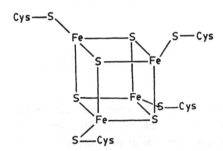

Fig. 5.2. Schematic representation of cubane-like structures found in 4-Fe sites in iron-sulfur proteins

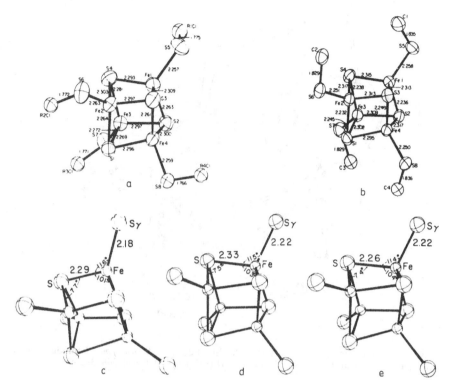

Fig. 5.3. Tetrairon cluster in the synthetic analogues $[Fe_4S_4(SPh)_4]^{2-}$ (**a**) and $[Fe_4S_4(SCH_2Ph)_4]^{2-}$ (**b**), and from active sites in *P. aerogenes* Fd_{ox} (**c**) *Chromatium* HP_{red} (**d**) and *Chromatium* HP_{ox}. Reproduced with permission from Holm RH, Ibers JA (1977) In: Lovenberg W (ed) Iron-sulfur proteins, vol 3, Academic, New York, chap 7

illustrated in Fig. 5.5 containing two tetrahedrally coordinated iron atoms bridged by sulfide. Mössbauer and ENDOR (Electron Nuclear Double Resonance) studies indicate that the active site in the reduced form of 2Fe–Fd contains two high-spin iron atoms, one Fe(II) and one Fe(III); meanwhile in the oxidized form it contains two Fe(III).

An interesting aspect in the structures of iron-sulfur proteins are some amino acid sequences which are approximately the same in a number of proteins. Thus, as observed in Fig. 5.6, there are also some sequences, e.g. Cys-X-X-Cys or Cys-X-X-Cys-X-X-Cys, in which the relative positions of the cysteine residues which are known to directly bind the Fe-S are absolutely conserved in proteins from different species. Since the coordinative environment of iron sulfur cluster in different species are practically the same, the specificity of sites in a given protein are fundamentally determined by the configuration of the rest of the polypeptidic chain. This feature, besides the wide distribution of ferredoxins in numerous different species, permits us to establish philogenetic relationships as that illustrated in Fig. 5.7.

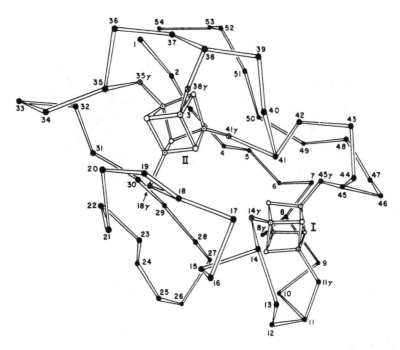

Fig. 5.4. Schematic diagram of main chain of the 8-Fe Fd$_{ox}$ from *M. aerogenes*. Reproduced with permission from Adman ET, Sieker LC, Jansen LH (1973) J. Biol. Chem. 248:3987

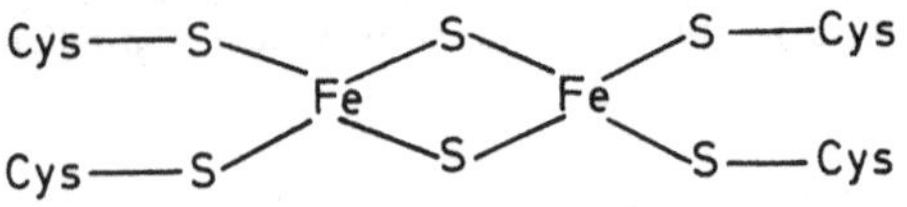

Fig. 5.5. Schematic structure of active sites in 2Fe-Fd

5.2 Synthetic Analogues of the Active Sites of Iron-Sulfur Proteins

The furthering of biological studies by chemical investigation related to the composition and stereochemistry of the biological coordination units constitutes an approach specially relevant for understanding protein properties. However such an approach is possible only by using relatively low molecular-weight complexes able to duplicate the biological unit in terms of composition, type of ligands, structure, redox properties, etc., i.e. small model molecules able to mimic the chemical behavior of protein active centers. Environmental and other effects arising from protein conformation are however not considered by such models. Nonetheless, since they reflect the intrinsic properties of the coordination unit unmodified by protein matrix effects, the latter can then be estimated by comparing both situations.

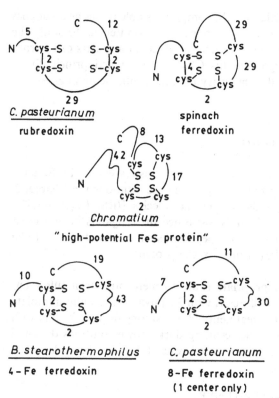

Fig. 5.6. Schematic representation of the relative positions of the cysteine groups in ferredoxins from different species. Numbers indicate the number of residues between the cysteine groups

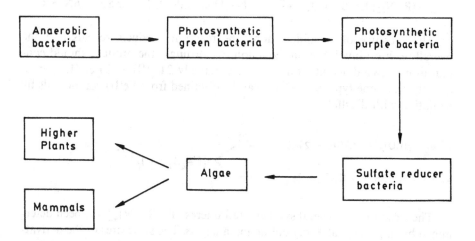

Fig. 5.7. Phylogenetic relationship deduced from the analysis of Ferredoxin distribution in different species

The synthesis of low molecular-weight compounds able to act as analogues of the prosthetic group in iron-sulfur proteins has permitted us to study the chemical properties of these centers as well as, indirectly, the matrix or environmental effects they are undergoing. This Chapter is devoted principally to analyzing pertinent aspects of the chemistry of some synthetic analogues.

5.2.1 Tetrameric Iron-Sulfur Clusters

Some aspects of the chemistry of transition metal clusters with a Fe_4S_4 core, specifically those related to their redox properties were mentioned in Chapter 2. To the same class of compounds belong the species $[(RS)_4Fe_4(\mu_3\text{-}S)_4]^{m-}$ (m = 1, 2, 3, 4) with mercaptide ligands SR. Since these ligands have a great similarity to the cysteinyl residues in polypeptides, such clusters turn out to be appropriate analogues of the Fe_4S_4 units in the protein.

Synthesis and Structure. An appropriate route to mercaptide-iron sulfur tetrameres is the anaerobic reaction of $FeCl_3$ with 3 equivalents of sodium thiolate in methanol solution followed by treatment of the resulting polymeric solid with alkali sulfide in methanol. From the resulting dark-brown solutions dark-red crystalline products like the oxidized ferredoxins, Fd_{ox}, can be isolated by addition of quaternary cations.

$$4FeCl_3\ 12NaSR \longrightarrow (4/n)[Fe(SR)_3]_n$$

$$4NaHS + 4NaOMe$$

$$\downarrow$$

$$(R_4N)_2[Fe_4S_4(SR)_4] \xleftarrow{R_4N^+} Na_2[Fe_4S_4(SR)_4] + RSSR + 6NaSR$$

The presence of disulfide among the reaction products reveals that the overall reaction involves iron reduction. Accordingly the product should contain iron in two different oxidation states namely $2Fe(II) + 2Fe(III)$. Consequently, the same type of products can be obtained from Fe(II) compounds by oxidation with disulfide.

$$2Fe_2S_2(CO)_6 + RSSR + 2RS^- \xrightarrow{-6\,CO}$$
$$\Big\rangle [Fe_4S_4(SR)_4]^{2-}$$
$$4FeS + RSSR + 2RS^-$$

The structures of several synthetic tetrameres $[Fe_4S_4(SR)_4]$ has been determined by single crystal X-ray diffraction analysis. The structures of the derivatives with R = CH_2Ph and Ph are shown in Fig. 5.3. The comparison of these structures with those of the prosthetic groups of *P. aerogenes* Fd_{ox} and

Chromatium $HP_{ox,red}$ also illustrated in the same Figure shows that from a structural point of view the synthetic cluster is an appropriate model of the active sites in these proteins.

In both cases the basic symmetry is the same, i.e. a tetrahedron of four Fe atoms interpenetrated by a second, larger tetrahedron of four sulfur atoms. Observed symmetry D_{2d}, which may be considered as resulting from a distortion of a more symmetric arrangement T_d, is similar for both synthetic and natural clusters. One of the very important features is that crystallographic data for all studied compounds do not show any difference between the iron atoms constituting the clusters, i.e. there are no localized Fe(II,III) valence states but mixed-valence species.

Spectroscopic Properties. Characteristic properties such as color, paramagnetism, sensitivity to the Mössbauer Effect etc. which tetrameric units $[Fe_4S_4(SR)_4]$ confer on both iron-sulfur proteins and their synthetic analogues make the investigation of these compounds by spectroscopic methods be very favorable. IR and UV-visible absorption spectroscopies, Raman resonance spectroscopy, nuclear and electron magnetic resonance, photoelectron spectroscopy, Mössbauer spectroscopy and magnetic susceptibility determinations are useful tools for characterizing specific compounds and especially for comparing the behavior of the proteins with that of their synthetic models. In the case of the iron-sulfur proteins it is specially interesting to determine if the $[Fe_4S_4(SR)_4]^{m-}$ anions are isoelectronic with the structurally similar units in Ferredoxins and High-Potential proteins.

UV-visible spectra of solutions of synthetic tetrameres $[Fe_4S_4(SR)_4]^{2-}$ (SR = alkyl thiolate or N-acetylcysteine-N'-methylacetamide) in organic solvents show absorptions in the range 413–420 nm which are in general not very sensitive to solvent effects: blue-shifts of 4–6 nm on going from pure solvents to 80% DMSO. In comparison, the spectra of proteins in aqueous solutions are relatively poorly resolved showing only a shallow peak or shoulder at 388–400 nm. However under partially denaturing conditions, as for instance aqueous-organic solvent mixtures, e.g. 80% DMSO, the maxima in the spectra of both the synthetic analogue and the protein are practically coincident.

Magnetic susceptibility measurements and nuclear magnetic spectra also show that both proteins and $[Fe_4S_4(SR)_4]^{2-}$ synthetic species have similar magnetic properties displaying magnetic moments per ion of about 1.0 BM at room temperature. These values are rather below those expected for iron(II) or iron(III) compounds. This feature, as well as variable temperature magnetic susceptibility determinations indicates behavior characteristic of intramolecular antiferromagnetic coupling. This effect is also observed in proton magnetic resonance spectra of both cysteine groups in the protein and alkyl thiolate groups in the synthetic analogue. Isotropic downfield shifts respect to the normal resonance of the ligands, caused by their contact interactions with the Fe_4S_4 paramagnetic core, of approximately the same magnitude are observed in both cases.

The equivalence of the oxidation state in both protein active sites and tetranuclear synthetic analogues is also clearly established by comparing the isomer shifts ^{57}Fe observed in the corresponding Mössbauer spectra. Isomer shifts may be indeed considered to be a measure of the electron density in vicinity of the iron atom in a given species relative to that in a reference compound. For iron compounds with equivalent coordination spheres, the isomer shifts correlate with iron formal oxidation states. Isomers shifts observed for both proteins and model compounds are consistent with the hypothesis that synthetic species $[Fe_4S_4(SR)_4]^{m-}$ are isoelectronic with the active sites in the 4Fe and 8-Fe-Ferredoxins and HP protein. Mössbauer spectra confirm moreover that all Fe-atoms in Fe_4S_4 units in synthetic and natural species are strictly equivalent in the time-scale of this technique (ca. 10^{-7} s) so they are, as mentioned above, electronically delocalized systems.

Oxidation-Reduction Reactions. As already mentioned, iron-sulfur proteins are the group of electron-transfer substances most numerous and widely distributed in biological organisms. The most important property of these proteins is therefore their ability to undergo reversible electron transfers between different oxidation states of the active sites.

Analogously to the tetranuclear cluster with Fe_4S_4 core analyzed in Chapter 2, the synthetic analogues $[Fe_4S_4(SR)_4]^{m-}$ show a clear tendency to participate in redox reactions. Indeed they have proved to undergo the electron transfer series summarized in Table 5.2.

One of the great advantages of the studies with synthetic analogues is the possibility of establishing a clear relationship of structural and analytical data with electrochemical properties and thus deducing the net charge per complex unit. Such a kind of information being essential for specifying the oxidation level of the active site is not possible to obtain directly from the protein it self. From the correlation between properties of both protein and synthetic analogues it is indeed possible to find univocally the correspondence between protein oxidation or reduction states and oxidation levels in the synthetic analogues which is

Table 5.2. Electron-transfer series in synthetic analogues $[Fe_4S_4(SR)_4]^x$ and their relationshiop with the oxidation states observed in iron-sulfur proteins

Synthetic analogues	$[Fe_4(SR)_4(\mu_3\text{-}S)_4]^{4-}$	$[Fe_4(SR)_4(\mu_3\text{-}S)_4]^{3-}$	$[Fe_4(SR)_4(\mu_3\text{-}S)_4]^{2-}$	$[Fe_4(SR)_4(\mu_3\text{-}S)_4]^{-}$
Proteins oxidation States		Fd_{red} $[HP_{s\text{-}red}]$	Fd_{ox} HP_{red}	$[Fd_{s\text{-}ox}]$ HP_{ox}
Formal oxidation States of Fe	4Fe(II)	3Fe(II) + Fe(III)	2Fe(II) + 2Fe(III)	Fe(II) + 3Fe(III)

Reference

De Pamphilis BV, Averill BA, Herskovitz T, Que Jr. L, Holm RH (1974) J. Am. Chem. Soc. 96:4159

also mentioned in Table 5.2 (vide infra). The starting-point for establishing such a relationship is the congruence of the active site properties in the proteins HP_{red} and Fd_{ox} with those of the dianions $[Fe_4S_4(SR)_4]^{2-}$ which reveals that these three species have the same oxidation level.

As shown in Table 5.2 the oxidation states of the tetrameres may also be formally rationalized considering they are built up of iron atoms with oxidation states $+3$ and $+2$ which are the normal oxidation states for iron coordinated tetrahedrically by weak-field ligands. However as emphasized above, the system as a whole must be considered for proteins as well as for synthetic analogues as a mixed-valence system and in no case as a trapped-valence one constituted by distinguishable iron atoms with different oxidation states. This feature that determine that these tetrameric iron-sulfur species are the first non classical system known in biology is supported by spectroscopic techniques able to detect electronic states with half-life times in a wide range $(10^{-4}$–10^{-16} s) and at temperatures as low as 1.5 K. All these studies completely corroborate the crystallographic results mentioned above, establishing that iron atoms in the tetrameres are always totally equivalent.

Non-bracketed forms of the protein shown in Table 5.2 are doubtless involved *in vivo* and can be also isolated or produced *in vitro*. It is observed that the oxidation levels of the synthetic analogues enclose all oxidation levels known for the proteins 8Fe- and 4Fe-Fd, suggesting moreover the existence of other levels as those mentioned in brackets in Table 5.2. One of these levels which, corresponding to the "super-reduced" form of the HP protein $HP_{s\text{-}red}$, can indeed be obtained by dithionite reduction of HP_{red} under denaturing conditions.

The equivalence of oxidation levels between analogues and proteins also provides a simple answer to the difference in the potential of about 0.8 V existing between the couples HP_{red}/HP_{ox} and $8Fe\text{-}Fd_{red}/Fd_{ox}$ which was originally considered paradoxical. Fd_{ox} and HP_{red} are indeed isoelectronic, whereas Fd_{red} and HP_{ox} differ by two electrons. The potentials originating this difference do not refer to the same redox couple and are therefore not directly comparable.

Although the comparison of the electrochemical behavior of analogue and protein is qualitatively excellent, there are some quantitative differences which can be attributed to medium or matrix effects. Thus for instance, by comparing the potentials of isoelectronic series of analogues and proteins, the former dissolved in organic solvents or aqueous solvent mixtures and the proteins in water, it is possible to observe as shown in Table 5.3 differences of about 0.2 Volt. Nevertheless, the potentials of the proteins in partially non-aqueous media tend to those of the analogues. The differences decrease with decreasing water content. Since non-aqueous media are often denaturing, observed differences in both media may be interpreted as matrix effects caused fundamentally by the protein tertiary structure. As mentioned above, these partially denaturing conditions permit the stabilization of the super-reduced form of HP-protein, i.e. the protein displays under these conditions a behavior similar to that of the analogues. The feature observed in the protein 8Fe–Fd from *A. vinelandii* which

Table 5.3. Oxidation-reduction potentials of iron-sulfur proteins and analogues in different media

Species	Medium	Couple[a]			Ref.
		4–/3–	3–/2–	2–/1	
One-Iron					
1-Fe-Proteins	H_2O, pH7	—	—	[− 0.60]	1
$[Fe(S_2\text{-}o\text{-Xyl})_2]^-$	DMF	—	—	− 1.03	2
Two-Iron					
2-Fe-Proteins	H_2O, pH7	—	[− 0.24– − 0.43]	—	1, 3
$[Fe_2S_2(S_2\text{-}o\text{-Xyl})_2]^{2-}$	DMF	− 1.73	− 1.49	—	4
$[Fe_2S_2(SPh)_4]^{2-}$	DMF	− 1.37	− 1.09	—	4
Four-Iron					
4-Fe-Ferredoxins	H_2O, pH7	—	[− 0.28– − 0.42]	—	5–7
HIPIP	H_2O, pH7	—	—	[+ 0.35]	1
HIPIP	DMSO-H_2O(4:1)	—	[≤ − 0.64)]	—	8
$[Fe_4S_4(S\text{-Cys(Ac)NHMe})_4]^{2-}$	H_2O, pH7	—	[− 0.49]	—	9
$[Fe_4S_4(S\text{-Cys(Ac)NHMe})_4]^{2-}$	DMSO(80%)	—	− 0.91	—	9
$[Fe_4S_4(SEt)_4]^{2-}$	DMSO(80%)	—	− 1.16	—	9
$[Fe_4S_4(SEt)_4]^{2-}$	DMF	− 2.04	1.33	—	10
$[Fe_4S_4(SPh)_4]^{2-}$	DMF	− 1.75	− 1.04	—	10
$[Fe_4S_4(S\text{-}t\text{-Bu})_4]^{2-}$	DMF	− 2.16	− 1.42	− 0.12	10

[a] Potentials in Volt versus Saturated Calomel Electrode, 25 °C for the analogues and versus hydrogen electrode at pH7 for proteins [in bracket].

References

1 Hall DO, Cammack R, Rao KK (1974) In: Jacobs A, Worwood M (eds) Iron in biochemistry and medicine, 8. Academic Press, New York
2 Lane RW, Ibers JA, Frankel RB, Holm RH (1975) Proc. Nalt. Acad. Sci. USA 72:2868
3 Estabrook RW, Suzuki K, Mason JI, Baron J, Taylor WE, Simpson ER, Purvis J, McCarthy J (1973) in: Lovenberg W (ed.) Iron-Sulfur Proteins, 1. Academic Press, New York
4 Mayerle JJ, Denmark SE, DePamphilis BV, Ibers JA, Holm RH (1975) J. Am. Chem. Soc. 97:1032
5 Zubieta JA, Mason R, Postgate JR (1973) Biochem. J. 133:851
6 Stombaugh NA, Burris RH, Orme-Johnson WH (1973) J. Biol. Chem. 248:7951
7 Mullinger RN, Cammack R, Rao KK, Hall DO, Dickson DPE, Johnson CE, Rush JD, Simopoulos A (1975) Biochem. J. 151:75
8 Cammack R (1973) Biochem. Biophys. Res. Commun. 54:548
9 Que Jr L, Anglin JR, Bobrik MA, Davison A, Holm RH (1974) J. Am. Chem. Soc. 96:6042
10 De Pamphilis BV, Averill BA, Herskovitz T, Que Jr L, Holm RH (1974) J. Am. Chem. Soc. 96:4159

contains two 4Fe-4S clusters with two different redox potentials, − 0.42 and + 0.35 V for the same − 1/ − 2 couple reveals that the polypeptide chain can apparently regulate the potential of the prosthetic group in a relatively wide range.

Substitution Reactions. Electron-transfer reactions analyzed above are not the only type of reaction in which tetranuclear active sites of iron-sulfur proteins and their synthetic analogues can participate retaining their cluster structures. Under anaerobic conditions they can also undergo ligand exchange processes as substitution reactions of the thiolate ligands.

Synthetic analogues described above are, in general, high spin paramagnetic species with metal centers coordinated tetrahedrally. Such kind of complexes are normally labile to ligand exchange. Thus for instance the tetrameric anions $[Fe_4S_4(SR)_4]^{2-}$ can be easily converted in the corresponding halides.

$$[Fe_4S_4(SCH_2Ph)_4]^{2-} + 4PhCOX$$

$$\rightarrow [Fe_4S_4X_4]^{2-} + 4PhCH_2SCOPh \qquad X = Cl, Br$$

The iodide derivative can be prepared by metathesis.

Structural studies of the halides indicate that the compounds maintain practically unaltered cubane-like structures. Also the antiferromagnetic coupling appears to be the same than that in the original clusters. However redox potentials are affected in some extend by ligand exchange. Thus the $E_{1/2}$ values for the couple 2-/3- in $[Fe_4S_4X_4]^{2-}$ are -1.25 V and -1.22 V (vs Fc^+/Fc) for the chloride and bromide respectively, meanwhile for the thiobenzolate derivative under similar conditions is -1.56 V.

Direct exchange reactions with nucleophiles are also possible. Thus, the water soluble species $[Fe_4S_4(SCH_2CH_2CO_2)_4]^{6-}$ undergoes in aqueous media a series of equilibria:

$$[Fe_4S_4(SR)_4]^{6-} + nL \longrightarrow [Fe_4S_4(SR)_{4-n}L_n]^{2-} + n(SR)^{2-}$$

where L can be OH^-, CN^- or X^- as well as H_2O. In this last case the product should be the solvated parent species $[Fe_4S_4]^{4+}$.

Thiolate exchange reactions are especially important for obtaining products which cannot be prepared directly as well as for extracting and exchanging protein prosthetic groups (vide infra). They allow the straightforward incorporation of new and often complex ligand systems around a preformed core to be made.

By addition of a thiol to non-aqueous solutions of a given tetrameric iron-sulfur derivative, fast equilibria are established.

$$[Fe_4S_4(SR)_4]^{2-} + nR'SH \rightarrow [Fe_4S_4(SR)_{4-n}(SR')_n]^{2-} + nSRH$$

These equilibria can be followed by spectroscopic techniques, specially NMR and UV-vis absorption spectroscopy. Although they often lead to near statistical product distributions, there is a certain influence of thiol acidity. Thus for the reaction

$$[Fe_4S_4(S-t-Bu)_4]^{2-} + 4RSH \rightarrow [Fe_4S_4(SR)_4]^{2-} + 4t-BuSH$$

the following series of affinities have been established.

alkyl < Ac-Cys-NHMe < aryl

There the substitution degree increases with increasing thiol acidity. In these

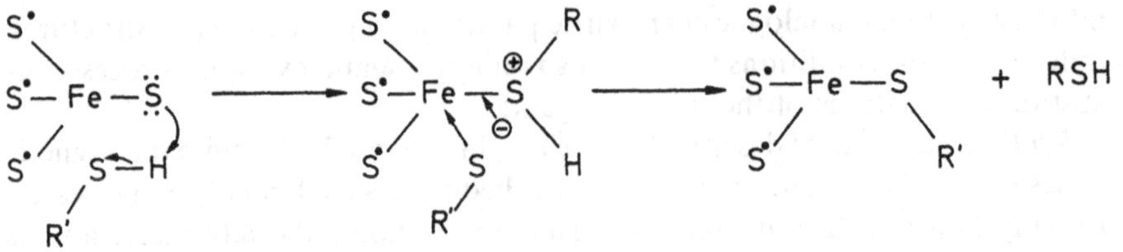

Fig. 5.8. Possible mechanism for ligand substitution in tetranuclear iron-sulfur $[Fe_4S_4SR_4]^{2-}$ species

equilibria, acid-base properties appear to be in general more important than structural factors.

Kinetic studies of thiolate substitution in alkyl (R = ethyl or tert-butyl) tetrameres dianions with arylthiols (R' = XC_6H_4SH with X = p-NH$_2$, p-CH$_3$, p-NO$_2$, o-NO$_2$) indicate that the rates are overall second order, first order in each tetramere and thiol. The most interesting feature of these kinetic studies is the dependence of rate constants on thiol acidity. This has led to the postulation of the mechanism illustrated in the scheme on Fig. 5.8. Here the slow protonation step should be followed by fast separation of the alkylthiol from and subsequent capturing of the generated arylthiolate anion by the metal center.

Among thiolate substitution reactions, those permitting the incorporation of oligopeptides are specially interesting. Three examples of this type of reactions are shown schematically in Fig. 5.9. Because of the volatility of *tert*-butylthiol it can be easily removed from DMSO reaction mixtures permitting practically quantitative product formation. The dianion $[Fe_4S_4(S\text{-}t\text{-}Bu)_4]^{2-}$ is therefore a very convenient starting material in the preparation of other thiolates by ligand exchange. The treatment of the peptide complexes with benzenethiol leads to their quantitative conversion to $[Fe_4S_4(S\text{-}t\text{-}Bu)_4]^{2-}$.

5.2.2 Dinuclear 2Fe-2S Analogues

Although there are three types of active sites in non-heme iron proteins, namely containing one, two, and four iron atoms per center, in the discussion above only tetrameric Fe-S species have been considered. This topic has been treated with some detail because these compounds meet properly the concept of cluster and are therefore clearly within the scope of this book. That is not the case for the di-iron species which form the 2Fe-2S active sites in the ferredoxins nor for the mononuclear center in the rubredoxins. However, in order to get a better understanding of the chemistry of the iron-sulfur proteins, an overview of the chemistry of the 2Fe-2S analogues is also outlined in this Section.

The relatively high stability of the species $[Fe_4S_4(SR)_4]^{2-}$ prevents, in most cases, the direct synthesis of the dimers $[Fe_2S_2(SR)_4]^{2-}$. However in some special cases, namely by using ligands as the 1,2-dithiols whose structures are

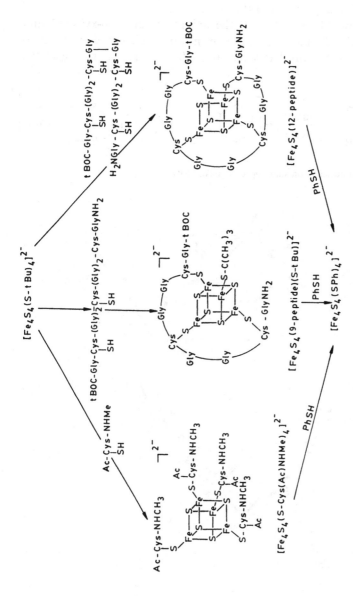

Fig. 5.9. Thiolate substitution reaction in $[Fe_4S_4SR_4]^{2-}$ species [Que Jr. L, Anglin JR, Bobrik MA, Davison A, Holm RH (1974) J. Am. Chem. Soc. 96: 6042]

not appropriate for compounds with cubane-like geometries, these dinuclear compounds may be obtained with good yields. The more widely used ligand for that purpose is the o-xylene-α, α'-dithiol. Other derivatives can be obtained straightforwardly by ligand exchange. The scheme in Fig. 5.10 illustrates some examples of compounds prepared in this way.

Figure 5.11 shows the structure of the anions $[Fe_2S_2(S_2\text{-}o\text{-xyl})_2]^{2-}$ and $[Fe_2S_2(S\text{-}p\text{-tolyl})_4]^{2-}$. In these structures iron atoms are coordinated tetrahedrically and the Fe_2S_2 core has a planar configuration. The nature of the

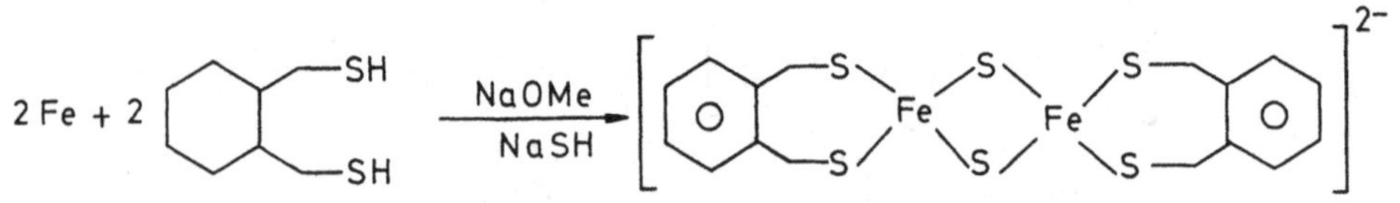

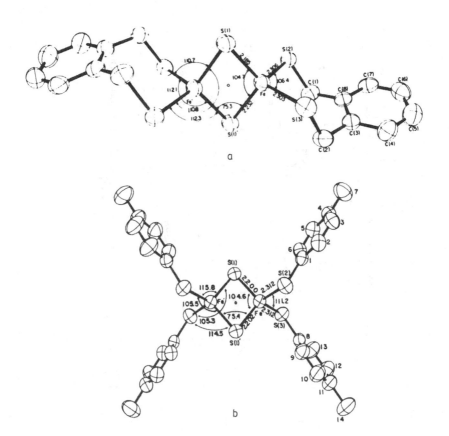

Fig. 5.10. Synthesis of dinuclear iron-sulfur species $[Fe_2S_2(SR)_4]^{2-}$

Fig. 5.11. Molecular structure of the anions $[Fe_2S_2(S_2\text{-}o\text{-xyl})_2]^{2-}$ and $[Fe_2S_2(S\text{-}p\text{-tolyl})_4]^{2-}$. Reproduced with permission from Maryerle J, Denmark, DePanphilis BV, Ibers JA, Holm RH. (1975) Am. Chem. Soc. 97:1032

ligand does not significantly affect the dimensions of the Fe–S core. Fe–Fe distance is relatively short (2.7 Å) and indicative of a stabilizing metal–metal interaction.

Optical spectra and magnetic susceptibility measurements of the o–xylyl dimer and 2Fe–Fd$_{ox}$ indicate that the Fe$_2$S$_2$ core in both the analogue and the proteins are isoelectronic and virtually isostructural. Moreover, electrochemical studies of the synthetic dimers reveal the presence of two quasi-reversible waves corresponding to the formation of the anions [Fe$_2$S$_2$(SR)$_4$]$^{3-}$ and [Fe$_2$S$_2$(SR)$_4$]$^{4-}$. The stabilization of these species is, however, not possible because of dimerization to [Fe$_4$S$_4$(SR)$_4$]$^{2-}$. Comparison of the analogue with 2Fe-Fd$_{red}$ is therefore difficult. Nevertheless, as shown by the following scheme, these synthetic dimer analogues possess, in principle, the required redox properties.

$$[Fe_2S_2(SR)_4]^{4-} \quad [Fe_2S_2(SR)_4]^{3-} \quad [Fe_2S_2(SR)_4]^{2-}$$

$$[Fd_{s\text{-}red}] \qquad\qquad Fd_{red} \qquad\qquad Fd_{ox}$$

$$2Fe(II) \qquad Fe(II) + Fe(III) \qquad 2Fe(III)$$

In spite of the similarity of the dianionic analogue with the oxidized form of 2Fe-Fd, there are some discrepancies with the reduced form. According to Mössbauer and ENDOR (Electron Nuclear Double Resonance) results, in 2Fe-Fd$_{red}$ the iron atoms are non-equivalent corresponding to Fe(II) and Fe(III) in a situation of trapped valence. That could not be observed in the synthetic analogue. Moreover, the second reduction potential of the analogue i.e. from tri to tetraanion, is only about 0.24–0.28 V more negative than the first one, meanwhile 2Fe-Fd$_{red}$ is apparently unable to accept a second electron.

5.2.3 Active-Site Core Extrusion Reactions

Iron-proteins can themselves participate in thiolate substitution reactions. There are indeed two types of this kind of process, namely active center core extrusion and protein reconstitution. Both are related to displacement of the Fe–S core between the protein and an appropriate organic thiol:

$$\text{Haloprotein} + 4RSH \longrightarrow [Fe_4S_4(SR)_4]^{2-} + \text{apoprotein}$$

$$[Fe_4S_4(SR)_4]^{2-} + \text{apoprotein} \longrightarrow \text{haloprotein} + 4RSH$$

In the displacements of the Fe–S core from the protein, high concentrations of organic solvent are required due principally to two reasons, namely, to achieve partial protein denaturation allowing the organic thiol access to the Fe–S core, and to maintain the integrity of displaced Fe–S centers. Thus it is possible, for instance, to cause the extrusion of about 95% of the active sites in 8Fe-Fd by using a solution of thiophenol in 80% DMSO-H$_2$O. Products are identical to the corresponding synthetic analogues.

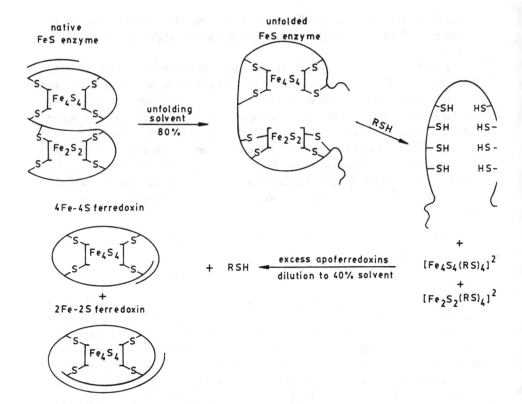

Fig. 5.12. Interprotein transfer of Fe-S centers

Protein extrusion reactions are effective only when the protein active site constitutes an integral substructure as that which occurs in the iron-sulfur proteins with Fe_nS_n core and in heme iron proteins where it is possible to extrude the iron-protoporphyrin IX prosthetic group.

Extrusion as well as reconstitution reactions appear to be an excellent tool in the identification of the active sites in numerous iron-sulfur proteins and enzymes in which the sites cannot be directly investigated. However there are some requirements for the use of this technique, e.g. protein solubility and stability under denaturing conditions, and distinctive spectral or other characteristics permitting the differentiation of the extrusion products from the halo-protein. Structural integrity of the core during and after extrusion also appears as an important condition. Indeed, in some cases it has been observed that an equilibrium dimer-tetramer can occur.

$$2[Fe_2S_2(SPh)_4]^{2-} \longrightarrow [Fe_4S_4(SR)_4]^{2-} + PhSSPh + 2PhS^-$$

This reaction is minimized by high thiophenol concentrations and by using the aqueous component buffered at pH over 7.5.

5.2.4 Interprotein Cluster Transfer

In the case of large, complex iron-sulfur enzymes, the application of the cluster displacement technique described above is not always effective. Among the difficulties often found in certain complex iron-sulfur proteins are, on the one hand, an extreme low solubility in the media required for accomplishing the displaced Fe-S centers and, on the other hand, the presence of other prosthetic groups that can obscure the absorption spectra of the Fe-S chromophores. For studying such complex systems an interesting alternative method, also based on the cluster displacement reactions with thiols, has been developed. It uses organic thiols to transfer Fe-S centers from a protein under investigation to low molecular-weight apoferredoxins. The method is described schematically in Fig. 5.12. There are however some restrictions that limit the application of this technique. Most of them are related with the opoproteins which to be used successfully must have great affinity for and high refolding rates around the Fe-S centers.

Subject Index

Scripts in Inorganic and Organometallic Chemistry

Ed.: Gmelin-Institut, Frankfurt am Main, FRG

Volume 1

W. Petz, Gmelin-Institut, Frankfurt am Main, FRG

Iron-Carbene Complexes

In collaboration with J. Faust and J. Füssel

1993. Approx. 180 pp. 23 figs. 14 tabs. ISBN 3-540-56258-3

E. O. Fischer received the Nobel prize in 1973 for the investigations of complexes with a formal metal atom-carbon double bond. Among these, the Iron-Carbene species is readily available and has proved to be a versatile reagent in organic syntheses. It is rather simple to tune the electronic properties of this Fischer Carbene and to control reactivity and stereospecificity of the reagent in, e.g., cyclopropanation reactions.

This first volume of the **Scripts in Inorganic and Organometallic Chemistry** addresses graduate students in the fields of coordination compounds and organic synthesis. It covers the chemistry and structural aspects of iron-carbon compounds with an iron-carbon double bond. The first part deals with the carbene moiety, the second with vinylidene ligands.

Springer

B3.04.040